VISIBLE NUMBERS:
ESSAYS ON THE HISTORY OF
STATISTICAL GRAPHICS

H0234825

Ashgate Studies in Technical Communication, Rhetoric, and Culture

Series Editor: Miles A. Kimball, Rensselaer Polytechnic Institute, USA

This series promotes innovative, interdisciplinary research in the theory and practice of technical communication, broadly conceived as including business, scientific, and health communication. Technical communication has an extensive impact on our world and our lives, yet the venues for long-format research in the field are few. This series serves as an outlet for scholars engaged with the theoretical, practical, rhetorical, and cultural implications of the field.

Also published in the series

Visible Numbers:
Essays on the History of Statistical Graphics

Edited by

MILES A. KIMBALL
Rensselaer Polytechnic Institute, USA

CHARLES KOSTELNICK
Iowa State University, USA

LONDON AND NEW YORK

Published 2016 by Routledge
4 Park Square, Milton Park, Abingdon, Oxon OX14 4RN
605 Third Avenue, New York, NY 10017

First issued in paperback 2017

Routledge is an imprint of the Taylor & Francis Group, an informa business

British Library Cataloguing in Publication Data
A catalogue record for this book is available from the British Library.

The Library of Congress has cataloged the printed edition as follows:
Visible numbers : essays on the history of statistical graphics / edited by Miles A. Kimball and Charles Kostelnick.
   pages cm. -- (Ashgate studies in technical communication, rhetoric, and culture)
  Includes bibliographical references and index.
  ISBN 978-1-4094-4875-4 (hardcover) 1. Mathematical statistics--Graphic methods. 2. Statistics--Graphic methods. 3. Charts, diagrams, etc. 4. Visualization--Technique. 5. Information visualization. I. Kimball, Miles A., editor. II. Kostelnick, Charles, editor.
  QA276.3.V57 2016
  519.5--dc23

2015024492

ISBN 13: 978-0-8153-8650-6 (pbk)
ISBN 13: 978-1-4094-4875-4 (hbk)

Contents

List of Figures

Plates

List of Tables

To my sweet wife and colleague, Ann Hawkins.
Miles A. Kimball

To my family—Clare, my wife; our children John, Matthew, Ann, Christine, and Peter and their spouses; and our grandchildren Isabelle, Isaac, Francie, Henry, August, and Olivia—for always giving me support, empathy, and perspective.
Charles Kostelnick

Acknowledgments

First, we wish to thank the authors who contributed to this volume for sharing their collective expertise on the history of data design. Because of their willingness to participate in this project, we were able to juxtapose in a single volume the rich, diverse perspectives of many accomplished scholars in rhetoric, history, and statistics. We also greatly appreciate our authors' diligence and patience as we constructed this volume, sometimes in fits and starts, as we made the long journey to publication.

We appreciate the expertise and encouragement of the many people who helped us along the way, beginning with Ann Donahue, Ashgate Commissioning Editor, who skillfully and patiently guided us as we worked through several drafts and reviews. We also thank the production staff at Ashgate, including Peter Stafford, who carefully copyedited the book, and Kayleigh Huelin, who adeptly composed the pages and images. Books like this one with such a large array of images require special care in their production, and Kayleigh handled this critical and daunting task superbly. We also thank Ann Hawkins for her expert bibliographic assistance, which she generously shared with us during a key step in the manuscript preparation process.

To the anonymous reviewers of the book manuscript, we wish to express our gratitude for sharing their expertise in the history of visual communication. They provided candid, insightful commentary on the chapters and numerous suggestions for revision that significantly strengthened the book.

We appreciate the support of our home institutions of Iowa State University and Rensselaer Polytechnic Institute, USA and the encouragement and insights of our colleagues and of the many scholars of visual rhetoric and data visualization in the profession at large. We also sincerely thank our students, who generously shared their insights as we introduced them to historical examples and graphical pioneers and the rhetorical, cultural, and social forces that shaped their work. Our many classroom discussions enabled us to scrutinize existing scholarship in the field and to recognize both the potential and the compelling need for additional historical studies, giving us confidence that we could attract an audience for an entire book on the subject.

In addition, we thank all of the libraries, public and private collections, archives, and copyright holders that own the visual artifacts that appear in this book, as well as the online digital collections that enabled ready retrieval of many of these materials. We are grateful to all of these sources for allowing our authors to access and reproduce these artifacts so that they could analyze and share them with our readers. Many of the images that appear in this book have very few copies still in existence, making the contributions of their owners or curators all the more critical to this collection.

Finally, we greatly appreciate the encouragement and support of our spouses, Clare Kostelnick and Ann Hawkins, and for enduring our many historical anecdotes, research exploits, and unanticipated detours as we developed this project over the course of five years. In so doing, they made this venture both possible and enjoyable.

Miles Kimball, Rensselaer Polytechnic Institute, USA
Charles Kostelnick, Iowa State University

Notes on Contributors

Lee E. Brasseur, Professor of English, Illinois State University. Brasseur is the author of *Visualizing Technical Information: A Cultural Critique*, as well as an influential article on Florence Nightingale's use of graphics to compare deaths from battle with deaths from disease in the Crimean War.

Rebecca E. Burnett, Endowed Professor of Literature, Media, and Communication, Georgia Institute of Technology. Burnett's areas of interest include professional and technical communication; collaboration, groups, and teams; communication assessment; communication in the disciplines and professions; intercultural/international communication; visual rhetoric, and risk communication.

Dianne Cook is a Fellow of the American Statistical Association and a Professor of Econometrics and Business Statistics at Monash University. Her research is in data visualization, exploratory data analysis, multivariate methods, data mining, and statistical computing.

Robert Cook, University of Massachusetts. Cook is a doctoral candidate in the Research and Evaluation Methods Program at the University of Massachusetts in Amherst. His background as a computer scientist and software developer includes building graphic user interfaces and data displays for commercial applications.

Michael Friendly, Professor in Psychology, York University. Friendly's research interests include the history of data visualization and the development of novel methods for statistical graphics. In addition to his five books and numerous articles, Friendly and his collaborators significantly expanded awareness of historical graphics with their website, Milestones in the History of Thematic Cartography, Statistical Graphics, and Data Visualization (http://www.datavis.ca/milestones).

Alan G. Gross, Professor of Communication Studies, University of Minnesota-Twin Cities. Gross is the author of *The Rhetoric of Science* and *Starring the Text*; co-author of *Chaim Perelman, Communicating Science, The Scientific Literature: A Guided Tour*, and *The Craft of Scientific Communication*; and co-editor of *Rhetorical Hermeneutics and Rereading Aristotle's Rhetoric*.

Derek Harnanansingh, Senior Developer, MediaCore. Harnanansingh's areas of interest are front-end web development and usability. Harnanansingh currently works for an education media startup company in Victoria, British Columbia, and previously ran his own web application freelance business.

Marguerite Helmers, Professor of English, University of Wisconsin Oshkosh. Helmers is the author of *The Elements of Visual Analysis*, and co-editor of *Defining Visual Rhetorics* and *The Traveling and Writing Self*. Her publications include books and journal articles on rhetoric, visual rhetoric, composition theory, and travel literature.

Miles A. Kimball, Professor of Communication, Rensselaer Polytechnic Institute. Kimball is the author of *Document Design: A Guide for Technical Communicators* and articles about the history of visualization.

Charles Kostelnick, Professor of English, Iowa State University. For several years, Kostelnick has written about visual rhetoric and communication. He is co-author of *Shaping Information: The Rhetoric of Visual Conventions* (with Michael Hassett) and *Designing Visual Language: Strategies for Professional Communicators* (with David Roberts).

Mark Monmonier, Distinguished Professor of Geography at the Maxwell School, Syracuse University. Monmonier is one of the best-known cartographic researchers today. He has written 18 books on cartography, including the influential and popular *How to Lie with Maps*, and is editor of *Cartography in the Twentieth Century* (Volume 6 of the *History of Cartography*).

Matthew J. Sigal, York University. Sigal is currently a doctoral candidate in the Quantitative Methods area of Psychology at York University. He is primarily interested in methods of data visualization and the use of interactive graphics in statistical pedagogy and influence diagnostics.

Howard Wainer, Distinguished Research Scientist, National Board of Medical Examiners. A well-known author and statistician, Wainer pioneered historical research in visualization in his essays with Ian Spence on William Playfair and culminating in their replica edition of Playfair's *Commercial and Political Atlas* (2005, Cambridge). Among Wainer's 20 published books are *Visual Revelations* (2000, Erlbaum), *Graphic Discovery* (2004, Princeton) and *Picturing the Uncertain World* (2009, Princeton). He was also principally responsible for the English translation of Jacques Bertin's 1972 classic *Semiologie Graphique* (1983, Wisconsin; 2010, ESRI). For the last 25 years he has authored the wildly popular "Visual Revelations" column in the statistics magazine *CHANCE*.

Candice A. Welhausen, Assistant Professor, University of Delaware. Welhausen teaches courses in professional and technical communication including biomedical communication and visual rhetoric/document design. Her research interests include visual rhetoric, rhetorical theory, and technical communication.

Kevin Van Winkle, Texas Tech University. Van Winkle is an Adjunct Professor of English at Colorado State University-Pueblo, where he teaches composition and literature. He is also a doctoral student in Texas Tech University's Technical Communication and Rhetoric program. His scholarship includes work on visual rhetoric, multimodal discourse, and popular culture.

Introduction

Charles Kostelnick and Miles Kimball

Today we are used to seeing quantitative data portrayed in a dizzying array of graphical forms. Virtually any quantified knowledge, from social and physical science to engineering and medicine, as well as any business, government, or personal activity, has been visualized in charts, graphs, thematic maps, and now data-rich interactive displays. Several terms have been employed to describe these many modes of showing numbers visually: data visualization, data design, data display, Edward Tufte's "visual display of quantitative information," or as we propose here, simply *visible numbers*. Thirty years ago Tufte (*Visual Display*, Intro.) estimated that hundreds of billions of data displays were generated annually, but with the emergence of digital design, that quantity has grown exponentially. As many common and controversial examples attest—from the "hockey stick" graph of global temperature change to the red state/blue state maps we see with every U.S. national election—the visual display of data has a profound effect on the way we see the world today.

Despite their ubiquity, however, these methods of making data visible—these *visible numbers*—are relatively modern innovations, most with their origins in the eighteenth and nineteenth centuries. To a large extent, data visualizations arose as a logical response to a growing desire to quantify everything. As scientists, doctors, engineers, merchants, and governments began to collect empirical observations of the physical and social world, they quickly grew inundated with a sea of numbers, usually arranged in tables—themselves a form of graphic communication.[1] But as William Playfair (1759–1824) pointed out in his 1786 *Commercial and Political Atlas*, "a man who has carefully investigated a printed table, finds, when done, that he has only a very faint and partial idea of what he has read; and that like a figure imprinted on sand, is soon totally erased and defaced" (3). Accordingly, innovators such as Playfair, Alexander von Humboldt, Heinrich Berghaus, Florence Nightingale, Francis Galton, and Charles-Joseph Minard began to develop graphic and geometric methods to make data more visible to the eye and memorable to the mind.

Particularly in the middle and end of the nineteenth century, Playfair's successors began to visualize data about health, poverty, population, industry, and a variety of other areas of social science—what contemporaries would call

---

[1] See Brasseur, *Visualizing*, Chapter 5, as well as her contribution to this volume, Chapter 2.

political economy or *moral statistics*. Scientists and engineers continued to develop more sophisticated methods for showing data, as well as for collecting it through standardized forms. In the twentieth century, data design became both increasingly specialized within new and existing disciplines—science, engineering, social science, medicine—while it also became more democratized, with new forms that made statistical, business, government, and other data more accessible to the public. At the close of the twentieth century and the beginning of the twenty-first, an explosion in digital data design has immensely increased our access to data, including big and complex data sets, as well as our control over their display, particularly online in interactive visualizations.

Unfortunately, as these new graphic forms of communication became common practice, their historical development largely faded from view. Only in the past 20 years has a growing body of scholarship revived our awareness of the development of the graphical tools we use today as a matter of course. Much of that scholarship has focused on drawing our attention to lost innovators such as Playfair, von Humboldt, Nightingale, Galton, and Minard—a valuable and necessary first step. But much remains to be done. We need to expand and deepen our knowledge of these pioneers and their milieu, developing a critical understanding firmly grounded in visual rhetoric, visual culture, and fully contextualized historical scholarship.

Approaching Visible Numbers

In this collection of essays, we bring together the work of a diverse group of scholars who critically examine visible numbers from many different perspectives and disciplines. This eclectic variety of approaches seeks to push the boundary of scholarship beyond straightforward biography toward more complex historical scholarship—not only on the individuals who undoubtedly had an impact on the development of statistical graphics, but on topics, realms of knowledge, and societies that benefited from that development.

In hindsight, the development of visible numbers may seem linear—a sequence of progressive innovations by brilliant people. Progress has obviously occurred: we know how to do things with visible numbers that we once did not. However, closer examination reveals that this history does not flow logically along a straight course; it more resembles a river with many fascinating streams, tributaries, currents, deep pools, and stagnant oxbows, often established by individual—sometimes idiosyncratic—people who turned to graphic methods to address specific problems in their own social and cultural milieu. This multifaceted story defies analysis from a single perspective.

This complexity is easy to overlook, in part because the geometry of visible numbers has a certain air of inevitability to it; surely someone would have invented the pie chart if Playfair had not done so. And individual innovators undoubtedly had a significant impact on the development of visible numbers.

Some of the earliest practitioners, such as Playfair, leaped at once from nothing to great sophistication, as he did by developing the histogram. Yet even some of the luminaries of early data design created graphics so unique as to counter this sense of linear inevitability. Consider one of the touchstones of this field of inquiry: Minard's famous "Carte Figurative" (Marey 72, Figure 37), which describes Napoleon's 1812 march to Moscow. While somebody may eventually have created this graphic if Minard had not, the fact remains that nobody did so in the half-century between 1812 and 1869, the year of its creation. And nobody else did so from 1869 until after 1983, when Edward Tufte brought this fascinating display to our attention (*Visual Display*, 40–41), inspiring many re-designs using modern tools and sensibilities.[2] What is it about Minard's rhetorical, social, and cultural context in 1869 that led him to create such a graphic? What are the implications for this field of study that a flow map that makes something complex seem so clear reveals a historical development that is anything but?

In addition, significant individual practitioners did not always expand the state of the art, as Miles A. Kimball discusses in his chapter on Michael G. Mulhall, who was well known for making popular—and often bad—data displays. As Kimball illustrates, when viewed in the thick cultural and national context of the late nineteenth century, Mulhall's missteps can be as instructive as Minard's triumphs. The Victorian and imperial values that infuse Mulhall's graphical inventions reveal the competitive political and economic landscape of a post-Darwinian era, visualized in pictographic and other comparative displays of wealth and power.

Given this complexity, in creating this volume we chose to cast our nets widely, bringing together scholarly essays that scrutinize individual practitioners throughout history; the dynamics of specific cases and social problems to which the methods of graphic statistics were applied; and the development of individual genres and forms. All told, this volume seeks to answer questions such as these:

- How have conventions for visualizing data emerged, developed, evolved, and mutated? How have readers come to understand and accept (or reject) these conventions?
- What cultural and social values are embedded in the visual rhetoric of data displays? How has data display both influenced and reflected social or political conditions?
- What rhetorical strategies for data design have been deployed historically? How, for example, do data displays create powerful visual arguments or narratives?
- What role have technology and production practices played in inventing, generating, and disseminating data displays and in how readers interact with those displays?

---

[2] See for example Dragga and Voss (270–71), as well as Friendly, "Visions."

- How have innovative designers, seminal researchers and theorists, and iconic designs influenced the course of data design practices?
- How should we theorize, methodize, and perform the historiography of data design?

The chapters in this collection respond to these questions using a variety of approaches: exploring the impact of data design on historical events, examining its role in solving social and health problems, assessing its political and cultural dimensions, rhetorically analyzing conventions and genres, and comparing contemporary and past design techniques and technologies.

Visible Numbers: Bodies, Nations, and Historiographies

Almost inevitably, many of the essays in this volume address graphical developments in nineteenth-century Europe and North America and two overarching concerns: the state of human bodies (and by extension the state of public health), and the development of nation states, their economies and their peoples, and their imperial aspirations. Throughout the nineteenth century, researchers, scholars, and enlightened amateurs strove to use visual methods to understand human bodies and human societies, especially within the framework of what Auguste Comte (1798–1857) first described in 1839 as "sociologie" (466). A seminal expression of these concerns is the work of Adolph Quetelet (1796–1874), the Belgian statistician who championed "l'homme moyen," or the average man (*Sur l'Homme*). Quetelet set out this concept as a measurement derived from bodies and behaviors, as for example when he examined the height of Belgian soldiers or the relationship between age and propensity to criminal activities. But his average man also became a measurement of nations and their people. Rather than the superlative individual standing as the epitome of France and the French, the average Frenchman, who was simultaneously all Frenchmen and none of them, became the measurement of the nation. And the average Frenchman therefore became available as a measurable construct that could be compared to the average Englishman, Belgian, German, or American.

This measurement of citizens and states formed an important locus of activity in all of the nations of Europe and North America in the nineteenth century. As many scholars have pointed out, the nineteenth century saw an explosion in the effort to quantify virtually everything.[3] Theodore Porter described in great detail this enthusiasm for "the social calculus"—the measurement of all things, as a means of understanding and controlling human life and societies through centralized administration and government. This enthusiasm was an inheritance of the eighteenth-century rationalism that inspired the founders of the United States of America to enshrine the gathering of statistics in the country's very

---

[3]   See Patriarca (2003) for a useful summary.

constitution, in the form of the U.S. Census. In Napoleonic France, the Bureau de la Statistique Générale, founded in 1800 by Jean Chaptal, became the first of a series of official statistical offices formed in that country (Faure 292). Similar offices were opened in other governments, such as the Statistical Department of the British Board of Trade in 1832. Knowledgeable societies began to form and hold meetings, such as the Statistical Society of London (later the Royal Statistical Society), founded in 1834. This interest also crossed national borders, as evidenced by the nine International Statistical Congresses held from 1853 to 1876. In all of these developments, nations were concerned with using numbers to understand their populations both in terms of their physical attributes, health, and disease, and in terms of their productions, commerce, transportation, and a myriad of other activities. With this explosion of numbers came an interest in finding new ways to understand numbers visually, engendering what Michael Friendly has called the "Golden Age of Statistical Graphics."

Accordingly, we have structured the volume in three sections, each prefaced by a short introduction. First, in *Visualizing Bodies*, we include essays that examine visualizations of disease, health, and the evolution of human and animal bodies. In this section, Alan Gross examines the use of numbers and geometry to the question of physical evolutionary development, or morphometrics, focusing especially on the contributions of D'Arcy Thompson. Lee Brasseur analyzes Florence Nightingale's development of statistical tables and forms (in themselves a species of visible numbers) for reporting information about patient health and hospital administration. And Candice Welhausen and Rebecca Burnett examine the use of visualization to track, locate, and correlate data about smallpox epidemics, and they analyze how those visual displays defined the concept of risk and fostered the evolving science of public health.

A second section, *Visualizing Nations*, offers essays that examine the use of visible numbers in the development of nations and in the imperial competition between them. This section begins with Robert Cook and Howard Wainer's chapter on the development of moral statistics and their representation in Joseph Fletcher's thematic maps. Mark Monmonier contributes his analysis and criticisms of graphical practice in nineteenth- and twentieth-century American statistical mapping. Kimball turns the biographical approach on its head by examining Mulhall, a well-known graphic statistician whose graphic productions were very popular but often visually deficient. And Marguerite Helmers examines the use of thematic mapping amid a horrendous national conflict in the form of trench maps from the First World War.

A third part, *Examining Visible Numbers*, features essays that examine genres, technologies, and the more recent history of visible numbers. Here Charles Kostelnick examines mosaic and area charts from the perspective of visual rhetoric as a complex, evolving genre. Dianne Cook draws on the Statistical Graphics Video Library of the American Statistical Association to chart the development of statistical graphics at the dawn of the computer age. And Michael Friendly, Matthew Sigal, and Derek Harnanansingh document the evolution of the online Milestones Project, applying the methods of statistical graphics to visualize data design history. Finally, Kevin Van Winkle provides a concise annotated bibliography of notable scholarship on the history—or perhaps we should say histories—of visible numbers.

Histories of Visible Numbers

If no single one-size-fits-all approach captures the complexity and richness of the history of visible numbers, what approaches have been employed, and what new approaches might lead to new understanding? To address these questions more deeply, in the remainder of this introduction we will turn our attention to several salient issues attendant to the aims and processes of studying data design history. We will probe its historiography, methods, and future prospects, and we will draw on chapters in this collection for examples and best practices.

Canonizing Data Design

At least as consumers of data design, many people have a stake in its history and the genealogy of its many forms—from the ubiquitous pie charts that appear on monthly utility bills, to hi/lo charts that show stock prices on investment websites, to dense scatterplots that appear in journal articles. Public audiences historically have acquired the visual literacy to interpret data design through several channels—national atlases, mass media, formal instruction in schools, disciplines and professions, and most recently through the Internet. Collectively, over the past few centuries, these sources have inculcated us in the design and interpretation of many forms of visible numbers we take for granted today.

This organic cultural absorption of graphical methods has recently benefited from a growing body of scholarship that has led to greater public awareness of the history of data design. As we suggested earlier, the history of data design for public audiences has entailed establishing a canon of innovators and pioneers, to whom the genealogy of display forms can be traced and of whom, presumably, all current designers and consumers are the heirs. H. Gray Funkhouser began that canonical process 75 years ago with his extensive history of the field and its pioneers, and this approach continues in the works of prominent scholars such as Tufte and Wainer, who celebrate historical figures throughout their work. Key historical figures are systematically documented on Michael Friendly and Daniel Denis's website *Milestones in the History of Thematic Cartography, Statistical Graphics, and Data Visualization*, the development of which is the focus of a chapter in this collection. Other chapters highlight the accomplishments of key figures. For example, Robert Cook and Howard Wainer focus on Fletcher, who in the mid-nineteenth century pioneered the use of thematic maps to visualize the relationship between crime and ignorance. As a result of these many invocations of their work, figures like Playfair, Minard, Fletcher, and Nightingale have achieved something like celebrity status in the burgeoning world of data visualization. Among the educated public, canonical works like Minard's chart of Napoleon's Russian campaign and John Snow's cholera map of London have become icons of modern visual culture. By identifying and popularizing canonical innovators and their works, the history of data design has slowly begun to seep into the public consciousness.

Establishing a canon of data design has a more immediate effect for contemporary consumers and practitioners by identifying exemplars that model best practices. Tufte's pantheon of heroes functions more than mere historical curiosity: Minard's Napoleon chart, Tufte declares, "may well be the best statistical graphic ever drawn" (*Visual Display* 40). In explaining why, Tufte draws lessons for contemporary designers about "Graphical excellence" (*Visual Display* 51). On the other hand, Tufte also identifies many deficient historical examples—the villains of the narrative—that he mocks as substandard because they violate principles of effective design. In vilifying historical (and contemporary) mistakes, Tufte follows the modus operandi of Willard Brinton, whose 1914 *Graphic Methods for Presenting Facts* similarly critiqued modern American examples.

The accomplishments of prominent innovators can be threaded together to create a reasonably cohesive story, a grand narrative of sorts, as data design moved from science in the seventeenth and eighteenth centuries to social science, engineering, health, commerce, and national statistics in the nineteenth century, culminating in the first "Golden Age," and then in the twentieth century broadening into the workplace, education, and popular media, followed by a second Golden Age with the advent of digital data design.

Several insights emerge in this collection that inform, and to an extent resist, such a sweeping narrative: that the development of graphical statistics has largely been the province of disciplines, which have served as laboratories for innovation; that this development often, but not necessarily, entails incremental progress, resulting in steadily improving methods; and that this development is driven and defined by cultural and rhetorical factors that often remain unrecognized or unremarked. This collection also raises epistemological and ontological issues about designing, interpreting, and studying data graphics. Let's explore each of these notions.

Disciplines as Laboratories for Innovation

In part, the history of data design resists a single narrative because it is an inherently disparate enterprise scattered among many disciplines—first and foremost statistics, because its own history is so deeply intertwined with data design. However, the heroes of data design history represent a wide array of fields: political economy (Playfair), engineering (Minard), physical science (Humboldt, Etienne-Jules Marey, Galton), mathematics (Johann Heinrich Lambert, Gaspard Monge), medicine (Nightingale, Snow), graphic design (Otto Neurath), and cartography (Bertin, Beck), to name only a few. Just a glance at

Friendly and Denis's *Milestones* website shows the wide array of disciplines that contributed to its litany of significant innovations. Thus, the history of data design can be studied as a series of disciplinary projects—for example, analyzing the evolution of data design in experimental science, medicine, or economics—as well as a lens through which to study the development of a given discipline—mainly, how its visualization techniques reveal its methods and its epistemology, as well as its affinities with other disciplines.

Most of the studies in this collection examine data design through such a disciplinary lens. In the realm of medicine, Welhausen and Burnett's chapter on smallpox visualizations show how several genres (tables, maps, graphs) were used to advance understanding of disease—its consequences, its deterrence, and its possible correlations with climate, location, and other diseases—and more broadly to advance the field of public health. Gross's chapter examines how advancements in visual modeling over the past century define and foster the emerging discipline of geometric morphometrics. In her chapter on trench maps, Helmers shows how military strategists struggled to visualize troop movements along a bloody war front. In the realm of political economy, Kimball shows how Mulhall's charts, in addition to their modest contributions to the emerging science of statistics, advanced the commercial and political interests of the late nineteenth-century British Empire. In each case, the formation of a disciplinary identity coincides with, and indeed is impelled by, the development of data displays, the absence of which would have deprived the discipline of an essential tool for exploration, analysis, and argument. So writing data design history often closely parallels writing the history of a given discipline.

These disciplinary histories, however, do not unfold in isolation but rather overlap and intertwine, creating an interdisciplinary genealogy of data design forms and practices. As the evolution of data display ensued, many of the key figures crossed disciplines: Playfair bridged social science and engineering (he once worked for James Watt, inventor of the steam engine; see Wainer, *Graphic* 21); Minard traversed engineering, commerce, and history; Marey was a scientist with interests in logistics and photography; Galton was a polymath whose work included statistics, genetics, and weather. Some of the benchmark projects in data display resulted from extensive interdisciplinary collaborations—most notably, Francis A. Walker's 1874 *Statistical Atlas of the United States*, which brought together members of several professions—statistics, engineering, cartography, the military—to visualize increasingly large and complex sets of census data.

The interdisciplinary history of visible numbers also entails appropriations of display forms for discipline-specific purposes. Playfair appropriated graphing techniques from science and engineering to visualize economic data. The wind rose, used by early meteorologists to chart wind direction, was appropriated by Nightingale and her successors in medicine to chart the monthly incidences of diseases. Engineers and geologists appropriated line graphs to create land profiles for railroads and to visualize cross-sections of strata and rock formations. As genres evolved within disciplines, they mutated into new forms, serving ever more specialized purposes as they evolved from one stage to the next. This cross-disciplinary interaction both enriches and complicates the narrative, leaving historians often to speculate on disciplinary influences on a given innovator and ultimately showing the collective rather than the individual nature of the enterprise.

Centuries of Progress

We have argued in this introduction that the history of visible numbers, when looked at in sufficient detail, does not suggest an overall linear progression from simpler, more primitive forms to more complex, refined, and efficient forms. But this assessment is not universal—not even among the contributors to this volume. As such, the question of historical determinism or teleology is central to discussions of the history all of the contributors wish to chart.

Most of the chapters in this collection at least implicitly address the issue of progress in graphical display, and for the most part, a progressive view of history prevails. Gross, for example, charts the progress in scientific visualization over nearly a century, as plotting techniques for the morphology of species became more precise and sophisticated, including 3-D modeling. Cook narrates innovations beginning with the digital revolution in the 1970s, with one innovation succeeding the next in a rapidly progressive and synergistic chain of technological and visual discoveries. Helmers documents the development of trench maps in charting danger on the Western front: progress is achieved as the maps become more prolific, precise, and local, but they inevitably lacked some vital data, thereby both empowering and endangering their users. Here, progress comes with a deadly price.

Drawing on contemporary technology, historical make-overs test the theory of progress by using digital tools to recreate iconic displays, raising a provocative question: do the make-overs surpass the originals, or do they forfeit the insight or graphical dexterity of their hand-formed antecedents? In their chapter, Friendly, Sigal, and Harnanansingh engage in a process they call "understanding through reproduction" by using digital tools to gain insight into Minard's methods, with (as they acknowledge) mixed results. Cook and Wainer recreate some of Fletcher's maps correlating crime and ignorance, using digital tools to display the same data in subtle gradations of color as well as transforming the bivariate data into scatterplots, a genre unavailable to Fletcher. Cook and Wainer's results argue rather convincingly for historical progress, as modern techniques far more clearly reveal the correlations (or lack of them) than Fletcher's grayscale maps. Likewise, as Kostelnick shows, progress can also be achieved when a traditional genre like the mosaic is rapidly rediscovered and its audiences and permutations greatly expand, here with the help of digital technology.

But progress can be complex, inconsistent, and sometimes painfully slow. In his chapter on displaying physical and human data in thematic maps, Monmonier shows how poor practices can become entrenched and additional ones develop, partly because maps are such a "robust instrument" and partly because the conventional status quo (that is, less than total transparency) may benefit those in power, resulting in visual "inertia." Brasseur's chapter on Nightingale's development of hospital forms shows that progress can come not in the form of a new invention—tables had long been used to record scientific data—but in the novel application of old techniques to new purposes: to gather information about patients, which in the long run spurred enormous progress in public health. Yet this innovation was initially rejected by some hospital administrators, who feared their institutional weaknesses and failures would be exposed. Similarly, visualizing smallpox, as Welhausen and Burnett show, did not necessary foster new graphical forms. Rather, those studying the disease appropriated existing forms to understand, analyze, respond to, and eventually eradicate one of the deadliest diseases in history—progress by anyone's standards.

Cultural Studies and Visual Rhetoric

Against these strains of progressive historiography run approaches by those who have taken the development of information visualization as an opportunity to examine not the teleological progress of innovation, but the variable dynamics of cultures and rhetorical exigencies. In this volume Kimball's chapter on Mulhall is the primary exemplar of such an approach. Though a Fellow of the Royal Statistical Society, Mulhall was out of step with the progressive development of sophistication in graphic display and came under great criticism for his work. Yet the overwhelming popularity of his 1884 *Dictionary of Statistics* spread awareness of visualization far beyond the disciplines in which many innovations were developed to a broad public familiarity. Viewed through the lens of cultural studies, then, Mulhall—though his work is imprecise and fails to follow even the best practices of his own time—is perhaps as significant in some ways as innovators such as Playfair, whose data designs were initially known to a relative few. The work of such figures as Mulhall does not illuminate the grand narrative of information graphic development, but it can tell us what cultural and social impact information graphics may have had on the cultures in which they were developed.

Considering the history of visible numbers through the lens of cultural studies finds a counterpart in studies that approach the subject through the lens of visual rhetoric. Data displays are communication tools that operate in rhetorical situations with audiences that designers try to engage, inform, and persuade. Analyzing historical artifacts rhetorically can be confounded by the inability to recover facts about important situational variables, leaving scholars to make inferences and hunches from what remains of the historical record. As we have noted, many laud Minard's display of Napoleon's Russian campaign for its effective design, but do not particularly analyze or even question its rhetorical context or impact. This lack is somewhat understandable, given what little we know about Minard's motivations or audience, much less how they responded to his display. But what we do know suggests a fascinating visual rhetoric. Created as Louis Napoleon (the nephew of Napoleon Bonaparte) considered going to war with Prussia, the display appears to make a civic argument against ambitious, futile military conquests, warning leaders to heed the lessons of this disaster. Given the timing of the display, it could also be viewed as an epideictic artifact, a memorial from an aging Minard to his fallen contemporaries half a century earlier during the Russian campaign.

The rhetoric of visible numbers extends beyond a single artifact designed and deployed at a given moment in time. Using mosaics and area charts as his historical reference points, Kostelnick's chapter explores how data displays embody socially constructed genres that discourse communities shape and modify. For example, genres are invoked for certain kinds of data (wind roses for meteorology) and then sometimes adapted for other purposes (health). Although the formal and rhetorical boundaries between data display genres can often be hard to distinguish, historical perspective can clarify the social and disciplinary networks that shaped their genealogies. Minard's Napoleon chart, for example, deploys several genre conventions familiar to readers at the time: a line graph showing temperature, a map showing location, and a transportation chart that gauges quantity of troop movements with the thickness of the lines (Minard himself pioneered this technique by mapping trade routes). Combining all of these conventions in the same chart, however, was Minard's stroke of rhetorical genius, in which the novelty of graphical hybridity exploits the socializing effects of previous data displays.

Epistemologies and Ontologies

Finally, the chapters of this volume address a basic epistemological and ontological question: Are the graphics described in this book devices of measurement or devices of rhetoric—or both? In other words, are these graphics transparent geometric representations of objective reality processed by eye and brain, or do they participate in the project of creating a social reality? If the former is true, then a history of visible numbers would probably take the form of recounting the development of particular genres, and this development would lean toward an inevitable, teleological progress, based on the limitations of geometry and human perception. The linking of quantity to height in a bar chart, for example, seems naturally complementary to human perception. As the quantity increases, so does the height, matching our common perception that "more" equals "bigger." In this historiography, individuals would play a role primarily as *discoverers* of graphic forms, rather than as innovators of graphic communication.

The alternative view, however, maintains that visible numbers are based not just on reality, but on a statistical accounting of observed reality. Statistics and observation are both fraught with epistemological and ontological problems. Who is the observer, and how does he or she make decisions about what and how to count—or even what's worth counting? What prior epistemologies and ontologies influence those decisions? Subsequently, when users become designers, how do they decide to employ the language of design—color, line, shape, geometry—to convey those observations to an audience? And to what end or purpose? Is it merely to record reality, or is it to convince an audience to see things in a particular way or to take a particular action? Intentions aside, what version of reality will the audience perceive through the representation formed by a chart, graph, or map? A historiography based on this version of history would focus less on the development of graphic forms than on the rhetorical and social circumstances in which visible numbers were developed. In her study of battlefield maps, for example, Helmers probes deeply into the grim and wretched conditions of serving on the front lines in order to reconstruct how maps were actually used in that context. In this kind of history, designers and their desire to communicate with an audience, as well as viewers with their desire to understand the world, would play a key role.

Many, but certainly not all, of the chapters in this book address these issues from a social constructionist perspective: that is, displays are socially mediated forms of communication that are profoundly influenced by the collective forces of things like conventions, genres, and disciplines. For example, Kostelnick's chapter on the conventions of the mosaic display and Monmonier's chapter on the often bad practices of map designers both emphasize the importance of social factors in the history of visible numbers. The countering epistemology—that graphical displays are tools of measurement of an objective reality—places greater stock in individual cognition, whereby designs and their interpretation can be explained through innate human perception. Thus the redrawings of historical graphics using modern tools, represented by Cook and Wainer's chapter on thematic maps and by Friendly, Sigal, and Harnanansingh's chapter on the Milestones Project, take the statistical data as the central object, which can be re-visioned with newer techniques that better match our human capabilities to see and understand visual displays.

This problem does not map directly onto the question of progressive or organic history. Researchers in this volume and elsewhere have celebrated the contributions of individual innovators, without necessarily addressing the social construction of reality. Conversely, researchers interested in a social construction of reality do not necessarily argue that graphics, because rhetorical, are suspect—rather, recognizing that displays are socially mediated forms of communication that are profoundly influenced by the collective forces of conventions, genres, and disciplines.

This volume does not resolve the tensions between differing epistemologies and ontologies—but recognizing this difference leads to fruitful and potentially profound ways of understanding our view of how visible numbers came to be. In this regard, these tensions are complementary and productive, giving us plenty to ponder over, both here and in future scholarship.

Methods and Problems

Although the difficulty of telling the story of data visualization arises from the strong disciplinary roots of those researchers interested in the subject, clearly these disciplinary foundations have supported a wealth of fascinating scholarship. However, in more practical terms, given this wide variety of approaches and perspectives, how do we—and how should we—study the history of data design? What tools, methods, and frameworks do we use? How have these changed recently, particularly in the theory and the technology that inform and enable analysis? These questions invite several answers, given that an eclectic mix of traditional and new methods now fosters scholarly inquiry.

Although the resources vary for reconstructing historical circumstances and for locating artifacts, the main research site for examining the history of graphic display is the archive, including physical as well as digital collections. An example of the former is JoAnne Yates's article that draws on DuPont's corporate archive of its "chart room," where for several decades DuPont managers used graphs for decision-making. Another example is Kimball's article on Charles Booth's maps of London poverty, which relied on the archives of the Booth survey held by the London School of Economics. Several of the studies in this collection also entailed archival research to recover original visual materials. No doubt similar, unexamined troves of archival materials can reveal many additional insights into the history of information visualization.

But increasingly, online archives of historical documents (sometimes characterized as the "digital humanities") provide invaluable tools for surveying graphical displays in published sources. For example, the Library of Congress website contains a digital collection of the three nineteenth-century U.S. Statistical Atlases, the Hathi Trust Digital Library's archive includes nineteenth-century European atlases and albums, and the Gallica website of the Bibliothèque Nationale de France includes a wealth of historical books and other materials. Likewise, Friendly and Denis's *Milestones* website provides a wide range of examples with links to other resources. Finally, Google Books (http://books.google.com) has made access to a large number of historical texts and page images convenient and inexpensive. The digital curation of images will undoubtedly increase in the future, offering scholars additional sources they can access online.

However, one of the most difficult challenges of studying historical graphics is that, for the most part, they must be found and indexed manually, one at a time. There is no universal convention among library catalogers of how to describe this material. What one cataloger calls a "chart," another calls a "map"; a "diagram" to one cataloger is a "plate" to another, or a "figure," or an "illustration." Unfortunately, few librarians have the expertise to distinguish between these terms, or they apply idiosyncratic or localized terminology that sows confusion more broadly. Finally, the amount of legacy information in library catalogs (that is, information entered before our recently increased interest in graphic display) is immense. This deeply entrenched inconsistency makes searching for graphics a relatively hit-or-miss affair, a matter of manual and time-consuming collection. As a result, scholars interested in the history of visible numbers are often unable to take advantage of many of the powerful search capabilities of computer and Internet technologies, which rely primarily on text strings. We and our predecessors have uncovered a wealth of examples, but we simply do not know what or how much we have yet to find, a circumstance both daunting and exciting.

But ironically, many of us doing research in this area lack the bibliographic, technical, and historical training of scholars in other fields like library science, art history, or the history of print. As a result, much of the research record on the history of data visualization lacks standard information about the material nature of the visual texts we examine, with even something as simple as the dimensions of graphics often missing from the scholarly record. Few scholars, for example, have delved into the question of production technologies and their impact on data design: What, for example, was the effect of the shift from copper plate to steel engravings on the number, quality, and type of graphics that appeared? How and when did lithography and chromolithography begin to have an impact? What about the effect of the plethora of other short-lived printing technologies that appeared in the nineteenth century: heliotypes, mezzoprints, aquatints? Similarly, we know little about the consumption of these visual texts in the public marketplace. Were they relatively esoteric and expensive publications, or widely available and cheap? How many copies were produced and sold? For what markets were they developed? Who read them, and for what purposes? Were they intended to be considered parts of larger publications (a book, an atlas), or did the producers expect that these graphics would be separated from their context—perhaps framed and hung on a wall? (Certainly this has happened to many visualizations regardless, as print sellers often break up illustrated books in order to sell individual plates for a higher price.) In short, while we often know a good amount about the designers of data visualizations, we know little about the context of their production or the context of their use.

Finally, our intensive effort to find the "firsts" of the history of data visualization—the first pie chart, the first isomap, the first histogram—has not helped us recognize the stunning pace of development in the more recent history of data visualization, what we might refer to as the Second Golden Age of Statistical Graphics, beginning perhaps with John W. Tukey's *Exploratory Data Analysis* and continuing through the advent of the desktop computer and the World Wide Web. That history has largely yet to be written, although Dianne Cook's essay in this volume begins to chart that untrodden ground. What issues face researchers of this Second Golden Age? Graphics that appear on the Web are easily accessible—partly because researchers can search for and interact with them digitally and partly because an increasing number of dedicated sites like Andrew Vande Moere's *Information Aesthetics*, Manuel Lima's *Visual Complexity*, and IBM's *Many Eyes* catalog them. However, web-based visualizations also pose unique challenges: displays that visualize time-sensitive data change periodically, sometimes by the minute, and often displays or their host sites disappear—a problem that will only magnify in the future. And of course, many proprietary visualizations (analytical displays, dashboards) remain invisible to the public because of their ownership and confidentiality. So researchers examining the current boom in data visualization have enormous opportunities—as well as a new set of challenges.

Finally, one of the most promising methods of research is to apply data design itself as a means for studying its history. In this volume, Friendly, Sigal, and Harnanansingh describe how they have graphed historical data about designers (lifespans, places of habitation) to reveal previously unrecognized patterns and relationships. Creating visualizations, including interactive displays, enables scholars not only to re-envision history and but also to make it accessible to larger audiences. This method parallels efforts in the digital humanities like Paula Findlen, Dan Edelstein, and Nicole Coleman's 2008 "Mapping the Republic of Letters, 1700–1750," an interactive visualization of correspondence among Enlightenment intellectuals.

What's Next?

Based on where the history of data design now stands, where do we go from here? What discoveries are still waiting to be made, what stories still need to be told, and which ones are emerging in our midst? Several avenues seem worth pursuing:

- *Expanding the canon.* What designers and designs will enter the canon, from the distant past, or more likely, from the Second Golden Age? Key innovators in information visualization like Tukey, Tufte, and Martin Wattenberg have already achieved that status; over time, scholars will have to sort out the contributions of digital designers, though the uncertain longevity (and long-term functionality) of online artifacts may pose some challenges.
- *Archival stories.* What graphical forms and stories are yet to be discovered, lingering in archives, special collections, and other sites? Like Yates's study of DuPont's "chart room," archives of private companies, government agencies, and other organizations await scholars to excavate their files and tell their graphical stories. Online archives may also provide some windows into the past that were previously unavailable to many scholars.
- *International scope.* Although scholars have noted traces of ancient displays outside of Western culture, virtually all of the studies conducted thus far focus exclusively on European and North American advancements in data display. To what extent do other, nonwestern cultures contribute to

this history, in the distant or recent past? More specifically, what role did national atlases play in cultivating visual literacy around the world?

- *Disciplinary narratives.* Tracing the evolution of display genres that a given discipline invents, appropriates, and develops can reveal its epistemological methods and values, as well as its relationships with other disciplines using cognate displays. How do disciplines shape their own visualizations, and how do theirs differ from/build on those of other disciplines?

- *Theory and methodology.* Most of the history of data display has been told by people who are not professional historians, but enthusiastic scholars from a variety of other disciplines. As a result, the theoretical and methodological frameworks behind studies of that history are mostly implicit. How can we make these theories and methods more explicit and formalized?

- *Data that shape events.* Like Minard's chart of Napoleon's Russian military campaign, graphs showing climate change, or maps recording presidential elections, data displays often visualize world events. However, they can also shape world events, as Helmers shows in her chapter on First World War trench maps. What other data designs have influenced the direction of world affairs and public policy?

This is hardly an exhaustive agenda for a history as complex, multidisciplinary, and prolific as data design. Still, like any history, the history of data design has many purposes—among them to recover what has been lost, to trace origins and foundations, to recognize achievements, to weigh cause and effect, and (perhaps most of all) to understand more clearly what we do in the present. The chapters in this collection achieve all of these ends, from many different angles and across different eras. At the same time, as the agenda above suggests, we realize that much more still needs to be done, that many more avenues await us to pursue. We hope that this collection walks farther down those paths and, equally important, points in directions that others can follow.

PART I
Visualizing Bodies:
Health, Disease, Evolution

In the nineteenth century a new and profound method for displaying data appeared: graphical displays that visualized people—their population, ethnicity, health, occupations, living conditions, physical characteristics, and a host of other criteria. Graphical innovators began charting these measurements in a variety of forms—thematic maps, bar graphs, wind roses, mosaics, illustrations—both to tell compelling "stories," as Tufte puts it (*Envisioning* 37, 108), about human beings and human behavior and to solve problems through visual exploration.

This part of *Visible Numbers* focuses primarily on the exploratory nature of data design in transforming health and science relating to the body. Particularly in the nineteenth century, data design became a valuable tool for defining and addressing compelling health problems and for arguing for improvements in health policy and medical treatment. Celebrated visualizations like John Snow's London cholera map and Florence Nightingale's rose charts on the casualties of the Crimean War have already received much-deserved attention, but much of the story of the burgeoning interest in visualizing health data remains to be told. In their chapter on smallpox, Candice Welhausen and Rebecca Burnett examine how visual displays, beginning in the seventeenth century, incrementally helped scientists, doctors, and public health officials to eradicate one of the deadliest diseases in history and one of the earliest to be charted. Welhausen and Burnett scrutinize those advances in public health in their social, cultural, and intellectual context, as data about smallpox outbreaks and experimental techniques for reducing risk were gradually made visible.

In her chapter on table design and forms, Lee Brasseur also examines the role of visualizing data in improving public health. Brasseur shows how Florence Nightingale designed and implemented tabular charting techniques that revolutionized the documentation of hospital patients, enabling medical professionals to analyze, compare, and identify patterns in the data and adapt treatments accordingly. Like many pioneers, Nightingale initially encountered resistance to her charting methods, but her innovation prevailed because it worked, dramatically improving the delivery of British health care.

In the second half of the nineteenth century, the theory of evolution transformed the scientific and intellectual landscape, fostering an interest in visualization techniques that illustrated the new paradigm. In his chapter, Alan Gross examines how visualizations of animal anatomy helped develop the discipline of geometric morphometrics, which plots evolutionary changes in the shape (e.g., bone structure) of mammals and reptiles. This process of precisely charting species' physical characteristics continued to unfold for nearly a century, most recently with 3-D digital modeling.

Chapter 1

Visualizing Evolution and Development: The Rise of Geometric Morphometrics

Alan G. Gross

So here and elsewhere an apparently infinite variety of form is defined by mathematical laws and theorems, and limited by the properties of space and number. And the whole matter is a running commentary on the cardinal fact that, under such *foedera Naturai* [laws of nature] as Lucretius recognised of old, there are things which are possible, and things which are impossible, even to Nature itself.

D'Arcy Thompson[1]

Although humans and Neanderthals share a common ancestor, they differ in skull formation. In Neanderthals, the rear of the cranium is characterized by a bun-shaped protrusion, the brows are heavy, the forehead slopes, and no chin is evident. Insofar as these differences are structural, they concern anatomy; insofar as they are functional, physiology. Changes in structure and function, however, also entail changes in shape, patterns of change beyond the disciplinary scope of anatomy, or physiology. Until the second decade of the twentieth century, there was no inkling of a science whose task was this subject, the shape of living things as they develop and evolve; until the last half-century, no science existed adequate to the precise analysis of changes in shape incident on development and evolution. In short, there was no geometric morphometrics, no happy combination of the mathematical and the visual. That at each stage this evolving science needed to be depicted as well as described gives me my subject, an analysis designed to deepen our understanding of scientific visuals as their practitioners struggle to live within, to exploit, and eventually to overcome the limitations of the printed page.

I begin my story in the second decade of the last century with D'Arcy Thompson, the first to intuit that evolution and development could be depicted by a fusion of the visual and the mathematical. This fusion he depicted in the form of vivid distortions in Cartesian grids. His idea was, however, an intuition only; because of the limitations of these grids, he was unable to turn his insight into a mathematically sound science of shape. Our story continues in the 1930s, a decade in which two separate attempts were made to create a science of shape. Both failed, though for very different reasons. Julian Huxley rightly insisted that in determining changes in shape real measurements only be employed. But his

---

[1] *On Growth and Form*, 1917, 740. All subsequent citations are to the 1917 first edition unless otherwise indicated.

mathematics fell far short of his ambition. He continued to graph changes in individual features, not realizing that these could not be isolated from the suite of which they were a part. Karl Pearson and G.M. Morant were more successful at their task. For the first time, to fix spatial relationships, they plotted landmarks onto shapes, realizing that only then could mathematical comparisons legitimately be made. Still, they failed to create a science, for two reasons: they lacked a true mathematical theory of shape, and they were forced by the limitations of the printed page to misrepresent a three-dimensional science in two dimensions. Nonetheless, theirs was a significant advance, one that lacked influence simply because it appeared in a specialist publication and seemed, in any case, to focus only on a minor historical puzzle concerning the identity of the Wilkinson Head. But Pearson and Morant managed to turn that puzzle into a major exploration of the basis required to create a science of shape that fuses the mathematical and the visual. Thus their work forms a bridge from Thompson and Huxley, from a morphemics that is potentially a quantitative science, to one that is fully.

Finally, I come to Fred Bookstein, F. James Rohlf, and their followers; I come to the present and to a true science, a quantitative endeavor that measures evolutionary and developmental changes in shape, employing to do so landmarks anchored firmly in three-dimensional mathematical space. This new science appears in two new visual forms: on the page, thin-plate splines, a far more sophisticated version of D'Arcy Thompson's grids, one that takes exact measurement into consideration. And more significantly, on the Internet, it appears in three-dimensional mathematical reconstructions, a visual apotheosis so vivid that it is easy to forget the wholly theoretical and mathematical origin of these images.

D'Arcy Thompson: Morphology Meets Geometry

There is no disagreement among the generations of his followers that the science of shape was inaugurated by D'Arcy Thompson in *On Growth and Form*, a book so important that it remains in print and is cited to this day, nearly a century after its first publication. This credit does not deny predecessors: Georges Cuvier is a brilliant and well-known example. But in a chapter virtually unaltered from the book's first to its final edition, Thompson provided one visual demonstration after another that dramatized the formidable power of geometry when applied to the changing shapes of growing and evolving creatures. Perhaps the most famous of his images—reproduced in Huxley and in Bookstein—superimposes a Cartesian grid on the shapes of two fish related by evolution, *Didion* and *Orthagoriscus mola* (Huxley 105; Bookstein, *The Measurement* 69). In Thompson's representation of *Didion* and *Orthagoriscus*, the vertical coordinates of *Didion* have become for *Orthagoriscus* a system of concentric circles, the horizontal coordinates a system of curves (see Figure 1.1). This transformation allows us to *see* evolution in geometrical terms.

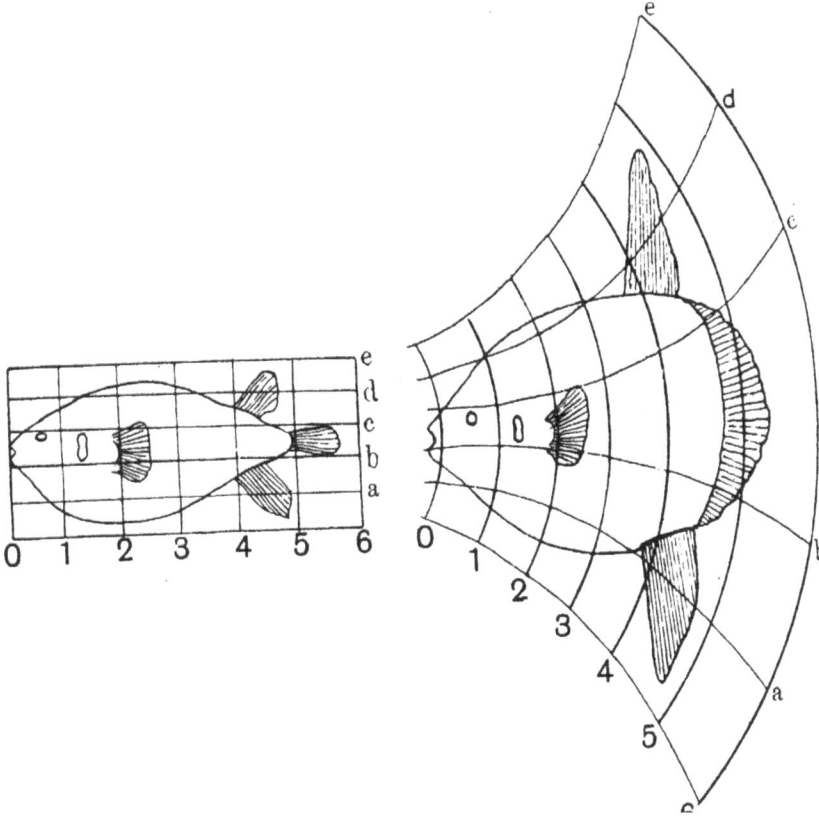

Figure 1.1   A Cartesian grid applied successively between *Didion* and
              Orthagoriscus mola
Source: Thompson, 1917, p. 751.

In his text, Thompson emphasized a crucial point about these images: development and evolution must be represented as integrated networks of change. For example, contemplating the differences in the shape of skulls of humans, chimpanzees, and baboons, Thompson noted "it becomes at once manifest that the modifications of jaws, brain-case, and the regions between are all portions of one continuous and integral process" (771). It is a process of progressive change analogous to the distortion of fossils under the influence of geophysical forces: "a simple and homogeneous transformation, such as would result from the application of some not very complicated stress" (761).

Thompson also applied these principles to evolutionary change. In Figure 1.2, reading from bottom to top, we see on the left conjectural reconstructions in the evolutionary tree of the horse; on the right we see the already discovered fossil heads of these horses. While in general the extent of agreement is remarkable,

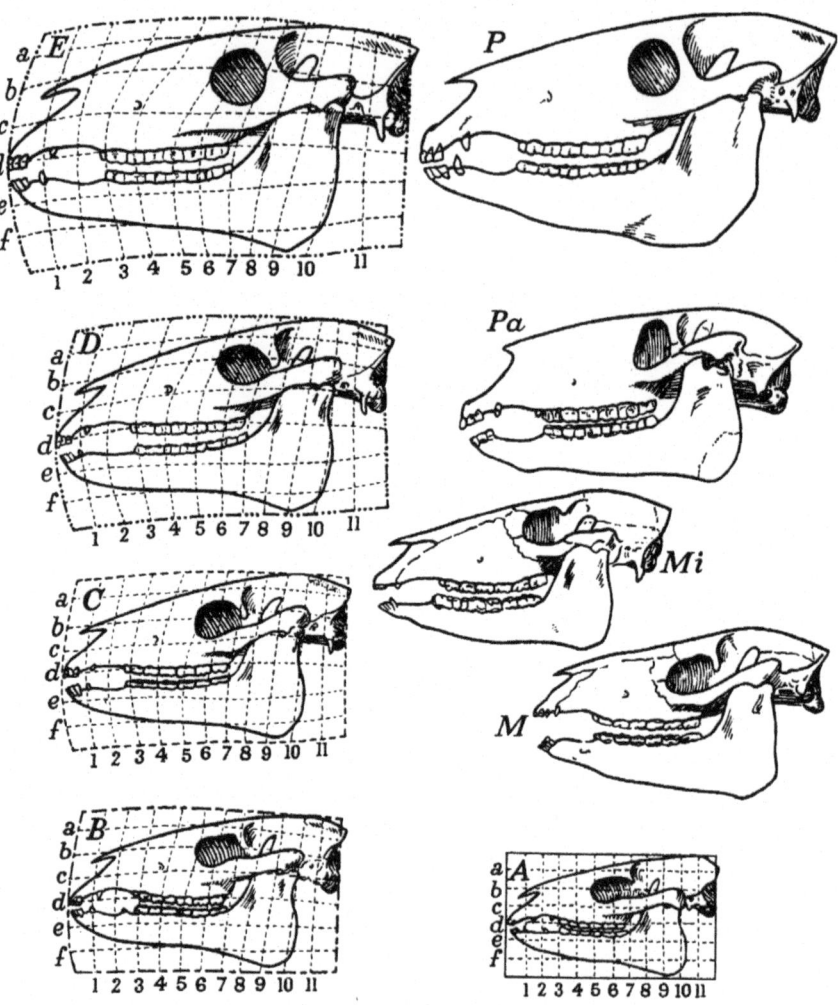

Figure 1.2   Conjectural fossil horse's heads in grids compared with
                  actual fossils
Source: Thompson, 767.

there is one case where it is not, *Pa*, *Parahippus*. But from this difference
between *Pa* and its contrasting conjectural image *D*, Thompson does not draw the
conclusion that his method has shortcomings. Rather, "though some writers have
placed *Parahippus* in the direct line of descent between *Equus* and *Eohippus*, we
see at once that there is no place for it there, and that it must, accordingly, represent
a somewhat divergent branch or offshoot of Equidae" (Thompson 768). This

conjecture has since been verified. Further emboldened, Thompson considered human ancestry. He found that as a consequence of his method there can be no common line of descent between *Homo sapiens* and *Neaderthals*: rather "among human and anthropoid types, recent and extinct, we have to do with a complex problem of divergent, rather than of continuous, variation" (Thompson 772). This conjecture has since been verified.

Thompson was well aware of some of the limitations of his method. First, it was not easily quantifiable. He pointed to:

> the simple fact that the developing organism is very far from being homogeneous and isotropic, or, in other words, does not behave like a perfect fluid. But though under such circumstances our co-ordinate systems may no longer be capable of strict mathematical analysis, they will still indicate *graphically* the relation of the new co-ordinate system to the old. (736)

He recognized further that even this result depends on the presence of a pattern of landmarks, a pattern of biologically significant locations he could not reliably supply, those that *fix* the degree of change over time. Even so, Thompson seriously underestimated the height of the barrier to the next logical step in his research program, the analytical unwieldiness that results when the method of coordinates is applied to actual measurements of networks of growth. It is this difficulty that a pair of images from Sokol and Sneath dramatize. It is clear from Figure 1.3 that the cat genus *Dasyurus* is incommensurable with the possum genus *Phalanger* (see also Bookstein, *The Measurement* 77). Thompson's simple geometric transformations simply cannot bridge the gap. His images inspired future generations of scientists by posing a problem in the form of a solution, a solution that required a mathematics far more sophisticated than he could muster.

Julian Huxley: Quantifying Morphology

Huxley's 1932 *Problems of Relative Growth* acknowledged the ground-breaking work of Thompson: "the coordinate method" is "of the utmost importance as affording a graphic and immediate proof of the need for postulating regularities in the distribution of growth throughout the body" (106).[2] Moreover, Huxley concurred that the recognition of his own innovation, growth gradient patterns proceeding from centers of growth, is "implicit in his [Thompson's] Cartesian transformations" ("Appendix" 292). Nevertheless, Huxley felt, Thompson's methods were undermined by a defect so serious that a new beginning is necessary, one that took seriously into consideration differential growth within bodies over time, "the change of relative proportions with absolute size" (106). Thompson's method "is static instead of dynamic, and substitutes the short cut of a

---

[2]   All citations of Huxley are to the 1932 edition unless otherwise noted.

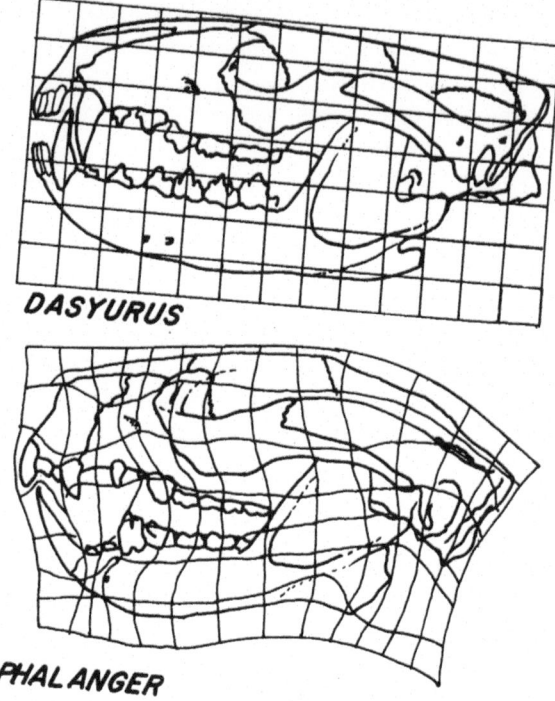

Figure 1.3   *Dasyurus* into *Phalanger*
Source: Bookstein, 77.

geometrical solution for the more complex realities actually underlying biological transformation" (106). Huxley was determined to map patterns of change of shape mathematically and to represent them graphically.

Figure 1.4 illustrates Huxley's point about the relative differential growth his mathematics is designed to capture. In Figure 1.4, the dashed lines represent the base-line of growth of the cephalothorax, the fused body and head of the hermit crab. The male gradients are marked by solid lines; the female, by diagonal filler. In the left side of the male, we can see a growth center at the third pereiopod (leg). We can also see that the male growth-coefficient diminishes both anteriorly and posteriorly. In the case of the first maxilliped (leg), the growth coefficients of the body and the leg virtually coincide. Anterior to this maxilliped, the growth-coefficient of the appendages is less than that of the body. In the case of the female, we can also see that the growth coefficient is less than that of the male, with the exception of the uropod, the final set of appendages.

The differential growth rates of the various organs of the hermit crab are typical in that they are described by the general formula: $y = bxk$, where y is the magnitude of the differentially growing organ; b and k are constants; b has

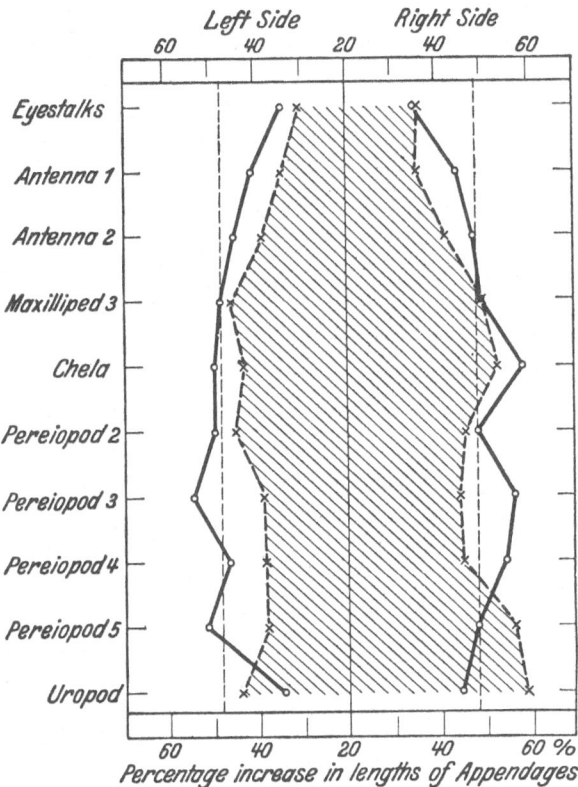

Figure 1.4   Growth profile of the hermit crab. Growth-gradients along the body of male and female hermit-crabs (*Euparugus prideauxi* [*Parugus prideaux* Leach 1815]). The male gradients are marked by solid lines; the female, by diagonal filler. The numbers give percentage increases of appendage length for a given percentage increase of body length, marked as dotted line

Source: Huxley 112.

no biological significance: it merely denotes the value of y when $x = 1$. But the constant k does have biological significance: "it implies that, for the range over which the formula holds, the ratio of the relative growth-rate of the organ to the relative growth-rate of the body remains constant" (Huxley 6). The formula can also be written log $y = $ log $b + k$ log x. In this form, the curve that y describes may be plotted as a line graph in which x and y are the abscissa and ordinate. Figure 1.5, for example, describes the relative growth of the chela or claw of the prawn *Palaemon malcomsoni*. Since the chelae increase in size relative the growth as whole, they are said to be "positively heterogonic" (Huxley 17).

To Huxley, it was this relative growth rate that has taxonomic and, therefore, evolutionary significance (208). According to this formula, we would expect to "find heterogonic organs varying in their relative size with absolute size of body in different-sized species of the same genus, or different-sized genera of the same family, just as in different-sized individuals of the single species. Our supposition is confirmed: this does occur. It is not universal or inevitable, but it does occur frequently" (Huxley 213).

This tendency is obvious in the four species of adult males of the genus *Golofa* depicted in Figure 1.6.

According to Huxley, these changes are controlled by rate-genes, a phenomenon we can trace in Figure 1.7 in the development of eye-pigmentation in *Gammarus chevreuxi*, a crustacean (228). On the left is a scale of eye-pigmentation from pure red up to black; at the bottom is a scale of weeks from the time of hatch, the dotted line, to and just beyond sexual maturity at eight weeks. At the center is a black-eyed male. The curve A-A is for the black-eyed *G. chevreuxi*, black-eyed from birth. Its pigmentation is controlled by the gene set *BBCCww*. Curves B-B and C-C are two forms of the red-eyed *G. chevreuxi*. In both, pigmentation is controlled by the gene set *bbCCww*, but rapidly darkening varieties, represented by B-B, also possess the dominant rate-gene *S*, while the slow-darkening mutants possess only the recessive rate-gene *s*.

We can see an analogous phenomenon in the case of human chest-girth as depicted in the graph in Figure 1.8. In it, the curve, a line coincident with the rear of the three-dimensional depiction, declines up to age 13, rises again, and then falls. Its "general shape … indicates that we are dealing with interrelated rate-factors" (Huxley 232). But the three-dimensional representation, illuminated directly from the top, shows something different, a multimodality in chest-girth: "multimodality shows that there exist a few main genetic types, each corresponding to one sort of body-build and determined by a particular gene or set of genes influencing relative growth of chest-girth and/or stature" (Huxley 232).

In his original publication, Huxley's confidence in his mathematical approach was high. He said of his "law" of growth that "[i]t may (like Boyle's law) prove only to be an approximation, and to be capable of modification in certain circumstances; yet (again like Boyle's law) it may remain fundamental" (31). Further, he said of Ivan Schmalhausen's formula combining growth-quotients and relative mass-factors that "it is clear … that if his general argument is correct, the two factors here recorded will serve to give a complete description of the facts concerning the relative growth of an organ from its first inception" (147).

But in an appendix, first published in 1945, this optimism proved unjustified. Haldane had pointed out that Huxley's formula could not apply to organs whose parts exhibit differential growth rates ("Appendix" 270). In addition, Huxley's "assumptions that growth is essentially multiplicative and that changes in the rate of self-multiplication affect all parts of the body equally" could not be sustained ("Appendix" 275). In the end, Huxley had to concede that attempts to add an adjustment factor to his formula had been unsuccessful and the obstacles

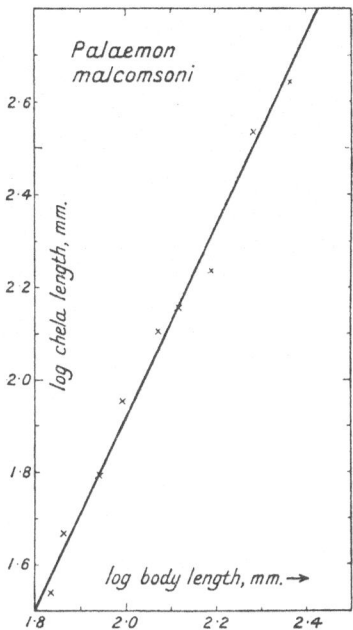

Figure 1.5   Relative growth of the chela (claw) in the prawn *Palaemon malcomsoni* [now *Macrobrachium malcomsonii* or Monsoon River prawn]. Logorithmic plotting
Source: Huxley *Problems* 17.

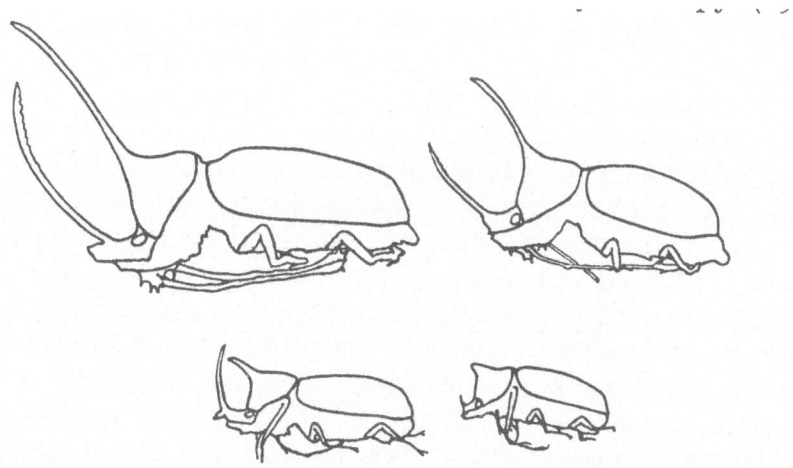

Figure 1.6   Above: *Golofa imperialis* and *Golofa caecum* [The name of this beetle is not present in contemporary taxonomies]. Below: two specimens of *Golofa porteri*
Source: Huxley, *Problems* 213.

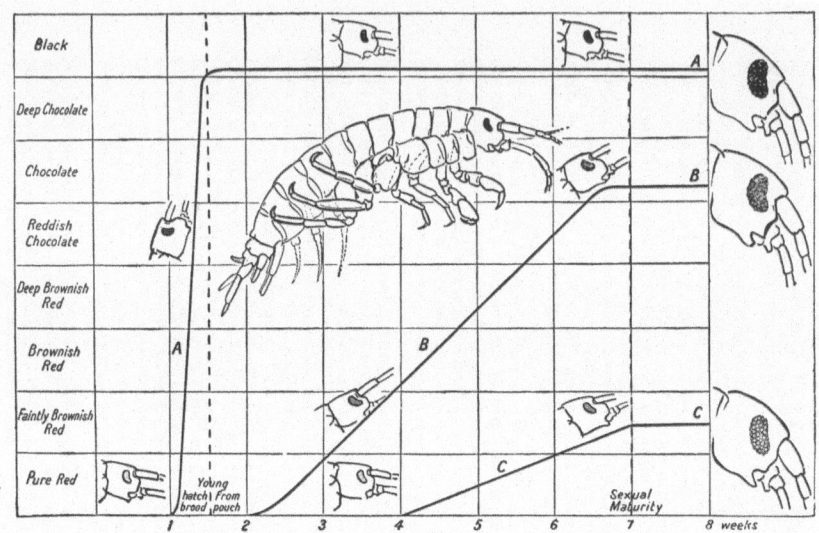

Figure 1.7   The effect of rate-genes on eye-pigmentation in
              Gammarus chevreauxi
Source: Huxley, *Problems* 228.

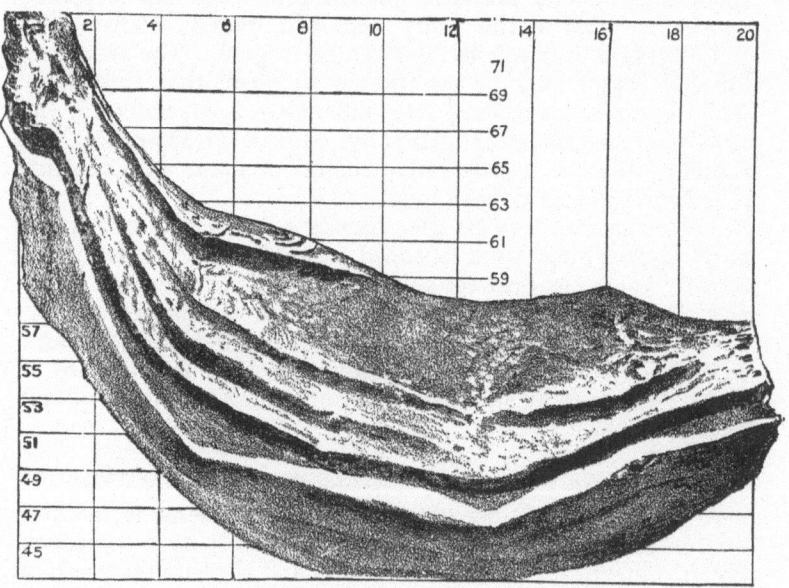

Figure 1.8   Ordinates mark human chest-girths as a percentage of stature; the
              abscissa gives the age in years
Source: Huxley, *Problems* 233.

encountered were unlikely to be overcome due to "statistical difficulties" ("Appendix" 299).

To these difficulties may be added those of measurement Huxley encountered in the absence of firmly established landmarks that would truly make patterns of differential growth exactly comparable. For example, in determining the relative growth of parts of the large claw of the fiddler crab, Huxley speaks of "the impossibility of assigning fixed points along the abscissa-axis to the several joints, since the very fact of their differential growth is causing their centres (or ends) to shift differentially with increase in absolute size" (81; see also 116).

The failure of Huxley's program is mirrored in the limitations of the graphs and images he selected to support his case. In the first place, despite his commitment to patterns of growth, the graphs we have seen show only relative growth in individual organs and body parts. A dramatic example of Huxley's problem is Figure 1.7: the quantification of a single trait is graphed, eye-pigmentation, while the drawings of Gammarus, however arresting, have no scientific value, as they add no new information by their presence. Indeed, in the article from which that graph is derived, no such images appear (Ford and Huxley, Figure 1, 69). Finally, though in numerous places Huxley acknowledged that growth occurs in three dimensions, only in a borrowed visual, reproduced in Figure 1.8, is a third dimension represented, a dimension that gives us additional information that the graph line occludes: chest-girth is multimodal; it clusters into distinct sub-types (96, 98, 103, 232). This is a graph that tries, but cannot overcome the limitations of the printed page. A new form of visualization that fuses images and their quantification had yet to be created. Karl Pearson and G.M. Morant will take the next step in the direction of a visual that represents the mathematization of patterns of change of shape.

Karl Pearson: The Beginnings of a Science of Shape

It is easy to see why Pearson and Morant's 1934 *Biometrika* article, "The Wilkinson Head of Oliver Cromwell and its Relationship to Busts, Masks, and Painted Portraits," should have had no influence on the development of the science that would eventually be called geometric morphometrics. The article's eccentric purpose, encapsulated in its title, seals its fate. This general neglect, however, says nothing about the level of insight Pearson and Morant manage to achieve into a science of shape fully capable of mathematization. When the article surfaced again in a 1998 review article, it received the highest praise: "Had Pearson gone on to supplement this splendid analysis with a distribution theory for such comparisons of vectors, instead of merely asserting their identity, he would have invented the entire core of contemporary morphometrics in 1935, and I would be out of a job" (Bookstein, "A Hundred" 16). This courteous hyperbole signals Bookstein's realization that, despite any appearance to the contrary, the article's purpose is far from eccentric. In fact, Pearson and Morant managed to turn a minor historical puzzle concerning the identity of the Wilkinson Head into a major exploration of the basis required to create a science of shape that fuses the mathematical and the visual.

What is the Wilkinson head? At the Restoration, Oliver Cromwell's body was disinterred and decapitated. Subsequently, his head was pierced with a spike and placed on public view along with the heads of two other leading regicides. The Wilkinson Head purports to be Cromwell's head. Preliminary testing yielded encouraging results: Cromwell's hat and helmet could have fitted such a head, and an examination of the skull indicated its age as roughly with that of Cromwell when he died. But the crucial tests that transformed possibility into probability were comparisons of the Head with the many existing portraits, busts, and death masks of the Protector. As it turned out, comparison with the portraits posed two difficulties. First, orienting the Head so that it was in the same position as that with which the painter viewed his subject turned out to be an almost insurmountable challenge. Second, "if a solid head and a mask or bust are seen from one position to be in corresponding aspects, they will appear to be so from other positions; but this is not so in the case of a solid head and a flat drawing or portrait" (Pearson and Morant 85). It is for these reasons that the attempts to fit the portraits to the Head are abandoned in favor of a comparison with busts and death masks. The comparison between the Head and the British Museum death mask in Figure 1.9, for example, is visually compelling.

Pearson and Morant summed up the results of their qualitative investigation: "we find that even without measurement, but simply by superposition, there exists a very remarkable accordance between the masks and busts of Cromwell and the Wilkinson Head" (100). This striking accord can be clearly seen in Figure 1.10: "the fit may ... be called good. The cincture comes, as it should, under the forehead bandage, the eyebrows, eyes, nose, upper lip and chin are in reasonable position; only the broken edge of the lower lip is too high. On the profile the limits of the embalmed flesh are good, but the allowance for flesh on the lips and chin is too great" (99).

It is important to see how well this purely visual accord survives the imposition of actual measurement, lest our eyes deceive us into giving simple coincidence more credence than it deserves. Figure 1.11 shows us eight "points," eight anchors in a network of vector measurements that will turn the subjective impression of coincidence into its mathematical equivalent, that will transform a pair of three-dimensional objects—figures 1.9 and 1.10, the Head and the British Museum death mask—into a pair of objects wholly constituted by mathematics and completely amenable to precise quantification (Pearson and Morant 97). Each of these eight points is exactly defined. Consider for example the following:

> *Glabella.* We use this term in a special sense. Let a plane be taken parallel to the Frankfurt horizontal plane and tangential to the upper border of the eyebrows. The point in which the trace of this plane on the forehead meets the 'midsagittal'

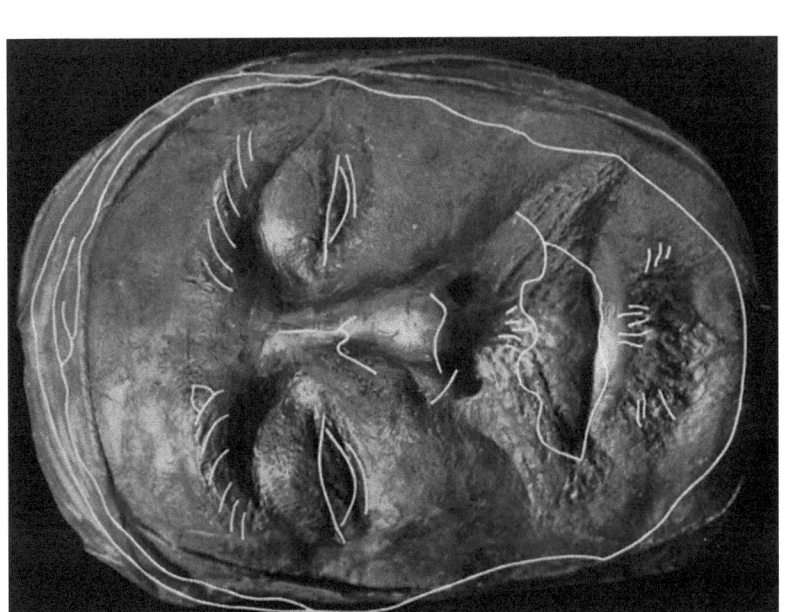

Figure 1.9   Full face of the British Museum Wax Death Mask fitted with the facial outlines of the Wilkinson Head

Source: Pearson and Morant, between 104 and 105.

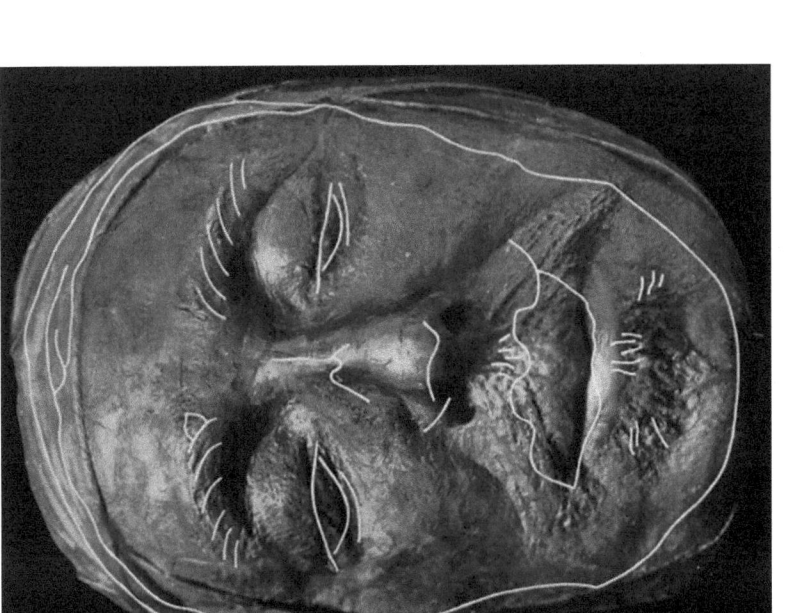

Figure 1.10   Profile of the British Museum Wax Death Mask fitted with the profile and the average flesh allowance of the Wilkinson Head

Source: Pearson and Morant, between 96 and 97.

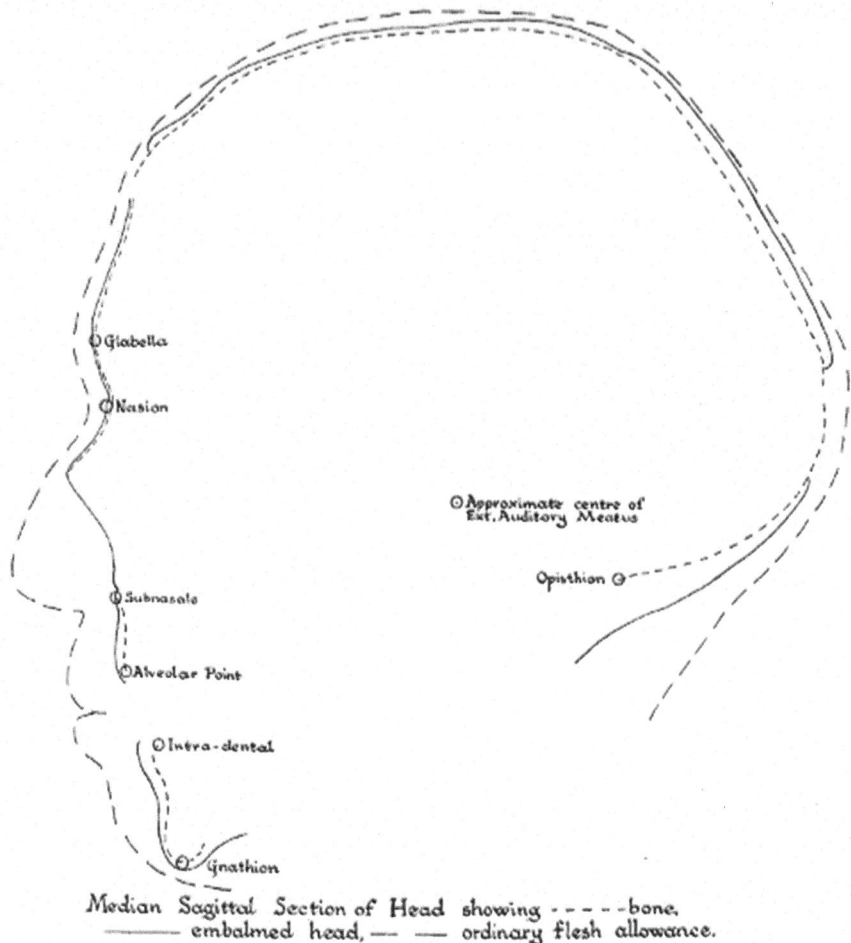

Median Sagittal Section of Head showing - - - - - - bone, —————— embalmed head, —— —— ordinary flesh allowance.

Figure 1.11   Median Saggital Section of Head showing bone, embalmed head, ordinary flesh allowance
Source: Pearson and Morant, between 104 and 105.

plane will here be termed the *glabella*. The point thus defined is easily determined on full face pictures or photographs. (Pearson and Morant 101)

While these points fall short of homology—while they are not taxonomically or evolutionarily significant landmarks—they give the comparison Pearson and Morant want to make a far firmer base than the measurement-averse grids of Thompson or the single-trait graphs and qualitatively differentiated images of Huxley. This assertion in no way minimizes the difficulty of pinpointing these "landmarks" with the desired accuracy. Indeed, Pearson and Morant rightly preferred to call their "values numerical appreciations rather than measurements" (103). Still, the Pearson and Morant appreciations issued in a summary table of remarkable coincidences force us to agree with them that "the accordance between the mean of the masks and the busts and the Wilkinson Head is astonishing," a conclusion that leads to the inference that "we appear to have what Bernoulli and Buffon would have termed a 'moral certainty' that the Wilkinson Head is the actual head of Oliver Cromwell" (105, 106; see Figure 1.12).

Characters	External Ocular Distance	External Orbital Distance	Length of Mouth	Nasion to Lip-line	Nasion to Wart Centre	Nasion to Gnathion	Nasion to lowest point of "beardlet" root	Nasion to Subnasal Point	Wart Centre to Right External Lid-meet	Wart Centre to Right External Orbital Margin
Mean Masks and Busts	98·7	113·75	58·9	75·2	28·05	122·3	95·9	53·8	48·9	52·4
Wilkinson Head	96·6	112·3	59·5	76·2	27·2	117·7	98·8	53·3	49·3	51·1

Characters	Wart Centre to Left External Lid-meet	Wart Centre to Left External Orbital Margin	Glabella to Subnasal Point	Glabella to Lip-line	Glabella to lowest point of "beardlet" root	Glabella to Gnathion	Subnasal Point to lowest point of "beardlet" root	Breadth of Nose, without alae	Interpupillary Distance
Mean Masks and Busts	74·9	81·2	71·9	96·65	121·9	139·6	48·25	32·2	73·55
Wilkinson Head	74·4	81·3	71·7	98·1	125·8	135·7	48·4	32·0	70·7

Figure 1.12   Final Comparison of the Wilkinson Head with Masks and Busts
Source: Pearson and Morant 105.

The significant conceptual advance this article represents in the calculation of comparative differences in shape cannot be matched by Pearson and Morant's visual realizations. Although their superimposed images powerfully convey the impression of likeness, they cannot convey its exact degree. Because Pearson and Morant had no way of conveying quantitative similarity visually, they must resort to tables. Additionally, while Pearson and Morant were the first seriously to address the problem that their objects of study were solid, that they existed in three dimensions, the limitations of the printed page forced them to represent those dimensions as two: in their case a view of the frontal and saggital planes, full face, and profile.

Geometric Morphometrics: The Shape of Development

"In ordinary language," Bookstein says, "the *shape* of an object is described by words or quantities that do not vary when the object is moved, rotated, enlarged, or reduced" ("A Hundred" 20). But a science of shape must go beyond ordinary language. Geometric morphometrics, pioneered by Bookstein and Rohlf, depends for its development on a mathematical theory. In the words of David Kendall, on whose mathematics of shape Bookstein replied, "if we are not interested in the location, orientation or scale of the resulting configuration, then we find ourselves working with a continuous stochastic process describing its change of shape" ("The Diffusion" 428). The application of this mathematical theory of shape to biology depends on the location of landmarks, biologically significant features, preferably homologous, that is, of taxonomic and evolutionary significance. It also involves the selection of so-called semi-landmarks, anchored in landmark locations, so that curves, necessarily free of landmarks, such as those on the skull, can be mapped. Changes in relative location of the entire network of landmarks and semi-landmarks track real change, that is, differences where any changes due to the orientation, enlargement, reduction, or movement of the object in question are factored out.

Geometric morphometrics completes the program Thompson initiated: "Differences in shape represented in this fashion are a mathematically rigorous realization of D'Arcy Thompson's (1917) idea of transformation grids, where one object is deformed or 'warped' into another" (Adams, Rohlf, and Slice 5). While Rohlf warned his fellow morphometricians about the difficulties and dangers of relying on morphometric analysis to make inferences of taxonomic distance, he was clear that work currently underway "represent[ed] a first step toward addressing what is a fundamental problem in morphometrics—the complete modeling of development and evolution" (Rohlf and Marcus 132; Adams, Rohlf, and Slice 13). The final section of this chapter focuses on the ways geometric morphometricians have learned to visualize changes in shape incident on this development and evolution. The first article with which I shall deal, "Fetal and Infant Growth Patterns of the Mandibular Symphysis in Modern Humans and Chimpanzees (*Pan trogolodytes*)" by Michelle Coquerelle, Fred L. Bookstein, José Braga, and Demetrios J. Halazonetis is a study in comparative development that focuses on a character peculiar to humans, the fusion of the two jaw-bones that form the chin. The article will "visualize how these two species [humans and chimpanzees] develop a vertical symphysis during fetal life and how this configuration changes towards a prominent chin in humans but towards an anteriorly inclined symphysis in chimpanzees" (Coquerelle et al. 509). The establishment of landmarks makes possible the creation of scattergrams and of 3-D representations of shape in print and in video.

Figure 1.13 consists of three representations of the right half of the human mandible age one year with landmarks and semi-landmarks designated over three dimensions. The landmarks are biologically significant locations; the curves are extrapolations from these landmarks. At the top is an outline diagram that displays all of the landmarks and curve semi-landmarks in X-ray view; it is as if a wire were bent in the form of a hemi-mandible. Below this are realistic depictions of the two sides of the hemi-mandible: the middle depiction turns this jaw bone 180° counterclockwise; the bottom depiction turns the middle one 180° further in the same direction.

In all of these cases the third dimension is an illusion. In the first case this illusion is created by overlapping; in the second by shading and foreshortening. Adams, Rohlf, and Slice make the essential point:

> A second, more fundamental, limitation is that of publication. Scientific papers are still largely limited to static, two-dimensional pieces of paper or computer screens that make the representation of volumetric changes challenging to say the least. [In these figures,] one can appreciate, but not fully comprehend the volumetric differences being represented. (11)

In these representations, information is omitted or distorted (Zelditch et al. 393). For example, in the top diagram in Figure 1.13 almost all information of shape is omitted. The middle representation occludes landmark 8; the bottom, landmarks 6 and 7. The middle and bottom representations also distort through foreshortening. Finally, all three representations omit most information concerning the shape and position of the deciduous teeth. Adams, Rohlf, and Slice foresaw the eventual solution: "As more journals go online one can anticipate the possibility of interactive publication graphics that would partially address this problem" (11). A problem for this chapter is that, although this possibility is being realized on the Internet, the evidence for this realization is hidden behind journal pay-walls.

Figure 1.14 is a scattergram in which three dimensions of the data are represented by principle components—PC1, PC2, and PC3—three measures of variance derived from the landmark data. The top line in the graph (in red in the online version) represents chimpanzees; the bottom line (in blue in the online version) represents humans. Along the horizontal axis, PC1, size is represented; along the vertical axis, PC2, certain aspects of shape, such as the lack of chin in chimpanzees.[3] The cluster of dots at the far end of the bottom line (blue in the

---

[3]   "PC1 (94.5% of the total Procrustes form variance [referring to the superimposition of mathematical matrices]) expresses overall size increase as well as allometric shape changes [the relationship between body size and shape], while PC2 (3.2% of the total form variance) separates the species. This second axis depicts some of the typical morphological differences between the species: the anterior corpus of the chimpanzee is longer and without a chin, relatively narrower anteriorly, and the condyles [referring to characteristics of the jaw bone] are relatively closer to each other than in the human mandible. PC3 accounted for only 0.5% of the total variance and hence is not visualized by surfaces here" (Coquerelle et al. 511).

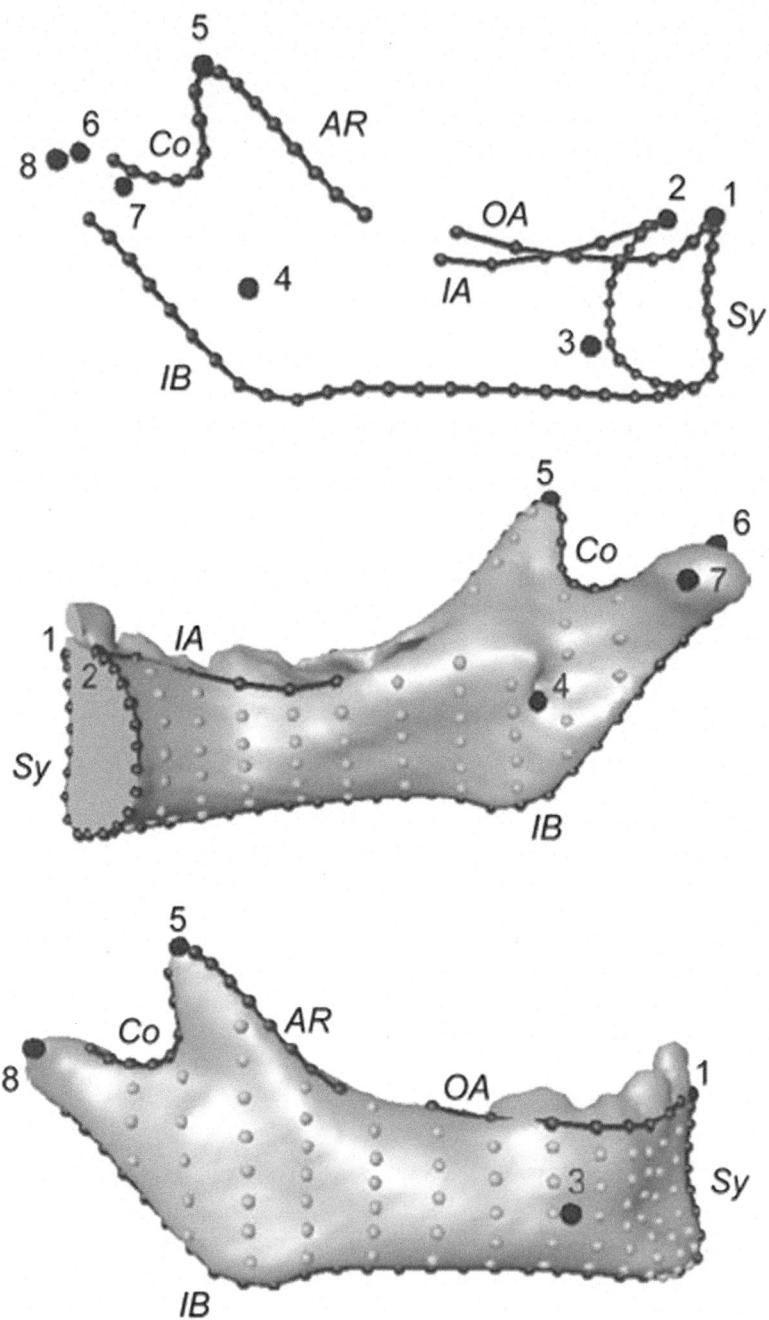

Figure 1.13   Three views of a human hemi-mandible, age 1 year
Source: Coquerelle et al. 511.

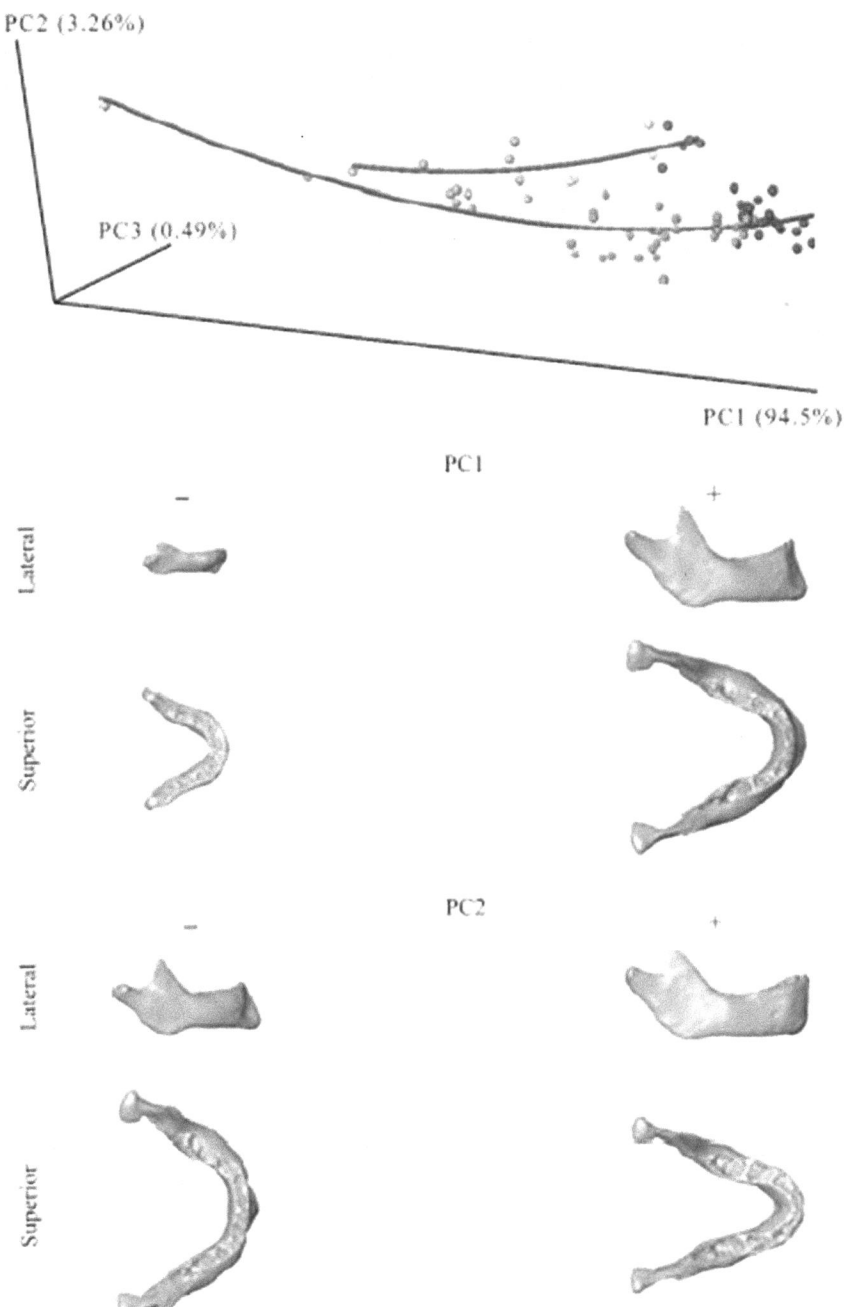

Figure 1.14   Scattergram comparing the development of the mandible in
humans and chimpanzees

Source: Coquerelle et al. 512.

online version) represent humans at DS2, dental stage two after the eruption of the M1, the first molar. These are further along PC1 than those of the chimpanzees and well below the chimpanzees in PC2. The representations below the scattergram are not those of actual chimpanzees and humans but of negative and positive extremes of PC1 and 2. These graphs differ from those of D'Arcy Thompson and Huxley in that they actually represent, though of course they cannot actually depict three-dimensional change.

Such representations are capable of correcting earlier work. In 1924, Bolk had published a study of human and chimpanzee mandibular development—of changes in the general shape of the human and chimpanzee jaw just before and after birth. Figure 1.15 consists of a series of 3-D representations of the development of the chimpanzee mandible. By means of these Coquerelle et al. were able to show that Bolk had been mistaken. Moreover, their results were not limited solely to analysis at the macro-level, the changing shape of the mandible as a whole. Notice the representation in lateral view at column D, row iv. The black line represents the floor of the tooth chamber, the arrows the direction of tooth growth at DS 1. Notice that the second tooth from the left, a canine, cannot erupt until the jaw lengthens at DS2.

In their representations of chimpanzee and human mandibles, Coquerelle et al. go one step further. Figure 1.16 is a screenshot from a 3-D video of the development of the human mandible, front and side view.

The verisimilitude of these 3-D representations is a striking realization of Thompson's dream, a coincidence between the mathematical and the biological. The achievement is indeed extraordinary, so extraordinary that we have to remind ourselves that figures 1.13 to 1.16 are *all* mathematical constructs, constructs hypothesized closely to resemble the real. In an email, Bookstein makes the point forcefully: "The heightened persuasiveness that often accompanies [their] increased visual clarity is often the scourge of the competent reviewer, whenever the power of the image overrides all the appropriate skepticism that we try to teach our students when facing this unfamiliar semiotics" (June 16, 2011).

Geometric Morphometrics: The Shape of Evolution

Up to now I have been dealing with development; now I move on to evolution. In "Cranial Shape Differentiation in Three Closely Related Delphinid Cetacean Species: Insights into Evolutionary History," Ana R. Amaral et al. apply geometric morphometrics to evolutionary problems. Amaral et al. use landmarks to create a graphic unique to geometric morphometrics, the thin-plate spline, which visualizes biological change as a deformation (see Zelditch et al. 25). Like the graphs in Michelle Coquerelle et al., these actually represent, though of course they cannot depict, three-dimensional change of shape (see figures 1.17 and 1.18).

Despite appearances, no third dimension is depicted in these thin-plate splines. The bulges inward and outward are visual metaphors for expansion and

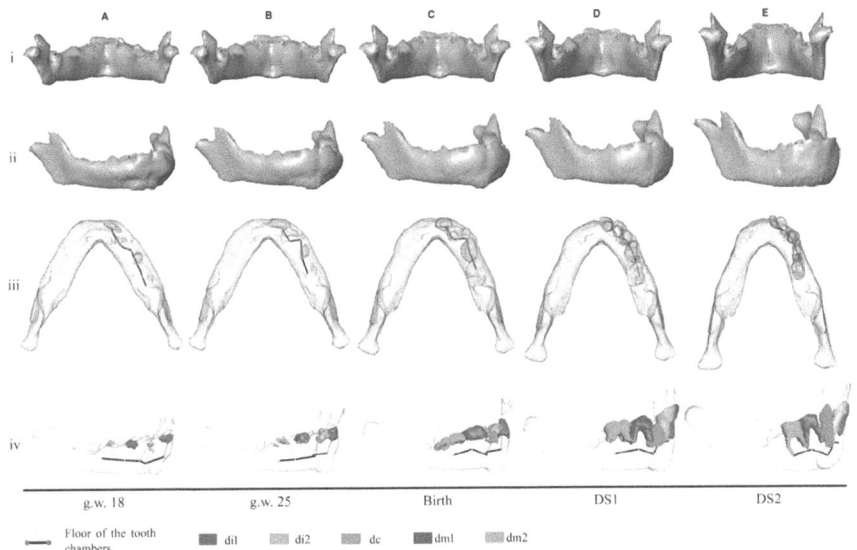

Figure 1.15   The development of the chimpanzee mandible from gestation week 18 to dental stage (DS) 2

Source: Coquerelle et al. 515.

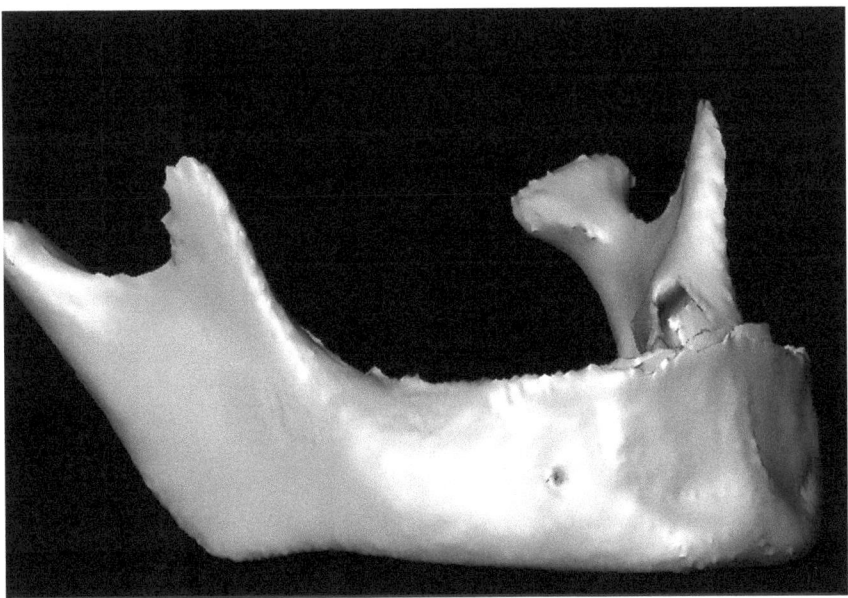

Figure 1.16   Screenshot from video of the development of the human mandible, front and side view

Source: Coquerelle et al., supplementary materials.

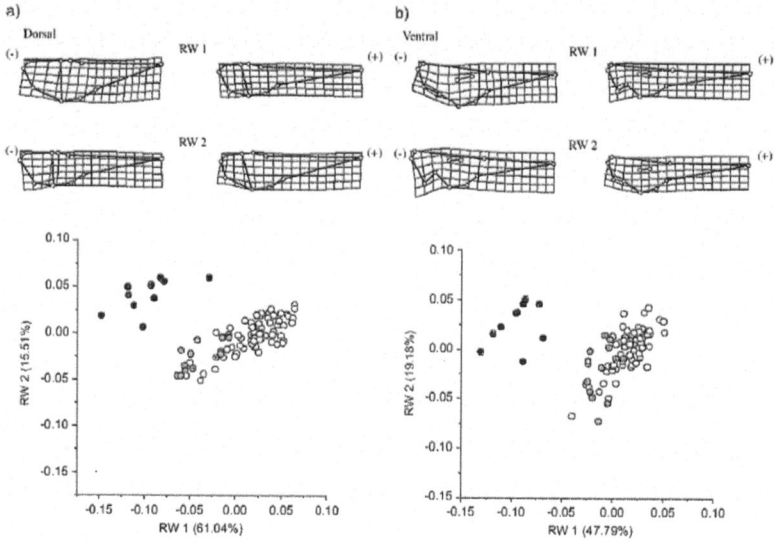

Figure 1.17   Thin-plate splines of the dolphin skull and the scattergrams on which
they are based. RW = relative warp analysis, a form of principle
component analysis "conducted in order to summarize the variation
in shape among the specimens in the sample" (Amaral et al. 40). In
each spline, landmarks are represented by open circles

Source: Amaral et al. 43.

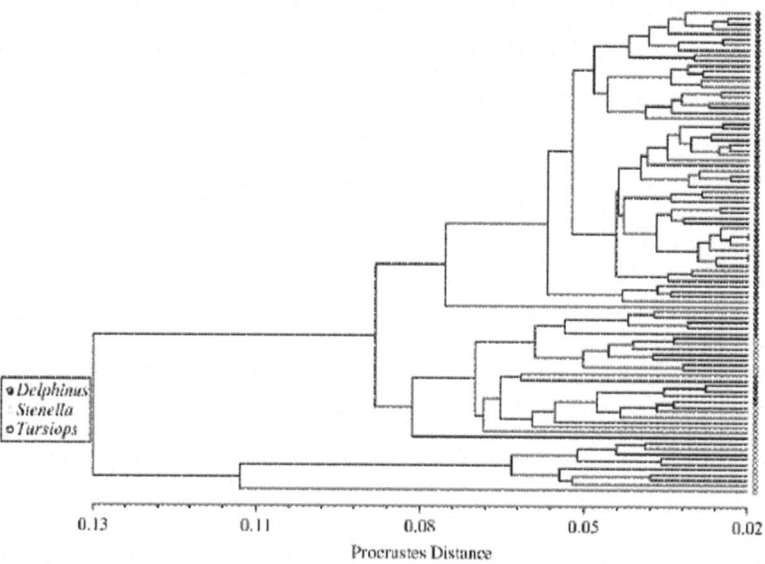

Figure 1.18   Phenogram of three dolphin species based on
geometric morphometrics

Source: Amaral et al. 43.

foreshortening in a plane. There are two spline sets for the dolphin skull: one for the dorsal view (from the back), the other for the ventral view (from the belly). In the case of each of these representations, the brain case is to the left, and the long rostrum or snout to the right. Within each set, the plus and minus views represent extreme values. The scattergrams show that one species, *Tursiops truncatus* represented by blackened circles, clusters toward the negative in RW1 and toward the positive in RW2. When we consult the relevant thin-plate splines, we see that this translates visually into a rostrum that is short compared to the cranium and a much larger brain case. When the dorsal results were plotted as a phenogram, a diagram depicting taxonomic relationships among organisms based on overall similarity, they "highlighted the close proximity of *D. delphis* and *S[tenella] coeruleoalba*, which appeared as sister taxon in a cluster separated from *T[ursiops]truncates*, which occupies a more dissimilar position" (Amaral et al. 41). Results from geometric morphometrics are largely corroborated by the findings of DNA analysis, suggesting that it is legitimate to make evolutionary inferences from morphometric data (Amaral et al. 44).

Similarly, in "Neanderthal Cranial Ontogeny and its Implications for Late Hominid Diversity," Marcia S. Ponce de León and Christoph P.E. Zollikofer compare the development of Neanderthal and human skulls to demonstrate that the two species belong to evolutionarily distinct lines of descent. In Plate 1, we see these differences dramatically presented, especially in *b* and *c*, where the left side in each case is Neanderthal, the right, human. In *b* and *c*, the lengthening and slimming of the human skull is vividly represented. In a video accompanying the article online, we see not only this lengthening and slimming, but also the emergence of the human chin accompanied by the disappearance of the Neaderthal "bun" at the back of the head.

Conclusion

This chapter has traced the visualization of biological shape from D'Arcy Thompson through Fred Bookstein and F. James Rohlf. In this development, there were two turning points: Thompson's insight that development and evolution could be realized geometrically in a visually striking manner, and the realization by Bookstein that Kendall's work on the theory of shape, a theory developed in the contexts of astronomy and archaeology, could apply to biology as well (Comment 222–6; Bookstein, "A Hundred" 19–20; see also Rohlf and Marcus). It is these turning points that led eventually to the intelligent and creative exploitation of the possibilities of representation in two- and three-dimensional space on the page and on the Internet: witness thin-plate splines and 3-D morphs of developmental and evolutionary trends. This exploitation, however, must be accompanied by a caveat: each of these visualizations has the peculiar character of looking like the real thing and of being at the same time a mathematical construct founded on a theory of shape.

Chapter 2

Florence Nightingale's Statistical Table for Hospitals: A Work of Utility and Art

Lee E. Brasseur

Florence Nightingale's hospital statistical table, which she designed to record data about patients in British hospitals, was a landmark in medical record-keeping. With this table, Nightingale hoped to help hospital medical professionals understand how best to treat sick and injured patients of varying ages. In fact, according to Charlotte M. Shell and Karen D. Dunlap, Nightingale, along with Dr. Ernest Codman, "may well deserve the credit for the birth of evidence-based medicine" (19–20). While tables had been around for centuries before Nightingale, her tabular "Hospital General Statistical Form" was unique in its day. Unlike her famous rose diagram, this table was not meant as a visual rhetorical tool, but rather as a visual *recordable* tool.

Nightingale's Rose Diagram

Most enthusiasts of the technical visual know Nightingale's most famous graphic—the rose diagram—that she designed from original tables of data to change the queen's and Parliament's actions regarding treatment of soldiers in wartime hospitals. Nightingale and William Farr's original report to this audience included both impassioned text and tables of data to focus attention on the appalling medical treatment of soldiers during the Crimean War. Unfortunately, this report languished with inaction for some three months. Consequently, Nightingale decided that the only way to get the queen's and Parliament's attention was to transform the tables into diagrams. These diagrams, shaped very much like the petals of a rose, showed how the size of the rose's "petals," representing deaths, decreased in size over time once Nightingale and her nurses instituted her cleanliness and health care measures. This transformation of data from a series of tables to a rose diagram resulted, finally, in real changes in the treatment of wounded and sick soldiers in wars. In this particular instance, Nightingale's skill at turning tabular data into data visualization resulted in real change for injured and ill soldiers (Spence, *Information* 6).

Nightingale's Statistical Table for Hospitals

In this chapter, however, I will not examine a visual device that Nightingale transformed from a table into a visualization, but rather her own design of a recordable table that doctors and nurses in British hospitals could use to record information about their patients: the "Hospital General Statistical Form" (see figures 2.1 and 2.2). Nightingale's own experience as a nurse in England and in the Crimea taught her the sad lesson that nobody regularly recorded information about patients, their diseases, deaths, recoveries, discharges and admittances. As a result, mistakes in treatment were made again and again.

To address this problem, Nightingale invented a tool for collecting data on patients in hospitals. It was not a visual tool that was intended for rhetorical effect, but rather a tabular form that doctors and nurses could fill out in order to understand frequencies and patterns of diseases and injuries, including their treatment. By using this form, medical personnel could better understand and apply consistent treatment patterns. Once in use, the table was designed to make it easier to note the number of patients with certain diseases or injuries at certain ages.

Though a table, this form did carry the same kind of visual power that her earlier rose diagram did, but it did it through its accumulation of marks. Nurses, doctors, and other health professionals would place these marks in a row within a certain column that was devoted to a particular disease or injury and associated with a patient's age. Readers could then study its marks and numbers and use this information to understand how to consistently treat similar kinds of disease and injury. With the information gathered in this way, hospital administrators, doctors, and nurses could more easily understand how and whether to treat different patients, what diseases were most diagnosed in which hospitals, and how to determine the best length of stay and treatment needed. Nightingale's hospital table, therefore, was not a rhetorical visual like the rose diagram but instead a visual *enabler*.

When Nightingale presented this table to the Royal Statistical Congress, she described it thus:

> Each Table is divided vertically into columns containing the ages in monthly and yearly periods from under 1 year to 5. Above 5 the ages are given quinquennially. The disease list is divided into two sections, one printed on the left hand, the other on the right of the sheet. The left-hand division contains the diseases most frequently admitted into hospital; the right-hand the rarer forms of disease.
> This arrangement is necessary for the purpose of limiting the size of the Tables. (*Notes* 160)

This new scheme, designed with the help of Farr, offered the basic categories of data that hospitals *should* collect: the number of patients in a hospital at the beginning and end of a year, the number of patients who had recovered or who had been either discharged as incurable or dismissed at their request, the number

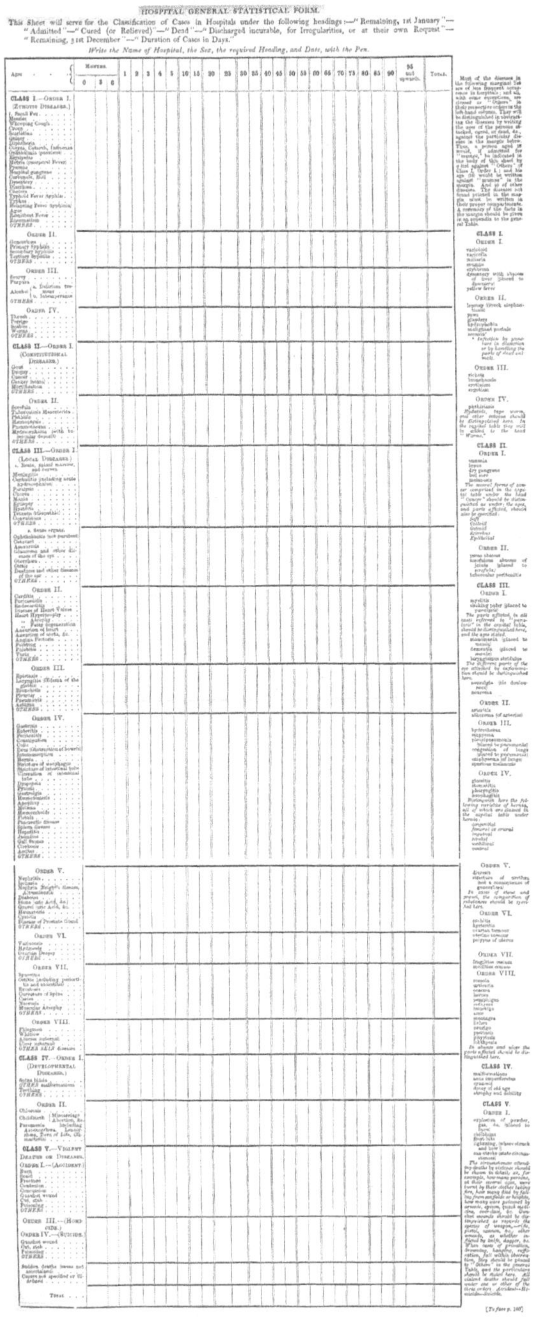

Figure 2.1   Hospital General Statistical Form designed by Nightingale
Source: Nightingale, *Notes* 161.

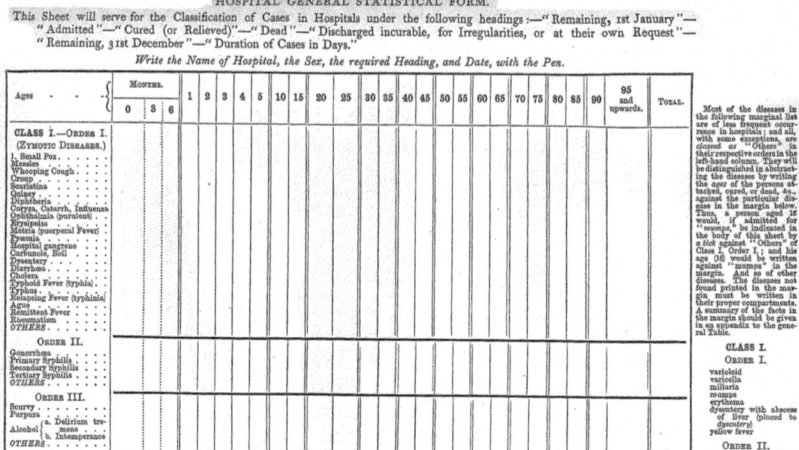

Figure 2.2   Close-up, Hospital General Statistical Form designed
                by Nightingale

Source: Nightingale, *Notes* 161.

of patients who had died, and the mean durations of hospital stays. The form contained 26 columns for entering the number of patients with specific conditions who were admitted, cured, dead, discharged incurable, or discharged at their own request, as well as the number of days hospitalized per case.

But before encouraging its use as a regular medical recording device, Nightingale had usability testing performed on it by having Guy's Hospital print a statistical analysis of patients from 1854 to 1861 and St. Thomas' Hospital from 1857 to 1860. She also asked St. Bartholomew's Hospital to print a table of its cases for 1860 alone (Street 105). Andrew Street notes that "Nightingale was aware of the technical problems of measurement and comparison, drawing attention to the complexities of risk adjustment, the possibility of data manipulation, the dangers of a narrow focus on a single outcome measure, and the potential perverse behavioral consequences" (105). This testing provided confidence that the form would work as intended.

When we consider the different kinds of information that could be entered in the columns and rows of Nightingale's hospital table by doctors and nurses, we can see why recording information in columns that were viewed as vertical and horizontal could create a more accessible means of collecting and gathering a considerable amount of key numerical information as it occurred. Without such devices, it would be difficult to learn how to diagnose, who to treat, what diseases and injuries to concentrate on, and other information hospital staff could learn simply from the ages and genders of previously admitted, sick, and dead patients. But during Nightingale's day, this level of regularity was unheard of. Thus, when submitting her paper to the Second Section of the Statistical Congress in 1860,

Nightingale wrote, "Up to the present time the statistics of hospitals have been kept on no uniform plan. Every hospital has followed its own nomenclature and classification of diseases, and there has been no reduction on any uniform model of the vast amount of observations which have been made in these establishments" (*Florence* 83). Accordingly, in her heading for the form, Nightingale explained that:

> The Sheet will serve for the Classification of Cases in Hospitals under the following headings:—'Remaining, 1st January'—'Admitted'—'Cured (or Relieved)'—'Dead'—'Discharged incurable, for Irregularities, or at their own Request'—'Remaining, 31st December'—'Duration of Cases in Days.' (Nightingale's "Hospital General Statistical Form," *Notes on Hospitals*, facing 160)

The categories she used were not, of course, chosen at random. It was well known that during the Crimean War, Nightingale would walk up and down the wards writing down the data as she saw it. Given the unfamiliarity of this practice of data recording, Nightingale had to provide explicit guidance on how to use the form. For example, she provided written directions that instructed hospital personnel to "*Write the Name of Hospital, the Sex, the required Heading, and Date, with the Pen*" since this was, at the time, not an accepted practice.

On the right-hand side of the form, she also provided a nomenclature for the names of the less frequently found diseases and injuries, subdivided into classes, and she gave instructions on how to record data on the less common conditions.

Most of the diseases and injuries in the following marginal list, then, are of less frequent occurrence in hospitals; and all, with some exceptions, are *classed as* "Others" in their respective orders in the left-hand column. As she instructed, they would be distinguished in abstracting the diseases by "writing the *ages* of the persons attacked, cured or dead, &e., against the particular disease in the margin below. Thus, a person aged 16 would, if admitted for '*mumps*,' have his condition indicated in the body of this sheet by *a tick* against 'Others' of Class I, Order I; and his age (16) would be written by 'mumps' in the margin, and so of other diseases. The diseases not found printed in the margin must be written in their proper compartments. A summary of the facts in the margin should be given in an appendix to the general Table" (Nightingale's "Hospital General Statistical Form," *Notes on Hospitals*, facing 160). Such an explanation is not unlike that of current health data systems. Indeed, although today standardization of disease and injury forms exists, some differences are still apparent because they are specific to a particular institution's or country's naming genres in *their* record collection.

Nightingale's Written Argument for the Table

In Nightingale's paper that was submitted to the Royal Statistical Hospital, which contained the first published version of the table discussed in this chapter, she argued its value based on the following points. First, Nightingale emphasized that, to that time, no British hospital had kept a uniform plan of statistics. As a result, each hospital had its own version, and it was difficult for medical personnel at different institutions to make wider conclusions. As Nightingale put it, "[t]he material exists, but it is inaccessible" (*Florence* 83). She then explained that the forms had already been tested at several hospitals at her urging, therefore enhancing her persuasive argument for the wider use of her recordable table throughout the country.

Nightingale went on to argue that the use of her tabular form "would enable us to ascertain the relative mortality in different hospitals, as well as of different diseases and injuries at the same and at different ages, the relative frequency of different diseases and injuries among the classes which enter hospitals in different countries, and in different districts of the same country" (*Florence* 83). Here, in particular, we see her precise understanding, based on her background as a nurse and trainer of nurses, of the kind of information most critical to collect. She backed up her argument by citing one of the results of her usability tests:

> For example, it was found that a very large proportion of the limited finances of one hospital was swallowed up by one preventable disease: rheumatism, to the exclusion of many important cases or other diseases from the benefits of hospital treatment. It has been shown most of the cases admitted to the hospitals, where the forms have been tried, belong to the productive ages of life, and not to the ages at the two extremes of existence. (*Florence* 83)

Here we see the scientist in Nightingale at work as well as the capable rhetorician. She understood her goal and knew that the table would need a supporting argument that members of the Royal Statistical Society could understand. She concluded her statement to the Society by writing, "The statistics of rare diseases and operations are still very imperfect, but by abstracting the results of such diseases and operations from the tables after a long term of years, trustworthy data could be obtained to guide future experience" (*Florence* 84). One of the most interesting words in this passage is "abstracting." Today, we do not typically think of data tables as abstracting data. Instead, we see them often as data in need of visualization. But, for the purposes of recording vast amounts of medical data, Nightingale's hospital tabular form was the kind of "abstraction" suitable for the situation. In other words, unlike a pie chart or a bar chart, the data were unknown at the beginning. It first had to be recorded, and her device—a table—was designed so that doctors and nurses could enter this information as it occurred. In a letter to Sidney Herbert from the Barracks Hospital in Scutari in 1855, Nightingale wrote,

> no statistics are kept as to between what ages most deaths occur, as to modes of treatment, appearances of the body after death, etc. etc. etc. & all the innumerable & more important points which contribute to making Therapeutics a means of saving life, & not, as it is here, a formal duty. Our registration generally is so

lamentably defective that often the only record kept is—*a man* [*sic*] *died*—on
such a day. (Goldie 97)

Here we see Nightingale, early on, recognizing the folly of treating patients and
not detailing a record about them. As each patient leaves—dead, alive, sick, or
healthy—medical personnel would not have the information to help best make
decisions about treatment of future patients who have similar complaints. As
Victor A. Ferraris and Suellen P. Ferraris point out,

> [a]lthough England had tracked mortality rates since the 1600s, the analysis of
> these rates was in its infancy during Nightingale's era ... Nightingale made the
> important observation that raw mortality rates were not an accurate reflection of
> outcome, since some patients were sicker when they presented to the hospital, and
> therefore would be expected to have a higher mortality. This was the beginning of
> risk adjustment based on severity of disease. (187)

As a result, in the right column of her 26-column hospital form, Nightingale
provided hospital personnel with an explanation for using the form. She specified
that the form allowed "for entering the number of patients with specific conditions
who were admitted, cured, dead, discharged incurable, or discharged at their own
request as well as the number of days hospitalized per case" (Dossey 224). She
aligned the table to aid hospital personnel so they could, for example, look up the
names of the diseases presented in the left-hand first row and then note the numbers
of people of a certain age who had any of those ailments. This information alone
could help hospital workers understand at what age certain ailments or injuries
were serious enough for hospital admission. As I described above, the form also
accommodated unusual conditions, as Nightingale had laid out in the explanation
on the right side of the form. Thus, much could be learned, such as at what ages
certain diseases were most likely to occur. As Barbara Dossey states,

> The development of these forms was based on Nightingale's belief that proper
> analysis of accurate statistical data could help identify the best treatments for
> certain illnesses and the optimal number of days needed for treatment as well as
> mortality rates from various causes in different hospitals. (224)

Nightingale also inserted her statistical table in her book *Notes on Hospitals* as
a three-page folded copy within its covers. Next, she strengthened her argument
by pointing out that "the authorities of St. Thomas's, University College, and St.
Mary's Hospitals" had used the forms and "accurately filled [them] up" (*Notes*
164). She then went on to explain why she had divided the diseases into two
columns—explaining that this was the direct result of her experience seeing how
actual hospital personnel worked in hospitals, especially when "filling up the
forms by different hospitals" (*Notes* 161). The results of this "user testing" were,
in fact, a considerable persuasive argument for the use of such forms, but the

embarrassment experienced by some hospital administrators at these "test sites," once the results were printed, made some hospital administrators reluctant to use them, for fear they might be looked at negatively. Nightingale, who was too ill at the time to attend the Statistical Society's meeting where her tabular form was presented, wrote that her forms "are intended solely for the tabulation of cases," but she also acknowledged the differing views in *Notes on Hospitals*, including the fact that the Statistical Congress thought that some other kinds of information should also be recorded (*Notes* 164). Thus, her scheme would ultimately add an annual tabular statement "not only of the total number of patients, but also of the total number of *cases of disease*, under the various heads," as well as other recommendations (Nightingale, *Notes* 164). In this way, more contextual information could be seen.

Nightingale considered this table to be part of "a series of model hospital forms that would ideally be used in all hospitals to provide uniform and consistent medical information about patients" (Dossey 224). This kind of medical information was gathered through Nightingale's considerable efforts. For example, before distributing her table to British hospitals for universal use, Nightingale wrote in a letter, likely to Sir James Paget, who was a member of the Royal College of Surgeons and a surgeon at St. Bartholomew's Hospital, to ask him what kind of information should be included in her form. As she wrote in a letter dated October 1, 1859, "I would ask you to do one great favour, viz., to send me a complete disease list, including surgical cases, such as would include all those who come into hospital at St. Bartholomew's" (*Florence* 80). Here we see Nightingale at her best: going to knowledgeable sources for the information she needed. Paget was well esteemed—a doctor to Queen Victoria and someone with whom she had previously corresponded. Nightingale continued to pursue this kind of careful research by revealing, in a letter to William Farr, dated January 31, 1860, that she had also requested disease statistics from St. Thomas Hospital, University College, and St. Mary's Hospital, in preparation for her own suggested hospital table (*Florence* 80).

Later, in a June 13, 1859 letter to Julius Mohl of the Woodward Biomedical Library at the University of British Columbia, she emphasized her continued call for creating forms that fit general standards. As Nightingale wrote, "Our Registrar-General means to draw up forms according to that note, and propose them at the next European Statistical Congress, which is to be held in London" (*Florence* 81). In other words, she wanted the forms to be as complete and usable as possible. This ease of use was paramount in her mind. We can see this in her letter to Mohl, where she stated her overall intent: "In order to make sure that the table would be useful, I wanted to make it simple to use, so that nurses and doctors understood easily how to enter facts on the forms" (*Florence* 81). When she finally finished the form, she wrote a paper to the Second Section of the International Statistical Congress in 1860, which was read by the Earl of Shaftesbury. In this letter she writes, in part:

> The proposed forms would enable the mortality in hospitals, and also the mortality from particular diseases, injuries and operations, to be ascertained with accuracy, and these facts, together with the duration of cases, would enable the value of particular methods of treatment and of special operations to be brought to statistical proof. (*Florence* 84)

As mentioned above, however, one of the major problems with the initial negative reaction from hospital administrators to her tabular forms was, of course, that they were published in the Statistical Society journal in a public forum and often reflected poorly on the hospitals, as is illustrated in her commentary where she wrote that her results "provoked opposition from those whose performance was the subject of public scrutiny" (Smith, *Florence* 105). According to Gillian Lancaster of The Royal Statistical Society, however, because Nightingale was aware of this reaction, as a result, when she wrote her final plans in her later book, *Notes on Hospitals*, she was careful to include wording that

> pointed out and recommended new information be gathered on the annual number of admissions, discharges, recoveries, as well as deaths, and the average duration of hospital treatment for different diseases by age and sex. She also demonstrated the greatly increased prevalence of communicable diseases among hospital nurses and attendants, compared to the normal population, and she advocated for the use of standard classifications of diseases and a set of hospital statistical forms for data collection, which were widely circulated. (Lancaster 103)

One of the major problems at the time was, according to Dossey, that "hospitals in the mid-1800s had no scientific or standardized approach to gathering data on admissions and discharges, and each hospital followed its own system of disease nomenclature and classification" (224). Obviously, this led to a great loss of information, as there was no consistent form used for the collection of different hospitals' data over time. Nightingale's table, then, was a first in providing a form that could record data in different hospitals all over Britain in the 1800s. As Lynn McDonald writes, Nightingale understood that the mortality in the war fields of Scutari alerted Nightingale to the possible harm that could be incurred, for example, by "bad hospital design or location" (*Florence* 75). According to Jocelyn Keith, who wrote about Nightingale's systematic classification of diseases, Nightingale did so with guidance from renowned statistician Adolphe Quetelet and, in collaboration with William Farr, had actually developed the first model for the systematic collection of hospital data, using a uniform classification of diseases and operations that was to form the basis of the ICD code used today (Keith 148). She goes on to state that Nightingale's form sets out the basic categories of data to be collected: the number of patients in hospital at the beginning and the end of each year, the number admitted during the year, the number who had recovered, the number discharged either as incurable or at their own request, the number who had died, and the mean duration of their hospital stay (Keith 148).

Today, when we think of a patient in a hospital, we know that a record is being kept of the patient's name, age, ailment, or complaint and that a detailed record of his or her treatment is regularly collected. Nightingale wanted hospital personnel to be able to collect, combine, and analyze data so they knew how to treat future patients. In some ways, the impact of Nightingale's tabular forms was not somewhat unlike the impact of John Snow's map of London showing the site of infectious water pumps in London in 1854. The information was there; it had just not been collected.

It is also true that while Nightingale was working on her hospital tabular forms that patients were laying in British hospitals with little standardized treatment. Since no records were kept, that meant that mistakes in treatment were made over and over. For twenty-first century audiences, it seems almost nonsensical not to record this kind of information. After all, how could hospital personnel possibly understand how to treat illnesses, wounds, and diseases for new patients if they didn't know the results of previous patients' history with the same illness, wound, or disease?

Unfortunately, in the 1800s in Britain, very few knew this information, other than through day-to-day experience and memory. In other places in Europe, however, primarily France, such record-keeping was more the norm, but not in Britain. It was important, then, for Nightingale to design a tabular form for hospitals that could serve as a record-keeping device and an important visual tool to record medical information as it occurs, in a form that then could later be studied and understood over time. This idea of systematizing data within a table, as Nightingale did, is reminiscent, I would argue, of certain artistic decision-making. For example, in Rudolf Arnheim's seminal work, *The Power of the Center*, he wrote about the interaction of "centricity" and "eccentricity," in art, but he also told me that he felt his ideas applied to technical graphic visuals as well (156; Personal). According to Arnheim, the first aspect of the power of the center is the function of the frame. Thus, surrounding Nightingale's center of data was a rectangular frame, resulting in a picture that developed its own power through its intersecting tabular framework of rows and columns enclosed in the rectangle. Indeed, if her table did not have a frame, doctors and nurses might have found it difficult to conceive of it as one recordable matrix and might have had trouble understanding the use of its columns and rows. As Arnheim wrote, a frame "means that the character of an object can be defined only in relation to the context in which it is considered ... The frame defines the picture as a closed entity, a center that exerts its dynamic effects upon its surroundings" (*Power* 42). Although not intended for artistic purposes, Nightingale's tabular frame did, in fact, define its whole, since lines of rows and columns crossed each other, and created a recordable arrangement in which nurses and doctors could enter data.

However, the instrumental purpose does not diminish its importance as a visual invention that mimics some of the same perceptual features we see in art. Nightingale's table is, I argue, in and of itself, a visual representation of what Arnheim refers to as "a concession to the gravitational coordinates of the outside world" (*Power* 56). In making this point, I seek to establish Nightingale's recordable table as a framed arrangement, which makes it easier to see as a whole work and within which tick marks, for example, could be inserted that represented numbers. Of course, as mentioned above, tables had existed beforehand, but the way in which Nightingale approached her design shows more than simple instrumental intentions.

When Arnheim wrote about technological pictures, for example, he made the point that "the better picture is one that omits unnecessary detail and chooses telling characteristics, but also that the relevant facts must be unambiguously conveyed to the eye" (*Art* 157). In this description, he included "simplicity of shape, orderly grouping, clear overlapping, (and) distinction of figure and ground" (*Art* 157). In addition, he wrote that "diagrams," like the "pocket map of subway lines" in London "give the needed information with utmost clarity" (*Art* 159).

It is certainly true, of course, that Nightingale's table was slightly different from other tabular structures in that it only defined the outer perimeters of the frame and the inner rows and columns with text. It is also true that the doctors and nurses filled in the columns and rows, but the table itself, as designed, holds a place in what should be considered "the domain of images," as defined by art critic James Elkins and exemplified by Arnheim. For Elkins, data tables are "schemata" (91). In this view, tables may not be considered a type of "writing or a picture," but "it is essential, I think, not to exclude images whose geometric order is latent or unmeasured" (Elkins 214). According to Elkins, "[l]ike picturing and writing, notating is a matter of degree, and a given image might only be a notation on balance, even while it is also a picture and a text" (214).

It is also interesting to note that Nightingale's creation of this kind of tabular reporting device was also evident in her youth. Cecil Smith writes that when Nightingale was young, in the 1830s, "With laborious patience, she kept a book in which she made a detailed comparison, in the form of a table, of the score, libretto, and performance of every opera she heard" (16).

What we are in effect seeing here, then, in a young Nightingale, is a natural interest in communicating information abstractly—using lines and words to indicate a more basic version. Thus, it is not surprising that, years later, as an adult, she would elect to use a tabular form to provide a logical and organized space for health workers to record information about their patients. It is also well known that, "in 1840, at the age of twenty, she announced her wish for further tuition in mathematics," although "her immediate family disapproved" (Diamond and Stone 67). Fortunately, she later was provided a tutor in mathematics and greatly benefited from it (Diamond and Stone 67). This, too, influenced her later recordable tabular forms.

History of Tables

Of course, tables have been around for centuries. But it is not the table itself—the visual form of columns and rows—that deserves our attention here, but rather the use of the table as a way to record and understand medical information and, then, make diagnostic decisions based upon it. Robert Horn dates the earliest use of lists and tables to 6000 BC in Mesopotamia, where people made extensive lists and tables of marks to represent inventory involved in trade (23). But the easiest, most recognizable use of tables in the way we use them today was in the second century in Egypt, where astronomical information was recorded to aid navigation, allowing for the organization of data and easy comparison (Horn 27). We can certainly understand today how helpful a tabular form is whenever we want to compare and contrast information carefully. However, in the computer-centered world we live in today, we often see tables only as starting points in spreadsheets, since we know that, once we enter data, a graphing program will automatically turn the numbers in the table into visualizations such as pie or bar charts, where we can sometimes see patterns easier.

But, in the history of tables, this is a relatively new development. Renowned statistician and visualization expert Jacques Bertin notes that "based on rational imagery, graphics differ from both figurative representation and mathematics" (*Semiology* 2). For Bertin, there were three functions of graphic representation: (1) to record information, (2) to communicate information, and (3) to process information (*Semiology* 12). For Nightingale's statistical tables, which were to be used in hospitals and warfront medical facilities, the purpose was twofold—to record and to communicate. Later, the information could be processed to aid in future actions and interventions. As Bertin acknowledged, "Modern research generally involves numerous components. It will constitute a unit, however, only if the information on which it is based can be thought of in the form of *one* double-entry table—albeit of very large dimensions" (*Semiology* 254). For Nightingale, those numerous components were, in fact, information about the wounded and dying soldiers in hospitals in a war zone that could be recorded and understood better if shown in an organized matrix.

Similarly, in modern day writing on information visualization, Stuart K. Card, Jock Mackinlay and Ben Shneiderman write that "graphical inventions" like tables "serve two related but quite distinct purposes. One purpose is for communicating an idea, for which it is sometimes said, 'A picture is worth ten thousand words.' ... The second purpose is to use graphical means to create or discover the idea itself" (1). In Nightingale's case in the Crimea, we see how the chaotic situation in the battlefield influenced her to create a standard recordable table for use in hospitals. In her book on hospital statistics, *Notes on Hospitals*, Nightingale notes that the health professional who fills out a medical record table can better help medical personnel in the future to understand "the *class* and *order* in which the case stands [for it] is then to be sought for in the left-hand column, and a mark or marks, as the case may be, are to be placed in the 'age' column in the line 'others'" (161). Such kinds of notation are quite consistent with both Bertin's as well as Card, Mackinlay and Shneiderman's points that people can use graphics, including numeric tables, to discover information that can impact future treatment. It is this end, according to Rafferty and Wall, in which Nightingale's table and explanatory text in her book *Notes on Hospitals* provides an insight into her "extraordinary research and rhetorical skills," revealing her as "an arch synthesiser and systematiser of data" (1063).

But Nightingale's hospital data tables not only serve as forms to be filled out. In fact, when it comes to Nightingale's hospital tables, the power of simply inserting tick marks to represent the number of patients with a certain disease or injury, or when patients were admitted, recovered, or died, can be seen not only as a utilitarian creation, but in some ways as an artistic creation that brings clarity to a cacophony of information. Antonio Damasio makes a cogent argument for the use of symbols as multi-powerful entities: "If those symbols were not imageable, we would not know them and would not be able to manipulate them consciously" (107). In fact, Damasio claims "that both words and arbitrary symbols are based on topographically organized representations and can become images" (106).

When the nurses and doctors at the hospitals or war fields in the Crimean War were inserting numbers and/or tick marks in a table, they were not only recording individual instances of a condition or event, their actions were also, in effect, acknowledging the truth of Rudolf Arnheim's statement in *Entropy and Art* that "[o]rder is a necessary condition for anything the human mind is to understand" (1). Thus, in this tabular form, Nightingale used the marks inserted by medical personnel to provide order so that the doctors, nurses, and administrators could more easily give it meaning. As Arnheim states, "[a]rrangements such as the layout of a city or building, a set of tools, a display of merchandise, the verbal exposition of facts or ideas, or a painting or piece of music are called orderly when an observer or listener can grasp their overall structure and the ramification of the structure in some detail" (*Entropy* 1). In a similar way, Nightingale's recordable table could allow doctors and nurses to grasp the overall medical situation in the Crimea and in British hospitals because they could then later use them to understand better how to treat patients.

However, it is also important to note that tables not only allowed for the doctors and nurses to reason and study based on a collection of numerous numbers or marks; they also, as Edward Tufte writes in his seminal book, *The Visual Display of Quantitative Information*, "work well when the data presentation requires many localized comparisons" (179). For medical doctors and nurses treating battle wounds and related physical illnesses, the recordable table matrix that Nightingale invented served both to communicate an understanding of current action as well as an opportunity to decide on future courses of medical action.

We might even consider that Nightingale's wartime tables were not only recordable devices, and, thus, served as intellectual visual devices; they even functioned as meaningful constructions that, according to Bertin's view of tables, are not only "one of the best ways to convey exact numerical values, [but] they are also more compact, assist in making comparisons, a convenient way of storing data for rapid reference, excellent for recording and communicating repetitive information and organiz[ing] information for which graphing would be inappropriate" (*Semiology* 387). Bertin also notes the importance of tables' value for their ability to be used for reference; to include grid lines that can be sorted and used for analysis. "They can allow for information to be ranked in ascending or descending order" (*Semiology* 388). In addition, tables are also helpful for "organiz[ing] information," "[assisting] the viewer in visually tracking data," and "emphasizing key information (*Semiology* 396). In fact, Elkins refers to tables as one of a group of graphical devices that he considers basic notational images "that have all the elements of emblems and are also based on geometric forms such as *reference lines*" (91; his emphasis). Elkins also cautions against dividing "word and image" and, interestingly, also argues that notation images (as I would argue tables are) can be problematic (83, 234). Thus, the ways in which they are made and the order in which they recorded are crucial. Elkins also notes that "tally marks" existed in prehistoric times and "therefore seem(s) to predate writing" (234). Of course, tally marks were also used when doctors and nurses recorded in Nightingale's tables in the Crimea.

For Nightingale, tables, if used throughout medical hospitals, were important records. Indeed, one of the most important findings that Nightingale discovered from the use of her tabular recordable forms in hospitals was that they were becoming forms that "have been prepared for recording, on one common form, all the facts of hospital experience" (*Florence* 83). Indeed, as mentioned above, Nightingale's main goal in creating a tabular record of illness and disease in the war front at Scutari was not simply to save the lives of current soldiers but also to allow transparency where disease and/or injury could be "ascertained with accuracy, and these facts, together with the duration of cases, would enable the value of particular methods of treatment and of special operations to brought to statistical proof" (*Florence* 84).

The impact of Nightingale's development of tables to record patient data in the wartime field cannot be emphasized enough. Still, Nightingale continued to lament the fact that other countries involved in the war were much better in not only treating their soldiers but also in recording their health information in the field. As she stated in a letter to Lady Canning, under the heading "Mode of Keeping Statistics":

> I think if you could see our *real* Statistics, you would think that I have been moderate in my Statements. In eight regiments in the front, of which the 46th actually lost more than its average strength from disease alone, we lost 73 per cent in seven months from disease alone. I am not aware that we can show any instance in our history of a similar disaster except in the Burmese War in '26. At Walcheren, which was called the 'ill-fated' expedition, we lost 10¼ per cent in 6 months from disease,—in the Peninsula 12 per cent in a year from disease. (Goldie 292)

Later, Goldie writes, "It is also important to note that, because of some of the critique Nightingale received when her tabular forms were published in the Statistical Congress' record, she had already learned from her London hospital usability test sites and, as a result, encouraged users to include new additional forms, including 'Hospital Administration and Discharge Book,' '1. In-Patients,' '2. Out-Patients' and '4. Sanitary Statistics of Wards'" in *Notes on Hospitals* (175). Nightingale also sought to rhetorically persuade her audience by writing that the form she provided "has been tried on a somewhat large scale and has been found to answer" (Goldie 175).

I would be remiss, however, not to point out that there have been critics of Nightingale's tabular form. Such critics felt that the proposed forms were too complex and that they included a system for the classification of diseases devised by William Farr—with which some pathologists strongly disagreed (Cohen 1984). In this view, her critics pointed out that it was Nightingale's error that she did not display the same understanding of the new germ theory of disease and its implications for the treatment of contagious diseases. Yet, at the same time, she does have her supporters on this point, who acknowledge that in her later years Nightingale did realize the rightness of germ theory. But this late realization does not deter from the fact that her innovations in treatment and recording in tabular forms, as well as her own sanitary reforms, saved lives.

Overall Advantages of Tabular Arrangement

Finally, we must not forget that Nightingale's interest in mathematics and drawings of tabular arrangements as a way to communicate information was not the norm for a woman of her time. Victorian England was steeped in the history of very traditional roles for women as wives and mothers. What the young Florence Nightingale did in her recording of musical performances helped her "see" an order and design, and this continued into her adulthood. It is also understood, as noted above, that many people tend to see tables as simply dry records of data that are in need of visualizing. And, indeed, readers often tend to skip over tables in books and articles because the study required to understand a table's points can be cumbersome. At the same time, the examples we are studying here were tables that were used to record information as it occurred. Later, they could be analyzed for patterns and deeper understanding and possibly better treatment.

In sum, for Nightingale, the uses that could be made of the ordered collection and recording of meaningful health numbers and their resulting arrangement was not only logical but also a highly held ideal, based particularly on her experience in the war zone. As editor Lynn McDonald writes, "The terrible mortality rate of the Barrack Hospital in Scutari had alerted her to the potential for harm by bad hospital design or inaction" (*Florence* 74–5).

Usability Testing

It is also important to note, as McDonald writes, that Nightingale's persuasive power was not only the result of her firsthand experience as a nurse in the Crimean War but also the result of the choices she made after her experience in the Crimea, and the voices she listened to as she created "a pre-test of her questionnaire on hospital statistics, a very proper procedure, but one scarcely undertaken then and not always now" (*Florence* 75). For example, she wrote in an 1859 letter to Sir James Paget, surgeon to Queen Victoria, "I have had a set of new forms prepared (with the Registrar-General's sanction) for hospital statistics. I should be very glad if St. Bartholomew's would be so good as to fill up a set on trial" (*Florence* 76). Then she included a form with rows and columns for "age," "sex," "disease," "admissions," "discharge," "deaths," "discharged incurable," and "duration of the cases" (*Florence* 76). Today, this type of activity might be referred to as essential in testing a form's usability since it allowed for both collection and analysis of data. In Britain, it was ahead of its time, but, as always, it illustrates that Nightingale was extremely thorough. For example, in October of that same year, she wrote another letter, likely to Paget, in which she suggested that "the easiest way of keeping his statistics would be to have seven separate nominal books, or at all events a ledger with columns for each separate subject: remaining, admitted, and discharged, etc., in which he could enter day by day the particulars from his day book" (*Florence* 80). She went on to ask for "a complete disease list, including surgical cases, such as would include all those who come into hospital at St. Bartholomew's" (*Florence* 80). In doing so, Nightingale was taking much of her inspiration from the French hospitals of Lariboisière and Vincennes, which she had earlier praised (*Florence* 81). She also was, of course, tremendously influenced by statistician Adolphe Quetelet's statistical analysis of social data, who, as noted above, was a strong believer in the power of the use of tables.

Nightingale and Visual Statistics

When we think of the pain of the loved ones of those who were related to or who served along with the dead soldiers in the Scutari War in which Nightingale led a contingent of nurses to the front, we may be reminded of how she had to watch as many soldiers died who should have lived. She knew that often it wasn't their wounds that resulted in their deaths, but rather the dirty, crowded, unsafe conditions of the field hospitals to which they were carried and the lack of recordable data that could influence future treatment. Nightingale and her nurses' efforts to keep the warfront hospitals clean and to offer normal, expected illness comforts such as blankets for injured soldiers changed not only the death rate of the soldiers but also became legend.

Not surprisingly, her rose diagram has been much written about, especially in recent times, because of the dramatic way in which it visualized the war dead and illustrated how efforts at cleaning the infirmary and attending to soldiers' basic needs helped save their lives. These efforts enhanced Nightingale's legacy as a pioneering nurse who was to become the first female member of the Royal Statistical Society and who carried on regular correspondence with famous statisticians Adolphe Quetelet and William Farr. However, Nightingale's tabular forms, as described above, also reveal her skill at understanding the importance of minute collections of ordered information in health centers. They also show her forward thinking in conducting early usability tests in hospitals that led to a number of visual recordable devices that improved future medical treatments and saved lives. It was another indication of the important impact that Florence Nightingale had on the lives of British citizens in the mid-1800s. "In the years immediately after her return home, she was to devote a great deal of time to improving the methods of record-keeping both in the army and in civil hospitals" (Goldie 271). Today a simple recording table—paper or electronic—can save lives because of its day-to-day tabulated collection of medical actions as well as patient health and illness—facts that even today, in many ways, we owe to Florence Nightingale's passionate call for transparency. After all, hundreds of soldiers dying unnecessarily in field hospitals should be made transparent—not just through journalistic records but also through an accepted and regular use of appropriately designed forms that can be studied, compared, and used to prevent death. Any attempt at making data about the reasons for deaths and injuries visible can help to address any resistance, political or medical, and to diminish faults and displace blame. In terms of the Crimean War, undoubtedly Nightingale's efforts to make men's injuries, deaths, recoveries, and illnesses transparent served to change how medical records were kept and used, both at home and in the war field. As a result, many lived that otherwise might have died.

Chapter 3

Visualizing Public Health Risks: Graphical Representations of Smallpox in the Seventeenth, Eighteenth, and Nineteenth Centuries

Candice A. Welhausen and Rebecca E. Burnett

You can smell smallpox before you enter the patient's room ... On at least two occasions, smell alone alerted me to the presence of smallpox. As I walked down a hospital hallway in India, the dead-animal odor stopped me in my tracks ...

William H. Foege[1]

Smallpox: "Severest Scourge of the Human Race!"[2]

In 1980, after decades of an aggressive, strategic international vaccination campaign, the World Health Organization (WHO) announced that the smallpox virus had been eradicated (World, *Factsheet*). However, its future remains uncertain. Samples of the virus collected during the eradication campaign were initially slated to be destroyed, but they are still stored in response to concerns about bioterrorism (McKay). As we write this chapter, less than five miles from the main Centers for Disease Control and Prevention campus in Atlanta, the storage facility in the U.S., the virus remains in cryogenic suspension, under heavy guard and in solitary confinement, another stay of execution granted (World, *Sixty-fourth*).[3]

This "Stone Age disease," as Jonathan B. Tucker characterizes smallpox, is unlike any other: virulent, painful, disfiguring, and often lethal, with a 25–30

---

[1]   American epidemiologist and physician William H. Foege served as Director of the Centers for Disease Control and Prevention and was chief of the Smallpox Eradication Program (Foege, *House* 3).

[2]   This phrase concludes Edward Jenner's 1801 *Inquiry* (182).

[3]   Given the 2014 incident in which vials containing the smallpox virus were found in an unsecure NIH facility, we can't be sure—despite international agreements—that samples of the virus are not being stored at other unapproved facilities. The vials were not intended for terrorism; instead, they were the result of careless protocols. However, no way exists to ensure that other vials are not unintentionally stored at unsecure facilities around the world.

percent mortality rate (102).[4] Further, it spans the entire course of recorded human history; "[t]he prince and the peasant are alike subject to its influence" (Bose 80). Ramses V, for instance, is believed to have died from the disease in the twelfth century BC. His mummified body, depicted in Figure 3.1, shows signs of infection on his cheek, nose, and upper lip.

Centuries later, many early medical statistics depicted smallpox incidence; seventeenth-century mortality tables showed death rates. Line graphs and bar charts created in the eighteenth and nineteenth centuries showed variolation,[5] the practice of deliberately infecting patients with smallpox to confer immunity, and vaccination data. Thematic maps created at the end of the nineteenth century revealed clusters of cases during outbreaks. Further, the astounding medical achievements of vaccination and eradication were only realized through the institution of public health policies to control this insidious disease, which included analyses of detailed tables, graphs, and maps representing collected data. As William H. Foege notes of Edward Jenner, who invented the smallpox vaccine: "Jenner's discovery of [vaccination as a] tool for preventing disease is one of the great breakthroughs in science. Indeed, the modern era of public health can be traced to Jenner's 1796 experiment" (10). Today, more than 30 years after the eradication of smallpox, scientists continue to investigate its properties, and public health decision-makers continue to debate its future.

Thus, given its longevity, its etiology, and its profound historical influence on the evolution of disease control and prevention efforts, smallpox offers a unique opportunity to explore the ways in which visual thinking influenced the creation of historical statistical graphics in medicine and public health. The idea that disease and health can be understood, managed, and controlled from a population-based perspective—that is, that particular variables can increase or decrease risks—seems obvious from our twenty-first century point of view. However, these contemporary ideas began with an epistemic shift toward scientific thinking in the early seventeenth century and helped shape the beginnings of the discipline now known as public health.

In this chapter, we analyze selected graphics depicting smallpox epidemics created in the seventeenth, eighteenth, and nineteenth centuries, examining some of the historical and cultural factors that influenced their creation. We argue that rather than adhering to established visual genres for representing

---

[4]   Death rates among populations with no prior exposure, such as indigenous people in the Americas, were even higher. See Foege, *House*, and Donald R. Hopkins.

[5]   Unlike modern vaccination, which uses dead or attenuated viral particles to prompt an immune response, variolation (also referred to as inoculation) exposed patients to the live smallpox virus, usually through applying the pus of an infected person to a purposely cut or abraded area or by inhaling a powder made from an infected person's scabs. The goal was to infect new patients with the biological material of former patients who had had mild cases and presumably less virulent strains of the virus. The vast majority recovered with a 1–2% mortality as compared to the usual 30% mortality rate (NIH).

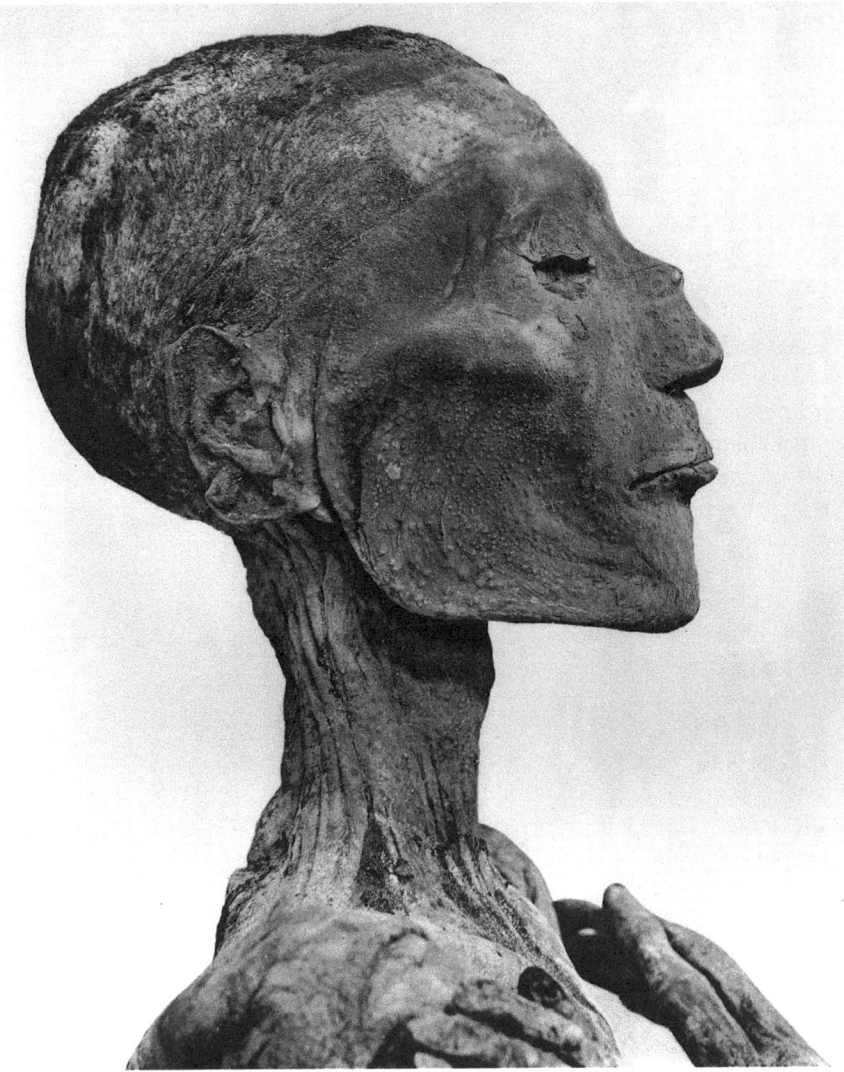

Figure 3.1   Mummified body of Ramses V, believed to have died from
smallpox in 1157 BC
Source: WHO Archives.

quantitative information, which only began to emerge in the early nineteenth
century,[6] the creators of these early depictions engaged in three rhetorically
driven tasks—correlation, location, and value—that facilitated and shaped

---

6   See Michael Friendly, "Golden."

risk perception in the field of public health and epidemiology. Specifically, *correlation* reveals a growing interest in seeking to better understand spatial and temporal relationships. *Location* signals space and place as important in such correlations. *Value* emphasizes selected categories of information (variables) such as demographics or incidence mapped onto locations, highlighting and bringing together the importance of correlation and location. Engaging in these interrelated rhetorical tasks allowed creators of early smallpox graphics not only to better understand the distribution and spread of this disease but also to visualize public health risks more generally.

We do not attempt a comprehensive historical account of epistemology, data visualization, the disciplines of medicine, public health, and/or epidemiology, the study of risk and risk analysis, or smallpox. Rather, we explore the historical intersections of these areas *through* selected visual representations of quantitative information about smallpox. We begin by briefly discussing the shift in epistemology during the seventeenth century, which led to an interest in managing and controlling disease from a population-based perspective. We continue by discussing risk perception and assessment and outline the origins of statistical graphics in medicine. We then propose our rhetorical tasks framework and analyze and discuss selected historic smallpox graphics. We conclude with three suggestions that extend the disciplinary conversation about visible numbers.[7]

Public Health, Risk Perception, and Early Statistical Graphics

Early statistical graphics were created in the fields of medicine, public health, and epidemiology, in part, in response to the push toward empiricism driven by the Scientific Revolution. Historically, illness and disease were often interpreted as punishments from God, with the *Bible* setting the precedent for the spread of plagues. However, a worldview characterized by superstition, folklore, and predestination gradually changed to one in which the causes of infectious disease could be explained through science. Prior to the seventeenth century, the actual causes of disease had not yet been identified, and thus addressing outbreaks of epidemic disease tended to focus on mitigating the symptoms of individual patients rather than attending to a modern-day public health perspective with its emphasis on disease prevention in populations. The late eighteenth and early nineteenth centuries were particularly crucial as attention gradually moved toward, as public

---

[7]   In our approach to this chapter, we have adapted Suzan Cozzens' argument—noted also in Laurence Smith et al., as "situated practices" (75) —which suggests that promising lines of investigation and inquiry can be conducted by examining specializations rather than a discipline as a whole; thus, we examine selected smallpox graphics as a way to illustrate the change in thinking from a focus on individual cases to a focus on large-scale public health efforts.

health historian George Rosen characterizes it, "maintaining and augmenting a healthy population ... for the political and economic strength of the state" (90).

Public health today is viewed as the study of disease and health in populations[8] through *surveillance*, "the ongoing and systematic collection, analysis, and interpretation of outcome-specific health data ..." (Stroup and Berkelman 5). Specifically, surveillance seeks to investigate and identify disease (and health-related) trends, patterns, and *risk factors*.[9] Risk assessment—"the characterization of potential adverse effects of exposures to hazards"—is fundamental to contemporary public health practice (National Research 321). Jonathan Samet, Robert Schnatter, and Herman Gibb explain that "a primary objective of epidemiologic research is to measure or assess the risk of disease in a group of individuals" as well as to "describe[s] exposures to populations and assess[es] dose-response relationships between exposure and risk, two components of risk assessment" (930). That is, risk assessment determines the potential causal relationships of the particular type, amount, and degree of *exposure*; the *dose*; the particular adverse health *outcome*; and the *response* (such as smoking and lung cancer). The relationship between the exposure and the likelihood of experiencing the adverse health outcome constitutes the risk.

Humans have long held a vested interest in mitigating risk, but the idea that risk could be understood and managed did not originate in modern statistical thinking. Rather, Blaise Pascal's discovery of probability theory in the mid-seventeenth century provided the early framework for the emergence of quantitative-based risk assessment, facilitating the means to assess risk mathematically and subsequently paving the way for modern medical statistics and its emphasis on inference (see Covello and Mumpower 104–5).

Prior to the seventeenth century, quantitative disease and health-related information had been collected primarily for descriptive purposes. This trend would change, however, in 1662, when John Graunt, an English merchant, published a series of life tables culled from the London weekly "bills of mortality" produced between 1604 and 1660 (Sutherland 545).[10] The history of medical statistics begins with Graunt's creation of visuals that reflect "the first recorded attempt to calculate empirical probabilities on any scale" (Covello and Mumpower 106). In these visuals, "Graunt develop[ed] some fundamental principles of public health

---

8   Stroup and Berkelman offered this definition: "While clinical medicine has the individual as its focus, *public health* is fundamentally concerned with preventing disease, disability, and premature death in the population or community" (1).

9   The term *risk factor* is a contemporary concept that originates in the longitudinal Framingham heart disease study (see Kannel et al.).

10   While we follow common practice and use the word "table" to refer to the visuals Graunt creates, in fact these representations bear only slight resemblance to contemporary statistical tables and are often visually similar to columns of lists. By the mid-sixteenth century, bills of mortality were compiled every week by London parishes; they routinely included burial counts as well as christenings (Sutherland 543).

surveillance, including disease-specific death counts, death rates, and the concept of disease patterns" (Thacker 2).

Graunt explained his interest in creating these tables, suggesting that "some other uses might be made" (quoted in Sutherland 542). Specifically, the information might be used to influence particular types of decisions (political, for instance), and his tables were used, for example, "to estimate the number of fighting men from London the king could rely on for future wars" (Klein 44). Sutherland notes: "John Graunt was the first person to whom it occurred that numerical information on human populations could be of more than ephemeral interest," with his work "represent[ing] the earliest attempts to estimate population size scientifically" (554, 550). Graunt's tables show the beginnings of careful, detailed, empirical inquiry, using quantitative data to speculate about populations affected by disease. For instance, he was the first to note high infant mortality rates (particularly in London), to realistically estimate the total population of London and the country, and to differentiate between endemic and epidemic disease in terms of cause of death (Bessant and MacPherson; Sutherland).

The significance of Graunt's work goes beyond the practice of merely reporting vital statistics to including projections that drew inferences. His work reflects crucial first steps not just in tracing the history of the emergence of quantitative graphics in medicine, public health, and epidemiology, but more important for our purposes, in documenting the beginnings of the epistemic shift that prompted new ways of thinking about disease, health, and risk. Graunt saw the bills as more than disparate pieces of information; instead, he saw *data* that could be collected, arranged, analyzed, and subsequently used to make particular types of decisions. He also saw that in order to facilitate these "other uses," data would need to be represented in ways that anticipated communicating information about correlations, locations, and values (quoted in Sutherland 542). Among other things, his life table grouped numbers of deaths visually, *correlating* potential temporal patterns of life expectancy. Some of Graunt's tables identify locations of disease, death, and burial sites. *Value* would later be demonstrated through public health decisions made as a result of better understanding disease and risks. The importance of Graunt's "political arithmetic" was soon recognized, adapted, and spread throughout Europe, illustrating the progression from descriptive reporting to statistical graphics as a visual analytical tool.[11]

Risk assessment today is often understood and evaluated from a narrowly focused quantitative perspective, "directed almost entirely to assessment of economic risks and risks to human health and life," with such evidence and analyses "expressed in dollar terms, days lost as a result of accidents or disease, and in disease rates or body counts" (Short 716). The conventions of statistical

---

[11]   See Friendly, "Golden" 505. For example, in 1669, Dutch mathematician Christiaan Huygens used Graunt's tables to produce rough survival line graphs; and in 1693 Edmund Halley drafted the first mortality tables from the Breslau Bills of Mortality (Wainer and Velleman 311; Friendly and Denis, *Milestones* 10).

graphics that communicate risk are now well established—for example, showing risk magnitude, relative risk, cumulative risk, uncertainty, and interactions (Lipkus and Hollands 150). This approach guides much contemporary public health risk assessment, but also reflects a twenty-first century perspective of risk that has developed and evolved over the past several hundred years with the integration of increasingly sophisticated statistical methodologies.

Yet this quantitative approach toward risk and risk assessment did not exist for creators of early statistical graphics in medicine, public health, and early epidemiology. Today the idea of *risk* refers to the likelihood—real or perceived—that an adverse event will occur. In a pre-Scientific Revolution worldview in which events were explained through acts of God, this modern-day concept of risk was largely irrelevant because people who believed their destiny was already controlled would have had no reason to be concerned with what we now refer to as risk assessment or risk management.

The smallpox graphics that we analyze later in this chapter illustrate a shift in thinking that ultimately leads to a contemporary understanding of risk and risk assessment, along with the evolution of risk as a visual, quantitative construct—that is, as a *thing* that could be quantified and then acted upon to mitigate the perceived outcomes. The creation of early statistical graphics in the fields of medicine, public health, and epidemiology demonstrates this progression and represents a problem-solving approach that responds to social, cultural, and political issues related to the spread of communicable and infectious disease, investigating smallpox epidemics, and communicating the risks of infection. The significance of the creation of these graphics is not what they actually depict (details about incidence of smallpox, for example). In fact, from a twenty-first century perspective, the designs may appear unremarkable. Rather, their importance is the visual thinking they represent.

The historic graphic representations of smallpox epidemics that we discuss are highly contextualized social constructions, showing the ways that an epidemic affected a particular population in a particular setting at a particular time. This criticality of context and perception has been widely recognized in contemporary scholarship. Woody Caan and Dawn Hillier observe that the "objective measure of risk does not tell the whole story and, in determining acceptability of any particular risk, perceived risk is likely to play a large role" (38), while scholars in technical and professional communication have argued that risk perception is a social and cultural issue (see Grabill and Simmons; see also Sauer). Further, the importance of social factors as well as the contextual framework in assessing risk is reinforced by Foege when he explains that "[t]he response to a plague or threat of a plague is in some ways dependent on the perception of risk held by individuals and decision-makers" ("Plagues" 12). In the next section, we apply our framework of rhetorical tasks to analyze selected graphics depicting smallpox epidemics. Our analysis accounts for this expanded understanding of risk by examining some of the social and cultural factors that shaped their creation.

Visual Genres and Historic Representations of Smallpox

Smallpox epidemics affected every aspect of a community—economic, social, medical—leaving people in that community with no sense of control. Historically, quarantine and isolation were commonly used to manage the risk of contracting communicable and infectious diseases. These methods of disease prevention and control also remained true for smallpox until the eighteenth century when two events occurred that would dramatically alter the risks of contracting the disease. First, variolation became widespread across Europe and the U.S., and second, the smallpox vaccine was invented. Prior to the eighteenth century, smallpox risk was primarily understood in terms of mortality from "natural" smallpox.[12] Later, after variolation increased and after it became accepted in Europe and the U.S., interest grew in assessing the risk of mortality from natural smallpox compared to the risk from variolation. Still later, at the end of the eighteenth century, mortality rates from natural smallpox were compared to mortality rates after vaccination. Graphics created at these different points in history reflect the changing reality and perception of smallpox risk.

Our discussion below follows this historical chronology of managing smallpox risk and considers the ways in which changes in risks were visually depicted. We do not focus on the history and evolution of the visual genres themselves (for example, when smallpox tables or particular types of graphics were first used). Instead we focus on the ways in which risks were visually represented, which allowed creators to organize information by performing the following rhetorical tasks:

- They *correlated* variables to show temporal and/or spatial relationships, visually hypothesizing about the relationships that may lead audiences to think of causation, whether verifiable or not.
- They *located* specific geographic patterns and trends in natural and built environments to emphasize the connections between two or more variables and those spaces/places.
- They assigned *value* to political, social, and cultural information to visually highlight the ways spatial and temporal variables might or could affect, for example, public health decisions. Value is reflected in the individual visuals we discuss later in the chapter as well as in terms of a broader value to these public health decisions.

These rhetorical tasks are not neat, nor are they exclusive; rather, they reflect the synergy, the overlap, and the messiness of a decision-making process that served, in part, as a precursor to the emergence of the visual conventions that have subsequently been defined and codified in contemporary graphical forms.

---

[12]   Smallpox can be contracted in two ways: through variolation, deliberately being exposed to the virus, or "naturally" by a person who is already infected.

Collecting, analyzing, and evaluating information about the ways a disease affects populations as a whole and over time reveal trends and patterns that suggest causal associations. The rhetorical tasks we have identified reflect the evolution from an emphasis on managing disease and health in individuals to managing disease and health in populations, which (along with other factors) shaped the ways in which risks are graphically represented in public health.

Depicting Risks: Mortality, Variolation, and Reduced Vaccination Rates

In this section of the chapter, we examine six graphics separated into three categories that reflect visual thinking about ways to depict smallpox outbreaks.

- Two graphics depict different ways to represent the *mortality risk from smallpox*: a diagram (what we now call a line graph) correlating death with disease duration, and a collection of two line graphs and two bar charts, correlating mortality with weather conditions.
- One graphic depicts a way to represent risks of mortality through *variolation*. (The table was used to persuade public health decision-makers to adopt variolation.)
- Three graphics represent the *mortality risks that accompanied decreased vaccination rates*: a street map noting incidence alone, a street map clustering the sequence of an outbreak, and a pair of bar graphs comparing the effectiveness of compulsory vaccination in two armies.

We analyze each graphic using our framework of rhetorical tasks (correlation, location, and value). We do not address each in equal detail, depending on available contextual information as well as the information included in the graphic itself. Further, our analysis is invariably articulated from our contemporary perspective, which necessarily shapes our understanding. In the twenty-first century, we not only define and use data differently than those in earlier periods, but we also have different views about what constitutes data.

Representing Mortality Risk from Smallpox

By the sixteenth century, the risk of dying from smallpox was a significant public health threat. The disease was globally widespread and feared (Rusnock 18); mortality rates began to be closely monitored in England during this period (Shryock 136). By 1629, the bills of mortality were routinely reporting smallpox deaths (Rosen 75); reporting such data became common in the seventeenth, eighteenth, and nineteenth centuries.

The first visual we examine, "Diagram representing the force of Mortality of Small Pox. Days 18–35," appears in the 1838 *British Medical Almanack*, depicting daily mortality over a 17-day period and most likely documenting the peak and decline of an epidemic (see Figure 3.2).

- *Correlations.* Beginning with a high of 271 deaths by day 18 (days 1 through 17 are not shown) to a low of 25 deaths by day 35, correlation is visually established among three variables: the *numerator* (deaths from smallpox), the *denominator* (the number of "constantly sick," which is 10,000), and *point in time* (day) in the life cycle of the epidemic. This line graph suggests a 35-day life cycle, documenting the most significant risk of mortality, "the force of Mortality" as the title suggests, as occurring about mid-cycle and then rapidly decreasing. Interestingly, the overall mortality rate is far below the expected 30 percent. In 1838, the year this graphic was published, the vaccine was presumably available and in use, yet the causes of this exact outbreak (for example, natural smallpox among an unvaccinated population or an outbreak resulting from variolation) are unknown.
- *Location.* No information is given about the location of the outbreak.
- *Values.* Understanding the relationship between time and risk of mortality allowed public health officials to determine the temporal progression of smallpox epidemics and subsequently enact measures such as quarantine and isolation that sought to mitigate exposures and minimize mortality risk.

In tracking the trajectory of the decline of this particular epidemic, this graphic visually shows a dramatically decreased risk of death after day 18 of the outbreak, with presumably a higher risk of death earlier in the cycle. Such a graphic today would probably show the entire time period. Thus, the creators of this graphic may have been attempting to focus only on the rate of decline. This representation also visually establishes temporal correlations by day within a 35-day period. Researchers may have used line graphs such as this to make predictions about how long an epidemic would last, when the greatest number of deaths was likely to occur, and when the epidemic might be expected to subside. This visual information might also been used to determine the risks of the epidemic spreading.

Our second visual also represents mortality risk—the "Synoptic Chart: Illustrating the Weekly Mortality by the Two Epidemics—Cerebro-Spinal Fever and Small Pox in the City of New York in the Year 1872" (see Plate 2).

- *Correlations.* This collection of graphs correlates mortality with the weather. The display presents two bar graphs representing number of deaths and two line graphs representing humidity (top panel) and temperature (second panel). A third bar graph is embedded along the top of the "Scale of Humidity" representing the "Gauge of Rain Fall" (see upper right-hand corner for label). This visual attempts to correlate death from smallpox (bottom panel) and death from cerebro-spinal fever (third panel; what

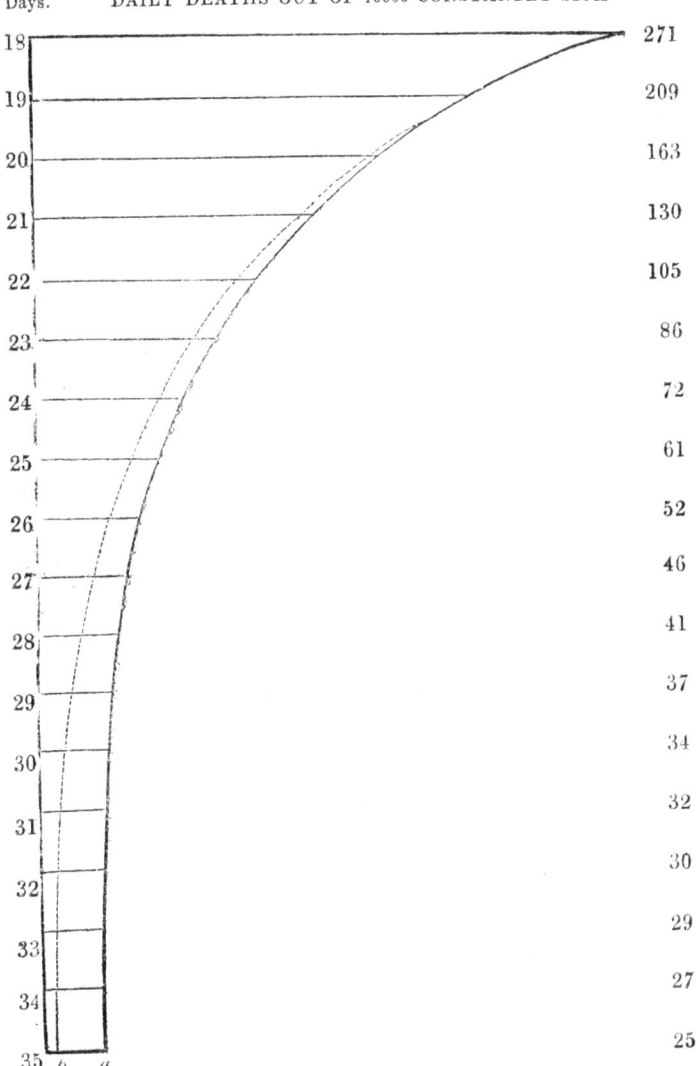

Diagram representing the force of Mortality in Small Pox.
Days 18—35.

Days.　　　DAILY DEATHS OUT OF 10000 CONSTANTLY SICK.

Days	
18	271
19	209
20	163
21	130
22	105
23	86
24	72
25	61
26	52
27	46
28	41
29	37
30	34
31	32
32	30
33	29
34	27
35	25

Figure 3.2　　Line graph that appeared in the 1838 *British Medical Almanack*
　　　　　　that traces daily smallpox deaths over a 17-day period
Source: Wellcome Library, London.

today we call meningitis) over a one-year period among three variables: humidity (which is also "sub" correlated with rainfall), temperature, and pressure of atmosphere (shown at the bottom). Correlating environmental factors with disease incidence and mortality has a long history (see also Rosen).[13] The purpose of these graphs, of course, is to identify parts of the natural environment as potential risk factors for contracting smallpox.

- *Locations.* These graphs display an impressive amount of data, clearly organized to lead viewers to see possible patterns in one spatial area (here in terms of a broad geographic area, New York City) and during one specific temporal period (one calendar year)—for example, a correlation illustrating an increase in deaths from smallpox with lower humidity. This collection of graphs demonstrates an effort to identify environmental causes so that appropriate public health decisions could be made to reduce the risk of mortality.

- *Values.* Grouping cerebro-spinal fever (meningitis) and smallpox in one visual presumed commonalities between the two diseases. In fact, grouping these two diseases suggests that the creators of this graphic may have believed that cerebro-spinal fever was sometimes a precursor to smallpox. This graphic might also simply reflect the two most significant public threats affecting this location at this time. Throughout history, officials were desperate to control disease outbreaks, so these graphs represent an attempt to better understand the risks of acquiring these diseases in this population in this particular location: New York City. No one presumed that weather could be controlled, but they did presume that populations could be moved to reduce exposure during high-risk times.

Plate 2 compares what we know today are two unrelated diseases—cerebro-spinal fever and smallpox—by week as well as simultaneously by year. This graphic visually demonstrates an increased risk of the two diseases in the earlier part of the year but with an unclear correlation between the two environmental factors under consideration (humidity and temperature). In the late-nineteenth century, miasma or 'bad air' was still a dominant theory of disease transmission; thus, these graphs attempt to visually assess the potential temporal and spatial associations between the two environmental factors represented and the potential risk of acquiring these two diseases—either simultaneously or separately.

Representing Risks of Mortality through Variolation

Before Jenner invented the smallpox vaccine, variolation presented the first viable alternative to quarantine and isolation for managing the spread of smallpox

---

[13]   Today such correlations are still common, with talk about "inversions" related, for example, to smog and dangers for at-risk individuals (infants, children, elderly, and those with respiratory diseases).

epidemics. Variolation had already long been used prophylactically by Asian and African cultures for centuries, but in the early eighteenth century it had yet to become widely accepted in England and the U.S. Smallpox cases caused by variolation usually proved to be much less severe than getting the disease the old-fashioned way and subsequently resulted in lower overall mortality rates. At the same time, the practice was highly controversial because it might "interfere with the 'natural' course" of the outbreak (see Hopkins, *Greatest*; Koplow 18). Since patients were deliberately infected, variolation also posed the risk that some individuals might still develop full-blown smallpox. Further, variolated patients could spawn new epidemics. In fact, in the mid-eighteenth century, the practice was initially prohibited in some American colonies for this reason (Hopkins, *Greatest* 255–6). Most patients survived a mild case, but for others it presented the same mortality risk as natural smallpox. The risk also existed that the variolation would be ineffective, and the patient would remain vulnerable to natural smallpox.

As the use of variolation grew in the mid-to-late eighteenth century, so did the need to assess the risks of natural smallpox in comparison (Shryock 136). In 1760 Swiss mathematician Daniel Bernoulli drafted and presented a report to the Royal Academy of Sciences of Paris, which included two tables that argued for variolation as a key preventative measure (see Bradley). Bernoulli's colleague, French mathematician Jean D'Alembert, disputed Bernoulli's calculations but ultimately agreed with his premise, and in 1761 D'Alembert created several hypothetical mortality line graphs, suggesting that variolation constituted a lower risk of death than natural smallpox (see Bradley; Funkhouser, "Historical" 296). Further in 1721, clergyman Cotton Mather also advocated the practice (which he learned about from his African slave), suggesting that the death rate from naturally occurring smallpox was 1 in 6 whereas only 1 in 60 died from variolation (Gehan and Lemak 9).

Our next visual appears at the beginning of a report by eighteenth-century British physician John Haygarth promoting variolation (see Figure 3.3). Haygarth argues in his 1801 report that the risk of "natural smallpox" outbreaks can be minimized by following the "Rules of Prevention," which included self-quarantine and "the utmost attention to cleanliness" (107–9). Through the visual information presented in this table, Haygarth argues in favor of variolation, a relatively new practice in England at the time, as well as provides statistical information used to persuade public health decision-makers—the Members of the Small-Pox Society in Chester—to adopt the practice in their communities.

- *Correlations.* Haygarth's table visually establishes relationships among the following variables: demographics (name and occupation), spatial (street of residence), and temporal (such as date of variolation, the date the patient began exhibiting symptoms of infection, and the date of death or the date when patients had recovered as indicated by "date of last scab"). The last two columns list whether patients "observed" or "transgressed" the Rules of Prevention, and whether they spread the virus to others, clearly showing a positive correlation between following the rules and the reduced risk of spreading the disease.

- *Locations.* Haygarth's table identifies geographic location of the 20 patients who had been variolated in the town of Chester in the late eighteenth century.
- *Values.* Prior to variolation each person signed a promissory note agreeing to follow the Rules of Prevention (described in more detail in the report) to prevent spreading the disease. Patients who followed these guidelines during recovery were paid 10 shillings (Haygarth 112).

Figure 3.3 is particularly important in demonstrating the epistemic shift toward new ways of thinking about disease and illness we discuss earlier because it visually shows the transition between an emphasis on investigating individual illness to an emphasis on a community-based perspective in attempting to bring together relevant quantitative information ("date of death" or "last scab"; "observed" or "transgressed" rules) and qualitative information (demographics, which using today's methodologies could be easily quantified). Hayworth's "data collection" includes gathering detailed demographic, spatial, and temporal information about individual patients and then visually collating his data, allowing him to make the argument to the Members of the Small-Pox Society in Chester that variolation was effective when the "rules" were observed. Creating this table allowed him to see information about controlling smallpox in different ways and provided a visual representation of the risks of not following "the rules."

Representing Risks of Mortality through Reduced Vaccination Rates

By 1810, vaccination was widely practiced in areas of Europe, Asia, and North and South America (Rusnock 18). Yet while the vaccine was instrumental in preventing and controlling epidemics, it did not entirely eliminate the risk of contracting smallpox. Historically, as well as today, the vaccine does not contain the actual virus (Jenner's original vaccine was made from cowpox); thus, getting the vaccine carried no risk of acquiring smallpox. However, unlike the lifelong immunity conferred through infection by natural smallpox or through variolation, and for reasons that are still not entirely understood scientifically today, the protection from the smallpox vaccine was not permanent. Thus, continued and routine vaccination in areas where the disease was still endemic was crucial for continuing to prevent future outbreaks. Our last three visuals, which were all created in the same 10-year period in the late nineteenth century, show the risks of less stringent vaccination rates (see figures 3.4, 3.5, and Plate 13).

Figure 3.4, "Smallpox outbreak in 1893 in Muncie, Indiana," shows an outbreak that occurred after vaccination rates had declined in the area following the last outbreak in 1876, illustrating the increased risks posed by declining vaccination rates. Routine vaccination as well as the compulsory vaccination laws instituted

REGISTER of the SMALL-POX in CHESTER, 1778.

i.		ii.	iii.	iv.	v.	vi.	vii.	viii.	ix.	x.	xi.
PATIENTS. Order. Name.	No.	Street.	Occupation.	Small-Pox Fever began.	Date of Inoculation.	Gentle Rules, or Promissory Notes.	Whence infected.	Date of Death, or Last Scab.	Washed and aired.	Infection communicated to	Rules observed or transgressed.
1. E. Bryly	2	Sty-lane.	Fisherman.	1778, Jan.	Jan. 30.	P. N. Jan. 30.				none.	observed.
2. M. Morris	2	Bridge-st.	Flour-dealer.	March 24.	April 3.	P. N. April 3.		April 25.	April 26.	Coleclough's 4th Family.	transgressed.
3. A. Collier	2	Northg.-st.	Bricklayer.	April 7.	April 14.	P. N. Ap. 14.		L. S. Ap. 29.	April 30.	none.	observed.
4. H. Coleclough	2	Bridge-st.	Labourer.	1st, Ap. 23. 2d, May 3.	April 26.	P. N. Ap. 26.	Morris's 2d F.			none.	observed.
5. Mr. Smith	1	Bridge-st.	Watch-mak.	April 22.	April 24.	G. R. Ap. 24.		L. S. May 4.	May 5.	none.	observed.
6. A. Singleton	2	Northg.-st.	Labourer.	May 6.	May 10.	P. N. May 10.	Liverpool.			Ashton's 7th F.	transgressed.
7. E. Ashton	2	Northg.-st.	Labourer.	1st May 30. 2d June 14.	June 4.	P. N. June 4.	Singleton's 6th F.	June 23.	June 24.	none.	observed.
8. H. Price	1	Gorse-St.	Shoemaker.	May 29.	June 8.	P. N. June 8.		July 6.	July 8.	none.	observed.
9. E. Evans	2	Bars.	Cobler.	May 30.	June 13.	P. N. June 13.	Croughton.	June 27.	June 27.	10, 11, 12, 13, Families.	transgressed.
10. A. Conolly	1	Bars.	Baker.	June 18.	June 23.	P. N. June 23.	Evans's 9th F.	July 10.	July 10.	none.	observed.
11. C. Jones	1	Bars.	Weaver.	June 22.	June 25.	P. N. June 25.	Evans's 9th F.	July 14.	July 15.	none.	observed.
12. H. Huxley	2	Bars.	Newsman.	June 20.	June 25.	P. N. June 25.	Evans's 9th F.	July 14.	July 15.	Smith's 17th F.	transgressed.
13. Mr. Jenkin	1	Bars.	Tanner.	June 16.	June 26.	G. R. June 26.	Evans's 9th F.	July 14.	July 15.	Morris's 15th F.	transgressed.
14. E. Alsop	1	Bars.	Soldier.	July 9.	July 11.	P. N. July 12.	10th, 11th, 12th, or 13th Families.	July 29.	July 29.	Downing 16 F.	transgressed.
15. M. Morris	1	Forest-st.	Shoemaker.	July 20.	July 23.	P. N. July 23.	Jenkins's 13th F.	August 3.	Aug. 4.	none.	observed.
16. A. Downing	1	Bars.	Sailor.	July 23.	July 27.	P. N. July 28.	Alsop 14th F.	August 2.	Aug. 4.	none.	observed.
17. A. Smith	2	Bunce-la.	Glazier.	July 22.	Aug. 6.	P. N. Aug. 6.	Huxley 12th F.	August 21.	Aug. 21.	none.	observed.
18. E. Tilston	1	Forest-st.	Shoemaker.	Sept. 10.	Sept. 26.	P. N. Sept. 26.		October 7.	Oct. 8.	none.	observed.
19. E. Johnson	1	Gorse-St.	Labourer.	October 4.	Oct. 5.	P. N. Oct. 5.		October 14.	Oct. 15.	none.	observed.
20. L. Bellis	3	Crooks-la.	Coachman.	1. Oct. 21. 2. Nov. 2. 3. — 3.	Oct. 29.	P. N. Oct. 30.		Nov. 27.	Nov. 21.	none.	observed.

Figure 3.3   Register of the small-pox in Chester, 1778 (Haygarth)
Source: Yale University, Harvey Cushing/John Hay Whitney Medical Library.

in many countries during the nineteenth century helped to reduce outbreaks. At the same time, because fewer outbreaks occurred, vaccination also became lax in many areas.

Figure 3.4 shows "infected houses" where one or more ill persons resided as well as the location of hospitals and other buildings. Edison's historical account of the outbreak explains that officials initially attempted to control the epidemic by quarantining the infected houses (shown on the map). However, this proved to be difficult to enforce, and on September 7, the State Board of Health quarantined the entire city; on October 8 infected patients were moved to the second hospital (Hospital No. 2 on the map) (Edison).

- *Correlations.* This map marks the "infected houses" street-by-street of the 1893 outbreak. The two variables shown are smallpox cases ("infected houses") and geographic space. "Infected houses" tend to be clustered, showing the spatial distribution of the disease and proposing a correlation between increased risk of contracting smallpox and space. The map does not provide temporal information beyond the year 1893.

- *Locations.* On the map, the natural environment includes a river, reserves, and a park (Heekin Park). The built environment includes streets and various buildings. The primary locations are the residential streets and the "infected houses" (marked with a red "x" on the map). Secondary locations include two hospitals, a school, and a courthouse. The map shows clear clustering of the "infected houses," with the greatest density north of the park, to the NNE and NNW, respectively.
- *Values.* The outbreak occurred as a result of lower vaccination rates in the area, with the last epidemic occurring 17 years earlier in 1876 (see *Epidemic*).

Individual cases are marked in Figure 3.4 showing the spread of disease in this 1893 outbreak. Identifying locations such as the factories, schools, and hospitals suggests that the map designers suspected that places where people were likely to congregate had some important connection with the transmission of the disease. The correlation of disease incidence and geographic space suggests that the map could be influential in persuading the city to resume a more consistent and inclusive vaccination program to reduce the spread of the disease.

Our next visual also maps an 1893 outbreak, this one in North Perth, Australia. See Figure 3.5.

- *Correlations.* This map shows three variables: the *location* of each case, possibly the place of initial diagnosis; the *order* in which each case was identified (case #1, 2, 3 and so on); and the *distance* of each case from the

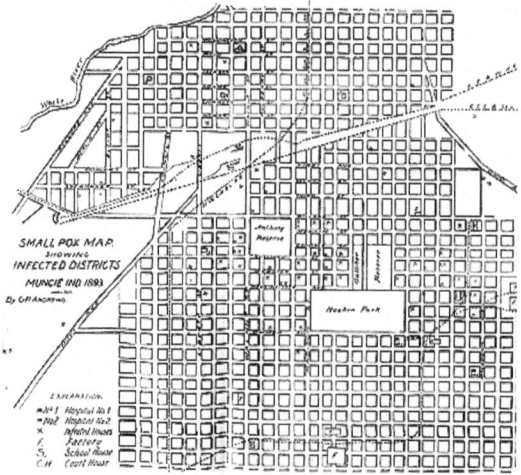

Figure 3.4   Smallpox outbreak in 1893 in Muncie, Indiana
Source: Courtesy of the Historical Medical Library of The College of Physicians of Philadelphia.

town hall. Distance is shown by marking the radius from the town hall in six quarter-mile increments. Visual relationships are hypothesized between *place* (location), *time* (the order of diagnosis via numeric sequence), and *space* (distance from the center of public and civic activities—the town hall). Cases tend to be clustered near each other, visually correlating these three variables and indicating an increased risk of contracting smallpox.

- *Locations.* This map shows the spread of this epidemic and, according to the report that accompanies the map, attempts to identify the location of patient zero—the initial case that presumably sparked the outbreak—to better understand the spatial and temporal distribution of the epidemic (Cumpston). Both natural and built structures are shown in great detail, with a number of built structures clearly labeled.
- *Values.* Smallpox was first recorded among non-native Australians in 1830, but native populations were affected as early as 1788 (Cumpston). The virus remained endemic in the country until its eradication in 1938 (New).

Much like Figure 3.4, Figure 3.5 marks individual cases in a way that shows a pattern of disease more explicitly rather than focusing on individual cases and demonstrating that the outbreak is not randomly or arbitrarily distributed in

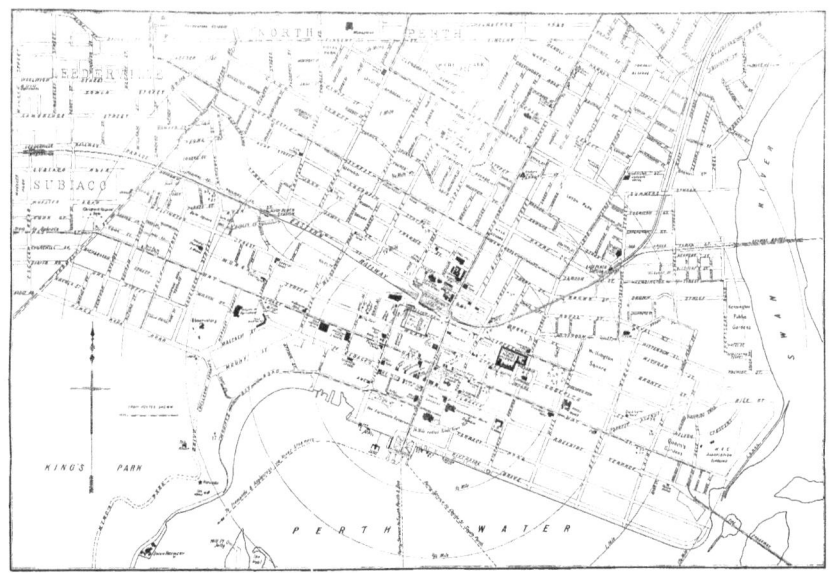

This Map shows the distribution of cases during the epidemic of 1893. The numbers correspond to the numbers given in the text.

Figure 3.5   Smallpox outbreak in Western Australia
Source: Yale University, Harvey Cushing/John Hay Whitney Medical Library.

this community; instead, the outbreak is clearly clustered and correlated with location—(#1, 3, 5, 6, 11, 16, 38), (#13, 24, 29, 25), (#4, 39, 40, 41, 42, 43), (#27, 45, 52), (#29, 32, 33, 36). Numbering the sequence in which each case occurred within the visual context of the outbreak shows an understanding of the infectious nature of the disease, a depiction of the development of the disease, and a sense that tracking these patterns might suggest ways to better manage and control epidemics.

Our final example shows the risks of non-compulsory vaccination by comparing smallpox incidence and mortality between the Prussian and Austrian armies. Plate 13 is included in a booklet entitled *Vaccination Law of April 8th 1874* (German Empire) that outlines the procedures for instituting mandatory vaccination in the general populace and includes detailed statistical graphics comparing smallpox mortality before and after the law was instituted. Text accompanying Plate 13, entitled "Table IV. Cases of illness and deaths from small-pox in the different armies from 1867–1901," which was originally published in 1883 and updated in 1901 for this booklet, explains that "the Prussian army enjoyed the advantages of careful revaccination, and the indirect protection of surroundings almost free from smallpox" whereas the Austrian armies were "incompletely vaccinated and [in their being] situated in the midst of incompletely vaccinated populations" (quoted in German Empire, 43).

- *Correlations.* The number of cases of smallpox is shown along with deaths for each year between 1867 and 1901 (displayed across the top of the graph). No data are available for the Austrian army from 1867 to 1870. The incidence and mortality are displayed across the bottom: blue numbers and bars show incidence, and the black numbers and bars identify mortality. The vertical scales display the possible numbers, from zero to 100 (actually a multiplier). Light grey bars show number of cases while the thinner black bars show deaths. Actual numbers are recorded along the x-axis for both.
- *Locations.* The exact location of the data collected is not noted on the graphs or in the accompanying text, but presumably smallpox incidence and mortality are shown for the entire Prussian and Austrian armies during the time period indicated.
- *Values.* The graph visually demonstrates the diminished risk of contracting smallpox if routine vaccination is practiced, noting that "No compulsory vaccinations before 1886; since then general vaccinations" (German Empire).

The graphs in Plate 13 directly correlate smallpox risk with vaccination as well as note the striking differences in incidence and mortality among the armies by stating, "It is impossible to adduce any other reason, than the effect of strictly carried out vaccination and revaccination for the remarkable differences in small-pox cases in the ... armies" (quoted in German Empire, 43). These bar graphs make explicit that the compulsory vaccination of the Prussian army in 1834 resulted in an immediate decline of deaths, advancing a strong and compelling argument in favor of mandatory vaccination for this particular population. Similarly, the compulsory vaccination of the Austrian army in 1886 also resulted in an immediate decline of deaths. The bar graphs show more dramatically than any textual explanation that vaccination reduced mortality caused by smallpox.

Conclusion

> The inmost kernel of every genuine and true understanding is visualization.
> (Schopenhauer, translated by and quoted in Funkhouser, "Historical" 269)

As we argue in this chapter, a shifting epistemic framework prompted creators of early statistical graphics in medicine, public health, and epidemiology to engage in visual thinking about information related to disease, health, and risks. The representations they created allowed them to tease out visual trends and patterns in disease occurrence, communicate risks visually, and subsequently make decisions that ultimately facilitated a broader, public health-based perspective. We illustrate these ideas with graphics that document smallpox epidemics in the seventeenth through the nineteenth centuries.

While our depictions focus on historic epidemic outbreaks of a specific disease, we suggest that our discussion may well be useful beyond this specific context. We have created an argument not by examining the chronological development of visual genres in medicine, public health, and epidemiology over this historic period but instead by identifying rhetorical tasks used to create visual representations that link problem-solving and visual thinking. Through our analysis, we also offer three contributions to the disciplinary conversation about the importance of visible numbers: discussions about correlations, locations, and values. Our tasks also serve as a lens for illustrating the move toward articulating relationships among variables that, in our examples, constituted smallpox risk and enabled public health officials to develop and enact public health policies that might mitigate those risks.

For us in the twenty-first century, visual representations of quantitative information are now often pre-defined and regularized, frequently applied formulaically to pre-existing data with minimal attention to the underlying rhetorical purposes that drive and subsequently inform their creation. In other words, genre conventions rather than rhetorical intent often dictate the creation of contemporary representations. At the same time, statistical graphics today are continuing to evolve in ways that embed and reflect contemporary social, political, and cultural values. These values are not only reflected individually in single representations, but holistically, highlighting the kinds of quantitative information that we as a culture have decided are important and are worth knowing and using to make particular kinds of decisions. In other words, these graphics do more than just "show" data; they organize, shape, and design information in ways that visually argue for particular interpretations, which then lead to certain types of decisions based upon this information.

Just as Graunt engaged in visual thinking by creating his tables and envisioning "some other uses," our contemporary ability (aided by new and emerging technologies) to find new shapes facilitates new visualizations. Indeed, engaging in productive, creative thinking now may result in new types of visual representation that appropriately respond to public health problems at the global health level. These new ways of seeing encourage us to envision relationships between and among quantitative information that have previously not been visible simply by looking only at the numbers.

PART II
Visualizing Nations:
Morality, War, Nationalism

As the nineteenth century unfolded, graphical displays told compelling stories about nation-states: their demographics, economic progress, moral condition, and imperial aspirations. Indeed, the term *statistics* originates linguistically and conceptually in data about "states." Statistical data opened new avenues for visualization, particularly later in the nineteenth century in the form of national atlases and albums that enabled the U.S., France, and other nation states to tell their stories in vast collections of graphical displays.

This part of *Visible Numbers* includes essays that explore the dynamic relationships between nation-states and their visualizations, charting how nations came to "see themselves" in a period of great innovation in data design. The moral and intellectual well-being of nation-states was partly measured by crime and education, and visualizing data about these indicators, and their correlation, began to shape public policy in the nineteenth century. Robert Cook and Howard Wainer examine Joseph Fletcher's graphical correlation of the two in England and Wales, as well as analyze data correlating education and economy in the U.S., both before and after the Civil War. In each case, Cook and Wainer use modern graphical techniques to re-envision the data and speculate on the implications for public policy.

Mapping practices profoundly shape the perception of nations, both their physical domains and the demographics and activities of their people. Comparing atmospheric mapping and census mapping over the past two centuries, Mark Monmonier analyzes how these two cartographic "domains" evolved along parallel paths in response to differing data sources and visualization problems, with designers sometimes resisting optimal methods, confounding the notion of cartographic "progress."

Nationalism also bred competition, economically and politically, and England stood squarely at the center of those rivalries, both domestically and internationally. Miles Kimball shows how English statistician Michael Mulhall used graphical techniques to visualize commerce comparatively across nations, positioning the English economy as a dominant imperial power. Deploying graphical displays to project power, Mulhall constructs a nationalist rhetoric to which many in his audience were readily receptive.

Nationalism also fostered imperial aspirations militarily, as Minard's iconic display of Napoleon's failed Russian campaign demonstrated, setting the stage for twentieth-century conflicts. In her chapter, Marguerite Helmers examines the consequences of nationalism in the trench warfare of the First World War and how visualizing those horrific conditions through increasingly sophisticated mapping techniques guided battlefield decisions, but with varying levels of success.

Chapter 4

Joseph Fletcher, Thematic Maps, Slavery, and the Worst Places to Live in the U.K. and the U.S.

Robert Cook and Howard Wainer

Joseph Fletcher

In 1847, Joseph Fletcher, a 30-year-old British barrister, published the first of several papers linking moral and educational statistics in England and Wales. Fletcher parlayed an amateur enthusiasm for collecting facts and figures about social and economic conditions into a career in the British government, first as secretary to different labor and health commissions, then as a school inspector for the British and Foreign School Society, a reformist agency dedicated to providing education to the poor of Great Britain. A key player in the early Victorian statistical movement—he served as editor to the journal of the Statistical Society of London from 1842 until his death in 1854—Fletcher created works reflecting the group's sentiments, reformist in their own right, and relied on the preponderance of statistics to make his positions clear (Alborn).

A theme running through much of Fletcher's work was the connection between ignorance and crime. This theme made its first appearance in a report to the commission of hand-loom weavers. He argued that weavers could transcend "the depravity" that accompanies rural upbringing if they could be concentrated in towns and educated in political economy and religion (so as to avoid socialist indoctrination) (Fletcher, "Reports"). Responsible statistician that he was, Fletcher was not one to fit the data to his hypothesis. In an 1843 survey of criminal convictions, he found the connection to be nuanced:

> Suppose, however, that we adopt the description of the state of instruction among the prison population contained in the gaol returns ... Here we find ... an excess of betrayed ignorance, which, considering the large proportion of juvenile offenders, the abstraction of many of the less poor on bail, and the absence of any nuptial pride or pretence to help out the ability of all, is by no means sufficient to dwell upon. The proportion ... [in prison] who can 'read and write well,' is as large as an extensive experience among our populations will lead any one to expect outside the prison walls. But granted that there is an excess of ignorance among the incarcerated, this only proves that the greater

number of them are derived from the poorer classes of society, upon whom, as such, are acting a thousand deteriorating influences ... (*Progress* 233)

Fletcher would continue to test the hypothesis in his later works, including four under the title "Moral and Educational Statistics of England and Wales," all reading as arguments for public education as a remedy against crime and other societal ills, such as pauperism and improvident marriage. In his first manuscript to take that title, in order to provide geographical context to the many tables of educational and moral statistics, Fletcher provided, almost as an afterthought, a key in the form of a single rudimentary map.

A Short History of Maps

Maps, whose invention dates back nearly 14,000 years, are the oldest form of graphic displays. This early development isn't surprising, for representing space as space is a natural metaphor. The oldest known map, engraved onto stone blocks and found in a cave in the Navarra region of Spain, depicted the surrounding landscape, including its mountains, rivers, and ponds, possible routes of travel, and on the side of one of the mountains, a herd of Ibex, a potential hunting target that may have provided the underlying purpose for the map's creation (Utrilla et al.).

Modern cartography can trace its roots more directly back to the ancient Mediterranean and Middle East. From this time and place come the first large-scale maps of the world and the cosmos; engineering plans for irrigation, drainage, gardens and building fortification; and maps of itinerary and military excursion. Mapmaking of the era may broadly (perhaps over-simplistically) be considered to emerge from two traditions: the Roman, emphasizing the practical in a way one might expect from a great and growing urban center with a bent toward imperialist expansion; and the Greek, focusing on larger theoretical aspects of geography, an emphasis true to the mathematical, philosophical, and scientific nature of ancient Greek thought (Dilke).

It is indeed the ancient Greeks to whom modern cartographers owe the greatest debt. In the period between Homer and Ptolemy, they added the concepts of a spherical earth, latitude, longitude, map projection, and orientation to the north to the science of cartography. After a medieval period that saw an emphasis in mapmaking toward the mystical and figurative, often seen as a regression given our pro-scientific bias, these lost scientific concepts were reintroduced into medieval mapmaking when Ptolemy's *Geography* was rediscovered, carried by a group of fourteenth-century Turkish refugees. This was a paradigm shift from figurative expression to scientific accuracy and was accelerated by the adaptation of printing techniques to mapmaking in the 1450s (Robinson, *Early*).

In the 200 years that followed, cartography saw evolutionary rather than revolutionary progress. It was a time of great cartographic activity, as it coincides nearly exactly with the Age of Exploration that saw European "discovery" of the Americas and circumnavigation of the globe, among other things. While undoubtedly facilitated by the advances in mapmaking of the previous century and a half, the era offered no profound cartographic innovations of its own, inclusion of the so-called New World to world maps notwithstanding. The mid-seventeenth century saw two developments in cartography that could be considered revolutionary: the addition of topography on general maps and the development of a map of an entirely new sort—the thematic map (Robinson, *Early*).

The Impending Crisis of the South

In 1857, Hinton Rowan Helper, a North Carolina native and son of a slave owner, called for the end of slavery in the United States, using every conceivable argument available to him to make the case that slavery was wrong for the South. To build what he believed was an unshakeable case that the institution of slavery must come to an end in the South, Helper used moral and fiscal arguments, balanced fact with opinion, and borrowed the words of prominent figures, historical and contemporary, political and religious, U.S. and foreign. And he used statistics. While clearly wishing that the weight of a passionate moral argument against slavery would be enough to compel the majority of Southerners—non-slave holders—toward abolition, Helper believed that a statistical case that the success of the South and the happiness of individual Southerners depended on the end of slavery would resonate more strongly:

> the political salvation of the South depends on the speedy and unconditional abolition of slavery. We will not ... rest the case exclusively on our own arguments but will ... appeal to incontrovertible facts and statistics to sustain us in our conclusions. (28)

Helper's treatise on the subject, *The Impending Crisis of the South: How to Meet It*, contained over 400 pages of textual argumentation. As compelling as his own words were on the subject, he relied strongly on the testimony of those whose words he felt would carry more weight, including five of the first six U.S. presidents (four of whom hailed from slave states), revolutionaries Patrick Henry and Benjamin Franklin, and great historical minds like William Shakespeare, John Locke, Plato, and Socrates. As many pages as he devoted to moral and social argument, he spent an equal number presenting statistical figures and their interpretation. His intent was clear:

> What we mean to do is simply this: to take a survey of the relative position and importance of the several states of this confederacy,[1] from the adoption of the national compact; and when, of two sections of the country starting under the same auspices, and with equal natural advantages, we find the one rising to a degree of almost unexampled power and eminence and the other sinking into a state of comparative imbecility and obscurity, it is our determination to trace out the causes which have led to the elevation of the former, and the depression of the latter, and to use our most earnest and honest endeavors to utterly extirpate whatever opposes the progress and prosperity of any portion of the union. (Helper 11)

Contrary to what is typically believed of abolitionists, Helper was a vehement racist whose sympathies lay with free white labor in the South rather than with the slave. He had a reputation for careless application of statistics. Though his intent in writing *The Impending Crisis* was to influence southern attitudes, the book was banned in the South while becoming a best-seller in the North. These facts complicate his premise, but they do not undermine its interest. And, like Fletcher in the U.K., Helper's focus on the relationship between social policy and societal quality and his reliance on geographical data to tell his story provide a useful tableau against which to tell ours.

Demographic Data Display

Scholars often date the gathering of what might be called social statistics from John Graunt's 1662 analysis of the London Bills of Mortality. Graunt analyzed and published birth and death data collected by London parishes. But Graunt included no graphs or maps in his collection. John Arbuthnot in 1710 famously used Graunt's data to illustrate his invention of null hypothesis testing, but he too did not include any graphs. In fact, it has been argued elsewhere that the clerical errors in Arbuthnot's paper provide strong evidence that he never graphed the data; had he prepared such a plot the errors would have stuck out (literally) like sore thumbs and been corrected.[2] That these errors remained uncorrected suggests that Arbuthnot never looked at Graunt's data visually.

Bits and pieces of graphic display of social data appeared in the late seventeenth and early eighteenth centuries, most notably Christiaan Huygens' drawing of a survival graph in a letter that he sent to his brother in 1669, but graphs of social (economic) data only entered the scientific mainstream in 1786 with the publication of William Playfair's *Commercial and Political Atlas*. Playfair prepared wonderful and elegant graphs of economic data, inventing the

---

[1]   Helper used this term to refer to the entire United States, a confederacy prior to the Civil War.

[2]   See Wainer, *Graphic*.

line graph, the bar graph and the pie chart in the process, but despite the book's purpose as a record of world economics, he included no maps.

Fletcher's Maps—Charting Moral and Educational Statistics

Fletcher's first map appeared in his 1847 manuscript on moral and educational statistics and was not thematic, but a general map of British counties that required the reader to apply the theme by connecting tabular data to it. Within the manuscript he explicitly rejected the use of thematic maps, suggesting that simply perusing the columns of numbers would suffice, avoiding the bother and expense of drafting and printing maps. He reversed himself completely two years later in two articles of the same title (but of much greater length) as his 1847 paper.[3] These included many shaded maps that were innovative in both content and format. Format first.

Fletcher, having a background in statistics, or at least what was thought of as statistics in the mid-nineteenth century, did not plot the raw numbers, but instead their deviations from the mean. And so the scale of tints utilized varied from the middle, an approach that is common now. For example, Andrew Gelman, in his remarkable book explaining the 2008 presidential election results, uses increasing saturations of red to show the proportion of voters that voted Republican in states where they were in the majority and increasing saturations of blue for those states that voted Democratic. The closer the vote came to a 50:50 split between the two parties, the closer the shading was to white regardless of the final outcome. But in 1849, the approach was an innovation. Centering all variables at zero allowed Fletcher to prepare a number of maps, on very different topics and very different scales, and place them side-by-side for comparison without worrying about the location of the scale. He also oriented the shading, as much as possible, in the same way. In his words, "In all the Maps it will be observed that the *darker* tints and the *lower* numbers are appropriated to the *unfavourable* end of the scale, whether of influences or results."[4] Of course, with variables like population density it is hard to know which end is favorable. He chose lesser population as more favorable perhaps as a reflection of Malthusian concerns about overpopulation still commonly expressed is his

---

[3]   Ordinarily it would be hard to know exactly what it was that compelled Fletcher to change his approach so completely, but he provides an evocative hint in his 1849 paper, when he states, "A set of shaded maps accompanies these tables, to illustrate the most important branches of the investigation, and I have endeavoured to supply the deficiency which H.R.H. Prince Albert was pleased to point out, of the want of more illustrations of this kind" (*Summary* 79).

[4]   Fletcher included this statement at the top of each table accompanying the plates. The section for these tables and plates appeared at the end of the book, following the last page of text (roughly 216).

day (e.g., Mill, *Principles*, Doubleday, *True Law*) or perhaps simply because it seemed to accompany favorable outcomes on the other variables.

However noteworthy Fletcher's innovations in format, the content he chose makes him truly special. He did not make maps of obvious physical variables like wind direction or altitude or even the distribution of the population (although he did make one such map, it was fundamentally for comparative purposes). No, he had more profound purposes in mind. Joseph Fletcher made maps of moral statistics, and opposed these maps with others to suggest causal interpretations to his viewers.

For example, juxtaposed next to his plot of *Ignorance in England and Wales*, he placed a map of the incidence of crime (see Figure 4.1 and Figure 4.2). He opted for this juxtaposition on the basis of detailed analysis of subsets of the data. He wrote, "We thus find that the decline in total ignorance to be slowest in the most criminal and the most ignorant districts" (*Summary* 44). He then tied this analysis into the phenomenon made observable through the juxtaposed maps, stating,

> the darkest tints of ignorance go with those of crime, from the more southern of the Midland Manufacturing Counties, through the South Midland and Eastern Agricultural Counties ... and it will be well to observe, as an example of their use, that all four of the tests of moral influences now employed are seen to be on the side of the more instructed districts. (*Summary* 91)

Looking into the phenomenon at a more microscopic level, he observed,

> The two least criminal regions are at the opposite extremes in this respect (the Celtic and the Scandinavian), with this important difference, that in the region where there is the greatest decline of absolute ignorance among the criminals (the Scandinavian), there is not one-half of the amount of it in the population at large which exists in the other. (*Summary* 44)

In addition to these maps, Fletcher produced parallel plots of *Bastardy in England and Wales*, of *Improvident Marriages*, of *Persons of Independent Means*, of *Pauperism*, and many other variables that could plausibly be thought of as either causes or effects of other variables. His goal in doing this was to generate and test hypotheses, which might then guide subsequent social action and policy. One such hypothesis he framed as follows:

> The practical resistance of the country or the people in these latter districts to any system of higher economy or better cultivation is as conspicuous on the map as it is on the face of the land itself; the regions occupied by the Celtic populations, forming a class apart, quite beyond all the others in the excess of their population in proportion the assessable value of the soil which it cultivates. A reference again to Map V. shows that these regions are equally *under* the average in observed delinquency and to Map IV., that they are equally in excess

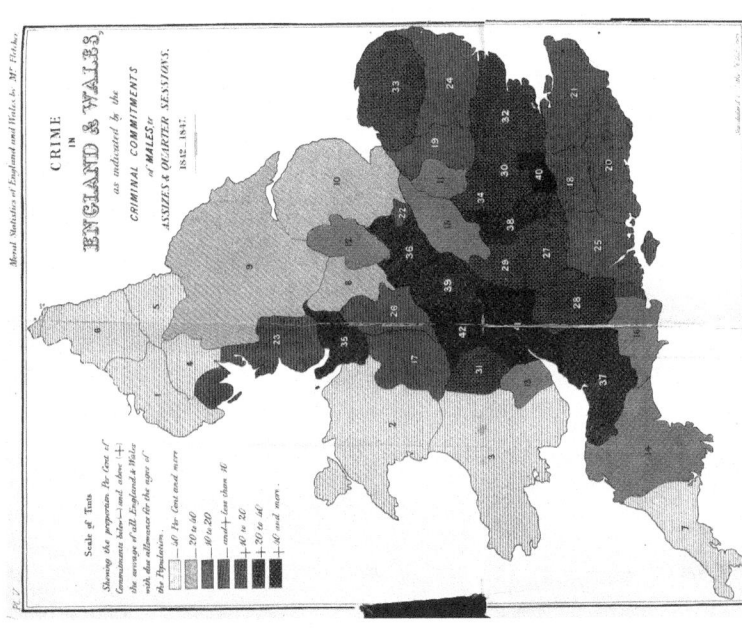

Figure 4.2   Joseph Fletcher's "Crime in England and Wales" ("Moral," 1849)

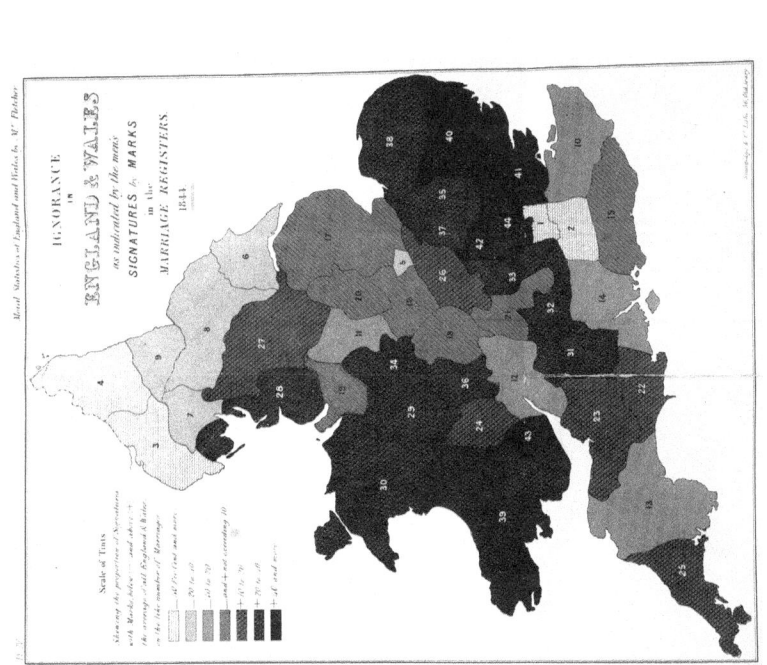

Figure 4.1   Joseph Fletcher's "Ignorance in England and Wales" ("Moral," 1849)

of ignorance; whence I am induced to draw the conclusion, to be tested by further experience, *that an ignorant people, engaged in rural industry, will exhibit a less amount of crime when that industry is organized on the plan of the small rather than of the large husbandry,* and, as a necessary corollary, from what has already been shown, that *the introduction of an improved economy into the organization of agricultural as of manufacturing labour, demands for its security and the general welfare a higher moral development among the whole population* than enabled society to exist in its ruder form. (*Summary* 90)

His epistemological viewpoint expressed here was both wise and surprisingly modern.

Early Thematic Maps

Adding a third variable onto a geographic background is a relatively recent innovation in cartography. Partially thematic maps appeared in China as early as 164 BC, but the additional variables plotted were geographic variables like altitude, which were ancillary to the main purpose of the map. As Robinson chronicles, thematic maps focusing primarily on the additional variable did not appear until much later:

> No map which is primarily thematic appears to have been made before the last half of the seventeenth century … It was not until some exceptional individuals became inquisitive about the earth, and later about its inhabitants and their activities, that their curiosity resulted in thematic maps. (Robinson, *Early* 17)

Robinson does an excellent job of defining and differentiating general maps and thematic maps (*Early*). Whereas it is inherent in maps to identify and locate geographical features, for the general map, that is its primary purpose; for the thematic map, such features serve only to give place to variables that traditionally fall outside the realm of general cartography:

> The thematic map concentrates on showing the geographical occurrence and variation of a single phenomenon, or at most a very few. Instead of having as its primary function the display of the relative locations of a variety of different features, the pure thematic map focuses on the differences from place to place of one class of feature, that class being the subject or 'theme' of the map. The number of possible themes is nearly unlimited and ranges over the whole gamut of man's interests in the present and past physical, social, and economic world, from geology to religion, and from population to disease. (Robinson, *Early* 16)

Initially, variables plotted on a geographic background were physical, such as magnetic phenomena, currents, and geology. The earliest published thematic map was produced by Athanasius Kircher in 1665, showing the horizontal circulation of ocean waters. Kircher conceived of a thematic map earlier than this point, describing in a 1641 publication how to make a map that showed distribution of magnetic declination over the earth, but the cost of actually producing the map was prohibitive (Robinson, *Early* 46). Edmond Halley, most famous for calculating the orbit of his eponymous comet, published what is believed to be the first meteorological chart in 1686. The first thematic geological maps were produced by Phillipe Buache in 1746 for a 1751 manuscript by Jean-Etienne Guettard. Guettard would be regarded as the father of geological maps for his emphasis of their importance. Thematic geological mapping became increasingly common and, in 1815, William Smith (1769–1839) prepared a map depicting the geology of England, showing that specific fossils were found in corresponding layers of rock; this severely conflicted with the biblical interpretation in which the rocks were laid down during the Great Flood. Smith's production has been called the "map that changed the world," instigating changes in thinking so that by the end of the nineteenth century, geology was, at least in terms of chronology and identification, advancing as a modern, evidence-based science (Winchester).

A lag between the development of the physical sciences and social sciences was mirrored in thematic mapping. Even the arrangement of numerical geographical data was a question until 1754, when A.F. Busching conceived of putting data points into rows and columns for the first time. So far as is known, very few maps of such social subjects as population, religion or industrial production appeared before 1820. The preparers of the earliest such maps merely wrote the statistic of interest on the map. An example of this practice is J. Wyld's 1815 map "Chart of the World Shewing the Religion, Population and Civilization of Each Country." Robinson notes that such maps only barely "qualify as thematic since they do not graphically portray the character of the distribution effectively" (*Early* 113). While putting numbers on a map to indicate the location where something is happening undoubtedly conveys more information than merely listing the data in a table, such a practice does not fully utilize the graphic medium.

More sophisticated use of thematic maps began on November 30, 1826, when Charles Dupin (1784–1873) gave an address on the topic of popular education and its relation to French prosperity. He used a map shaded according to the proportion of male children in school relative to the size of the population in that département. This graphic approach was improved in 1830 when Frére de Montizon produced a map showing the population of France in which he represented the population by dots, each dot representing 10,000 people. Although no one remarked on it for the better part of a century, it remains perhaps the most important conceptual breakthrough in thematic mapping. It was the direct linear antecedent of John Snow's famous map of the 1854 cholera epidemic in London. Snow's map, presented in Figure 4.3, is generally referred to as the beginning of modern epidemiology.

After Montizon's map, a number of others between 1830 and 1845 used shading to convey population. Often these maps were roughly drawn and clearly

Figure 4.3   John Snow's 1855 map showing the location of each death during
            the cholera epidemic that struck central London in September
            of 1854. Snow deduced the cause of the epidemic from their
            proximity to the Broad Street water pump

meant to be of only secondary importance to the prose in the document in which
they appeared. But a few were clever, remarkably polished displays. Exemplary
among these were population maps, designed and prepared by Henry Drury
Harness (1804–1883), a young lieutenant in the Royal Engineers, that appeared in
an 1837 atlas as accompaniment to a report prepared by railway commissioners.

     Maps of moral statistics started appearing at about the same time; these
typically focused on various aspects of crime. Most widely known were those of
Adriano Balbi (1782–1848) and André-Michel Guerry (1802–1866), whose 1829
map pairing the popularity of instruction with the incidence of crimes was but the
start of their continuing collaboration. Their work was paralleled by the Belgian
Adolphe Quetelet (1796–1874) who, in 1831, produced a marvelous map of crimes
against property in France, in which the shading was continuous across internal

boundaries rather than being uniform within a département. Guerry expanded his investigation in the 1830s to add three other moral variables (illegitimacy, charity, and suicides) to crime and instruction. Thus, despite graphical display's English birth, its early childhood was spent on the continent.

Applying Modern Techniques to Fletcher's Maps

Fletcher reluctantly brought thematic mapping of social data back across the channel, and in so doing added his innovative statistical approach to create new shading schemes to his variables. One can't help but wonder how much more effective his maps might have been had he access to modern robust statistics and color plotting techniques. Were he constructing these plots today, he could have looked at standard deviations rather than proportional deviations from the mean. He could have centered on the median instead of the mean for severely skewed variables. With affordable access to color printing, Fletcher would likely have followed Gelman's lead in plotting points above the median in varying saturations of red and points below the median in varying saturations of blue (or vice versa). And had he access to modern plotting software and laser printers, he likely would have abandoned the practice of categorizing the variables in favor of using a continuous method of shading. As Bertin has taught us, the representation of a variable should match the variable itself (*Semiology*). Thus, if a variable is continuous so too should be its graphical representation.

Plates 3 and 4 are versions, redone using modern variables and techniques, of Fletcher's plots of ignorance and crime previously shown as figures 4.1 and 4.2. Careful study will reveal the advantages of the modern approach. Most strikingly, the bicolor shading scheme renders his suggested interpretations all the more readily observable. Where comparison of Fletcher's maps may involve careful scrutiny of shading and frequent reference to the legend, at least before thorough familiarization, the modern versions of the same are more easily interpretable, needing only to have red matched to red, and blue to blue, to more quickly get an idea of how the proportion of crime within counties is related to the proportion of ignorance.

Mapping Hinton Helper's Moral and Educational Statistics

Hinton Helper, despite including a preponderance of tables of geographical data in *The Impending Crisis*, did not include even one map, nor did he make any attempt at graphical display that might aid in our interpretation of his data. The United States' own rich history of thematic mapping blossomed into the social arena a decade too late for him to utilize. Although some excellent thematic maps were produced prior to this point, their emphasis was on climatology and epidemiology. It was the census of 1860 that seemed to spark a burst of social thematic map activity with many maps of census data including themes of interest to Helper like agricultural production and distribution of the slave population. Helper missed an opportunity to be an innovator in this area and perhaps more compellingly make his arguments with strong visuals.

Given Helper's ability as a wordsmith and his reputation for statistical laziness, it becomes interesting to see if his story holds up to visual scrutiny. If we can apply modern techniques to Fletcher's maps, so too can we apply those techniques to build maps based on Helper's data. First we will simplify his argument so that we can evaluate it without production of maps for the multitudes of tables he produced (as persuasive as the North/South production of flaxseed may be to his case). Simply put, Helper's hypothesis is this: slavery is at the root of Southern states' disadvantage. We know better than to mistake causation for correlation, but we can at least evaluate the extent to which slavery is associated with hardship among the general southern populace. To do so, we constructed maps of four state-level variables from Helper's tables that get at the essence of the problem: proportion of population that is made up of slaves, illiteracy rate, wealth per capita, and death rate. (See plates 5, 6, 7, and 8 respectively.)

If Helper's premise is true, then the colors on the map showing the slave population (Plate 5) should correspond strongly with the other three maps (reverse coding wealth, since negative association is with lower values for this variable but with higher values in the other three; see plates 6,7, and 8). The only disadvantage conferred by slavery that is made apparent by comparing maps is that of illiteracy. Illiteracy figures included only white adults in both slave and free states, so the correlation cannot be blamed on the disproportionate illiteracy among slaves. In fact, the relationship only holds up when slave ownership is treated dichotomously state to state. Slave states are at a considerable disadvantage, but within the slave states this disadvantage does not seem to be strongly related to proportion of the population that is made up of slaves. For instance, South Carolina and Alabama are among the highest in slave proportion while being among those Southern states that compete favorably with the North in illiteracy. Arkansas and Tennessee have lower proportions of slaves to non-slaves, but are among the worst in illiteracy.

Southern states appeared generally to have more wealth per capita than the North at this time. In fact, wealth in the slave states corresponds almost exactly with proportion of the population that is slave, with South Carolina, Louisiana, and Mississippi the top three in both categories. It is unsurprising that if an economy treats human beings as property, those with the most wealth of this sort are also among those with the most wealth in general. Interestingly, Massachusetts, the wealthiest of the free states, was also the heart of the textile industry, making its wealth off of the same cotton that was making the South rich.

Death rates correlate to other social and economic variables, but proportion enslaved is not one of them. Treating slavery dichotomously favors the South on death rate, Louisiana being a notable exception. Massachusetts ranks second in death rate, a figure perhaps also attributable to its prominence in the textile industry, an industry notorious for both its pollutants and its unsafe work environment.

The Worst American State

As we can compare two variables by juxtaposition of thematic maps, we can look at one variable at different points in time in the same way. A more nuanced take on Helper's thesis would be that slavery was not so much responsible for any immediately observable shortcomings of the South as it was for perpetuating Southern decline and Northern advantage.

In 1931, renowned American essayist H.L. Mencken and Charles Angoff, his assistant editor at *The American Mercury*, published a three-part series in the magazine compiling state-level data on numerous variables in an effort to rank the different states in the U.S. The series was appropriately but pessimistically titled "The Worst American State." Acknowledging the conceit that "Every man tries to measure the level of a given culture by his own yardstick," Angoff and Mencken made their case that a general agreement could be reached by borrowing the 1928 words of Dr. A.J. Todd: "We shall have to agree that life on the whole is better than death, that health is better than sickness, that freedom is better than slavery, that control over fate is better than ignorance" (quoted in Angoff and Mencken 1). Among the indices selected for measuring which state was "worst," Angoff and Mencken included illiteracy rates, death rates, and taxable property per capita (i.e., wealth). As with Helper, the authors felt that a preponderance of tables was adequate for telling their story, but we will make our analysis and comparisons with maps of the three variables, shown in plates 9, 10, and 11.

Illiteracy, the most clear indicator of Southern disadvantage in the 1850s, demonstrates the same, only more strikingly. A third category of state enters the picture with the 1930s data: the newcomer state. Seventeen states joined the union in the time between Helper's book and the "Worst American State" series. For the most part, these newcomers are among the more literate states, with two notable exceptions: Arizona and New Mexico. These are, coincidentally, the two southernmost of the new states. Founded as part of the same non-slave territory, their illiteracy may stem more from a population made up of miners, railroad laborers, and non-English speakers.

Wealth, actually a slave-state advantage in the 1850s data, is now as clearly in favor of the free states as is literacy. Newcomer states likely win the category, with Montana, Arizona, and Nevada leading the way. If we consider only free and slave states, with the exception of Vermont, the poorest of the remaining free states is wealthier than the wealthiest of the southern states. Death rate confers an advantage only on the northern newcomer states, with a map of just the free and slave states remaining a murky picture and bearing no resemblance to the 1850s map. Arizona and New Mexico, the two least literate states in 1930, are also the most deadly.

While the North/South dichotomy is made clear in two of the three measures, how closely do those measures correspond to proportional slave population in the 1850s? Death rates are made no clearer by this consideration. South Carolina and Mississippi, two of the highest in terms of their slave proportion, are also among the top in death rate in 1930, but so are Delaware and Maryland, among the lowest in proportion of slaves among those that had them. Wealth in 1930 is nearly an inverse of the slavery map, Mississippi and South Carolina going from richest to poorest while the three slave states with the fewest slaves per capita—Delaware, Maryland, and Missouri—were the wealthiest by 1930. These were also the northernmost of the slave states and none of them seceded, so to attribute their relative success solely to their non-dependence on slavery would be ill considered. On the other hand, tracing the freeing of slaves to poverty in the South involves some unsavory but reasonable math. If one takes away a substantial portion of a population's wealth (their slaves), that is going to severely impact the wealth of the state. Reintroducing the freed slaves into the state in the form of an impoverished new citizenry hits the state's wealth figures a second time. The map of illiteracy in the South of 1930 looks nearly identical to the map of slave proportion in 1850. It is easy to imagine how slaves becoming citizens and having extreme educational and economic disadvantage conferred upon them could result in this outcome.

Slavery as the primary cause of southern decline is difficult to infer in this case, with 70 years between measures, numerous other plausible hypotheses, and the Civil War abruptly bringing an end to slavery rather than allowing it to run its course without such intervention. Drawing those lines of causality through the Civil War, while reasonable, would be disingenuous in making Helper's argument. Freedom for the slaves may very well explain the wealth and literacy disadvantages for the slave states, but that is at least in part a byproduct of the freedom for which Helper advocated. Helper's concern was for the impact of slavery on the white population, and it is difficult looking at these numbers to tell whether freeing the slaves was helpful or detrimental in that regard. On the whole, freeing the slaves seemed to create an overall shift in wealth, but this too is likely just a shift in definition, as these people existed without property prior to being freed and were simply not counted fairly. Fiscal poverty was a reality—though also a gross minimization of the condition—for the slaves even if it was not considered as such.

Putting causality aside, Helper's predictions for the South were realized by 1930. The three variables we chose for analysis from all of Angoff and Mencken's data closely reflected their final findings in regards to the Worst American State. From best to worst, they placed Massachusetts at the top and a slave state was not to be found on the list until numbers 25 to 27 (Delaware, Missouri, and Maryland), the northernmost and least slave-dependent of the slave states. At the bottom were Mississippi, Alabama, South Carolina, and Georgia, the four states with the heaviest proportion of slaves to free men. A non-slave state doesn't appear from the bottom up until number 39, newcomer New Mexico, and a free state doesn't appear until 23, Indiana. A map of Angoff and Mencken's overall rankings, with worst tinted red and best tinted blue, is shown in Plate 12. With the exception of a good part of the southwest, it doesn't read very differently than a modern map of the popular presidential vote, shown in Plate 14.

The Worst British County? Recontextualizing Fletcher in Modern Times

These thematic mapping techniques can also be employed to see how 150 years of social programs in the U.K., the arguments for many of which inspired Fletcher's thematic maps, have changed the faces of crime and ignorance on the landscape. To this end, let us look at revisions of plates 3 and 4 using modern data, shown as plates 15 and 16. With a careful look, we can see that the relationship between ignorance and crime seems more tenuous. The relationship does appear to hold in the area of highest crime, which is perhaps not coincidentally home to the better part of London, the biggest urban center in England. Overall, though, the lack of a strong link between ignorance and crime in the modern data suggests a possible interpretation that effective social programs have reduced the need of the ignorant to turn to crime in order to survive.

The maps can also be juxtaposed to see if geographical patterns of crime and ignorance have changed over time. As noted, crime rates are highest in the modern data in counties with major urban centers. This holds historically, as well. The 1845 crime map's two darkest red counties are the homes to London, its largest city, and Birmingham, its industrial center and "the first manufacturing town in the world" (Hopkins, *Birmingham*). Beyond that, understanding the geographical shifts of crime in the U.K. would involve deeper anthropological investigation. Counties with highest rates of ignorance were found primarily in the central latitudes of England and Wales in 1845, whereas the line of ignorance in the modern map is drawn to create a fairly strict North-South dichotomy.

Proceeding with Caution—Addressing the Modifiable Areal Unit Problem and Ecological Fallacy

David Unwin cautions against arriving at inferences based on thematic maps. He highlights two sources of potential misinterpretation: the modifiable areal unit problem and ecological fallacy. The modifiable areal unit problem recognizes that relationships may be a byproduct of the units of area selected to define the data. We can, for instance, observe a connection between slave populations in 1850 and the decline of wealth and literacy in the slave states by 1930. However, it could be that the wealthiest and most literate free states are dominated by large urban centers and were we to separate those out, the large rural sections of these free states are no more wealthy or literate than their rural slaveholding counterparts.

Defining our data at the state level obscures this plausible interpretation. Unwin notes that this may also be the actual driver for any relationship between crime and ignorance in Fletcher's data.

Ecological fallacy is the mistake of inferring that relationships between variables that may exist in certain areas also exist for individuals within those areas. If, for instance, we note a relationship between slaveholding and illiteracy at the state level, we should be cautious in inferring that slaveholders are illiterate. It is highly likely that the illiterate population of the slave states were largely found among the poor whites who owned no slaves. Unwin notes that the relationship between criminality and ignorance at the county level in England and Wales does not necessarily mean that the criminals are the ignorant in those counties. While that may be the case, one cannot draw that conclusion from looking at a county-level map.

Indeed, while this chapter is intended in part to demonstrate the superiority of the thematic map over simple tabular display of geographic data, its superiority lies in improved visualization of the information; it confers no additional causality. These sources of misinterpretation are, however, issues of data definition, not of data display. Clarity of presentation is only responsible for error in the sense that a failure to recognize a relationship leaves no potential to misinterpret it.

Scatter and Stem-and-Leaf Plots—Augmenting Map Display to Tell the Whole Story

Though understanding of the relationship between variables as well as their change over time through the comparison of thematic maps improves considerably upon sole reliance on tabular presentation, the process is neither easy nor precise—even using modern analysis and shading techniques. Fletcher's use of thematic maps was innovative but perhaps not the ideal tool available at the time, at least by itself. Had he an interest in astronomy, Fletcher may have come upon a more useful tool for depicting the relationship between these two datasets: the scatterplot. In his "investigation of the orbits of revolving double stars," published in 1833, British astronomer John Frederick William Herschel plotted the positions of stars on one axis against the observation year on the other and drew a smoothed curve between the points to represent the relationship between the two variables and, in so doing, unknowingly staked his claim as inventor of the scatterplot (Friendly and Denis, "Early" 116). Such a representation of Fletcher's 1845 crime and ignorance data more clearly illustrates the relationship between the two variables and is presented here in figures 4.4 and 4.5. In Figure 4.4 we show the relationship between ignorance and crime in the United Kingdom in the 1840s, and in Figure 4.5 we show the relationship between the same variables in contemporary Britain. The positive relationship Fletcher showed via his maps is made clear in the scatterplot based on his data, but perhaps not with the strength he would care to demonstrate.

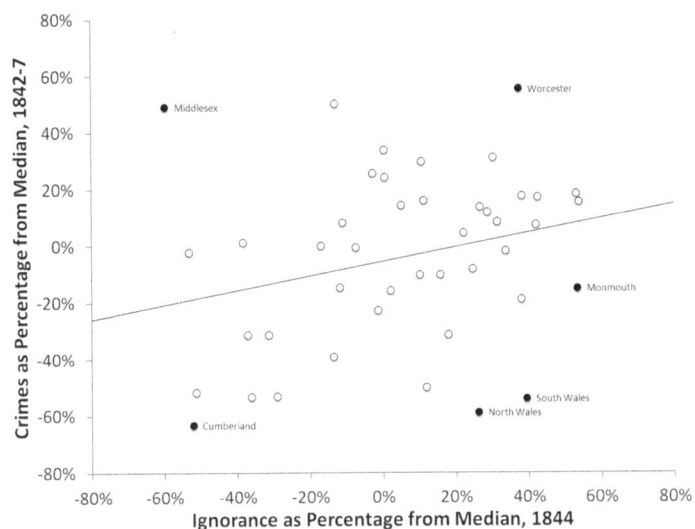

Figure 4.4   A scatterplot of ignorance versus crime for the nineteenth-century data vividly shows the positive relation that Fletcher discusses. It also highlights the unusual character of the county of Middlesex (home of greater London), as well as the similarity of the Celtic counties (Monmouth, north and south Wales)

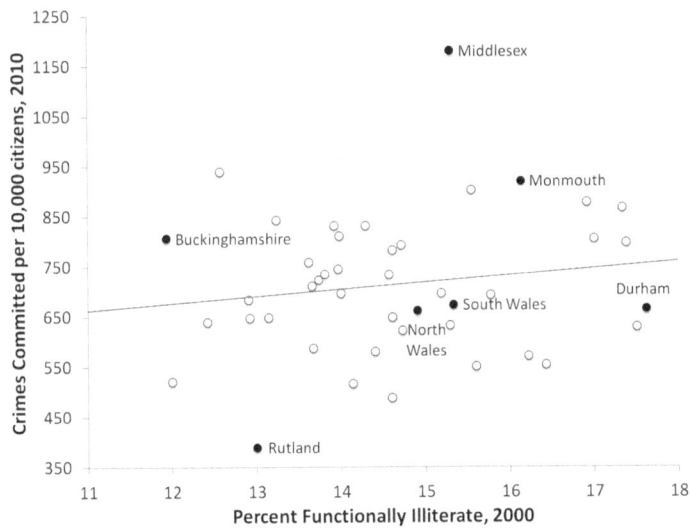

Figure 4.5   A scatterplot of ignorance versus crime using modern data maintains the overall historical relationship between the variables, though the relationship does not hold for most individual counties

The same positive relationship exists for the contemporary data, though the positions of the individual counties on these variables bear little resemblance to their historical positions. It is interesting that the exceptional relationship between high crime and low ignorance that Fletcher observed in the Celtic counties is considerably weakened, but the positions of the counties relative to one another is nearly perfectly preserved.

The bivariate scatterplot has the power to illuminate other relationships as well. In Figure 4.6 we compare the distribution of ignorance in the U.K. in 1844 with that of the present; in Figure 4.7 we do the same thing with crime. Ignorance is shown to have reversed such that the least ignorant counties in 1844 are now the most ignorant, and vice versa. There is no such reversal in the distribution of crime between then and now but the relationship is weak at best, with Middlesex being the most notable exception.

That there is a better way to represent the relationship between crime and ignorance in the U.K. than by juxtaposition of monochromatic maps does not diminish the ingenuity of Fletcher's approach in his day. Had he more sophisticated statistical tools, convenient color data mapping techniques, and knowledge of the scatterplot, he would no doubt have taken advantage of them in presenting his moral statistics. Given the tools available to him, his use of thematic maps in making his case was ingenious. Because of the tools we now have at our access, the use of thematic maps remains an excellent method for representing the geographic distribution of social data. For representing the geographic relationship between two variables, the addition of a scatterplot to make clear the exact nature of the relationship is invaluable.

In supporting Helper's claims, we noted the relationship between slaveholding and overall state rank by 1930, a relationship made clear in Plate 12 (assuming one can recall which are the free and slave states) but powerfully demonstrated with an even simpler device: the stem-and-leaf plot. Figure 4.8 is a back-to-back stem-and-leaf display with slave states on the left and free states on the right, ordered from worst state to best.

Applying Modern Display Techniques to the Modern U.S.

In the United States we saw Helper's prophecy realized by the 1930s. Well into the twenty-first century, do the trends of Southern decline and Northern advantage hold? Do the newcomers still largely outpace the slave states while not quite competing with the pre-Civil War free states? Again, we can apply our modern techniques to make a visual determination.

Rather than use illiteracy numbers, we chose a more modern conception of ignorance: eighth-grade NAEP (National Assessment of Educational Programs) reading scores (See Plate 17). There is a strong North/South dichotomy that

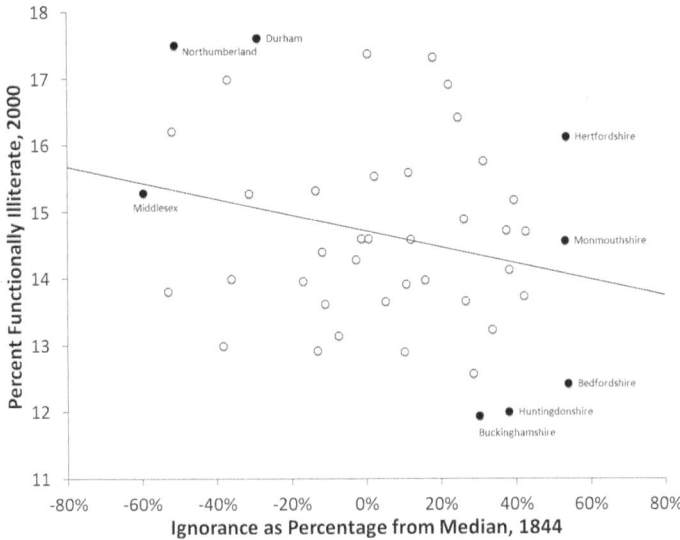

Figure 4.6   A scatterplot of ignorance for the two time periods shows a
               remarkable shift; the least ignorant counties in the nineteenth
               century tend to be the most ignorant now and vice versa

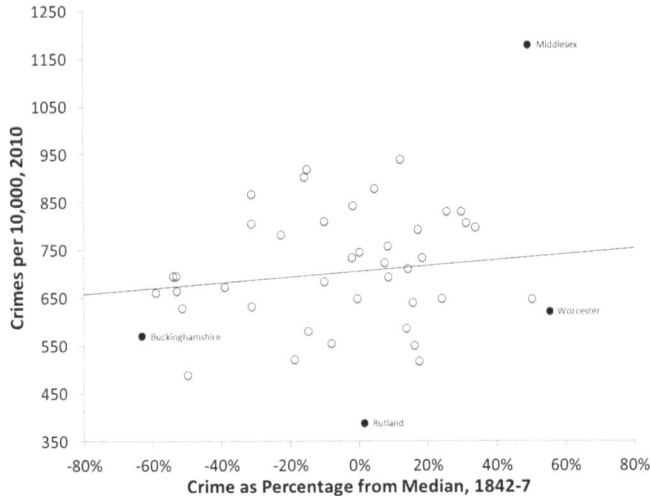

Figure 4.7   A scatterplot of crime for the two time periods shows only a very
               modest relationship between the level of criminality in a county
               over the two time periods depicted. The principal exception to this
               lack of trend is the county of Middlesex

	Slave States		Free States
		Worst	
SC AL MS	46-48		
LA NC TN AR GA	41-45		
FL VA TX KY	36-40		
WV	31-35		
MO MD	26-30		
DE	21-25	VT IN	
	16-20	OH PA	
	11-15	MI ME WI NH	
	6-10	IA IL RI	
	1-5	MA CT NY NJ CA	
	Best		

Figure 4.8   Stem-and-leaf plot of free and slave states' ranking on Angoff and Mencken "Worst American States"

	Slave States		Free States
		Worst	
TN AR MS AL SC	46-50		
GA WV LA	41-45		
TX NC	36-40	IN MI	
MO KY	31-35	OH	
DE FL	26-30		
	21-25	IL	
MD	16-20	ME WI IA RI CA	
VA	11-15	PA	
	6-10	NY VT	
	1-5	NJ CT NH MA	
	Best		

Figure 4.9   Stem-and-leaf plot of free and slave states' ranking on updated worst American States using modern data

transcends the old free/slave boundaries. One can nearly draw a straight line across the country from east to west starting with the Virginia/North Carolina border and successfully divide the high scorers from the low. California and Nevada extend to the north of this line, but they are at least partially in the southern portion with which they share lower test scores. Slave states Missouri, Kentucky, Virginia, Maryland, and Delaware all fit more with the northern states that they border than with the South. West Virginia, part of slave Virginia before the Civil War, at which point it seceded from Virginia rather than the union, appears out of place, surrounded by higher achievers.

Wealth per capita gets replaced by per capita income, and this map is not so clear cut (see Plate 18). With the exception of Wyoming, the states where income is the highest appear to share proximity to New York City or Washington, DC:

Connecticut, New York, Massachusetts, New Jersey, Maryland, and Virginia. There also appears to be a general trend toward low incomes in the former slave states. Our color scale is normalized, creating the impression that income disparity is equivalent to wealth disparity in the 1930s. Even ignoring a couple of extreme outliers in the 1930s, wealth disparity was such that the greatest wealth was about ten times that of the least, whereas income disparity today is such that the worst is still more than half that of the best. Of course, income disparity over time is unlikely to translate linearly to wealth disparity, but even considering that, this likely represents a narrowing of the gap.

For death rate, no updated statistic is necessary. Deaths per capita is still the statistic of choice, though it has been age-adjusted so as to better represent life expectancy rather than age demographics. For this, the map in Plate 19 shows a strong tilt toward the South, the five states with the reddest hues all former slave states. Even the free states appearing red on the map compare favorably to most of the slave states, only finding themselves on the wrong side of the mean because of the superior death rates found in the newcomer states in the West. This represents a considerable shift compared to 1930 and 1850, a change that may merely reflect that Helper's and Angoff and Mencken's data were not age-adjusted.

We use these three measures as a greatly simplified set of indices on which to determine today's Worst American State, presented on the map in Plate 20. What we see is that the worst states are predominantly former slave states. The four best states are all free states, with Minnesota and South Dakota, both post-slavery (and extreme northern) entrants into the union, tying for the number five spot. Only two former slave states appear in the top half of state rankings, and the eight worst states were all slaveholding. A new back-to-back stem-and-leaf plot of free and slave states in 2010 shows some improvement for the slave states compared to 1930, but the advantage is still largely to the free (See Figure 4.9).

In an effort to determine the extent to which Helper's hypothesis is supported by modern data and at least determine the relationship between slavery and Southern disadvantage today, scatterplots of slave proportion to each of the three modern variables are presented in figures 4.10, 4.11, and 4.12. No real relationship can be observed in any of these plots. There is a strong dichotomous free/slave relationship to overall state rank, but it is not related to the proportion of the population that was slave in 1850. This suggests something else is likely at play here. Use of the scatterplot avoids misinterpretation due to the modifiable areal unit problem. Had the plots agreed with what the maps tell us alone, we would need to rely on our understanding of the limitations of our data.

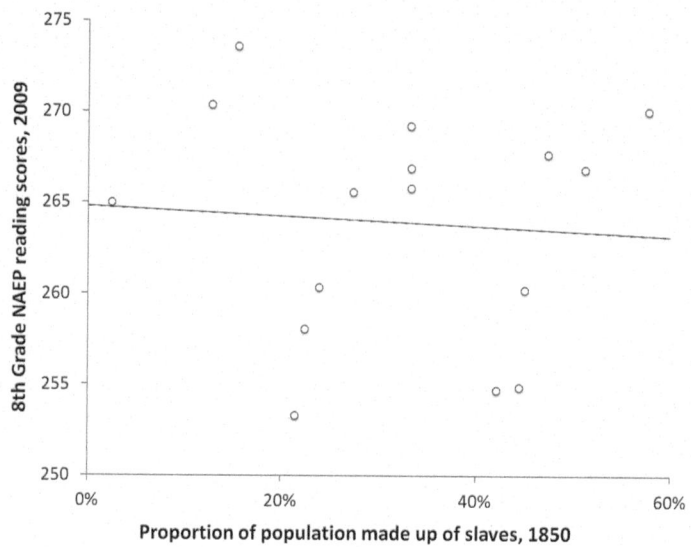

Figure 4.10   Scatterplot of relationship between 1850 proportion of population
that is slave and 2009 NAEP reading scores

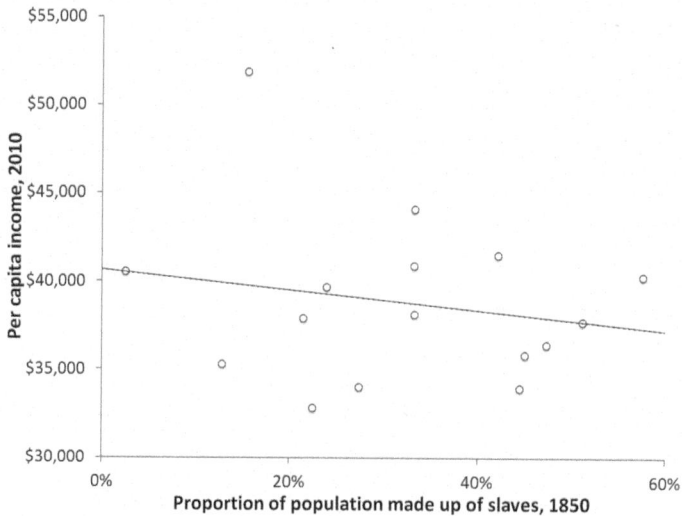

Figure 4.11   Scatterplot of relationship between 1850 proportion of population
that is slave and 2010 per capita income

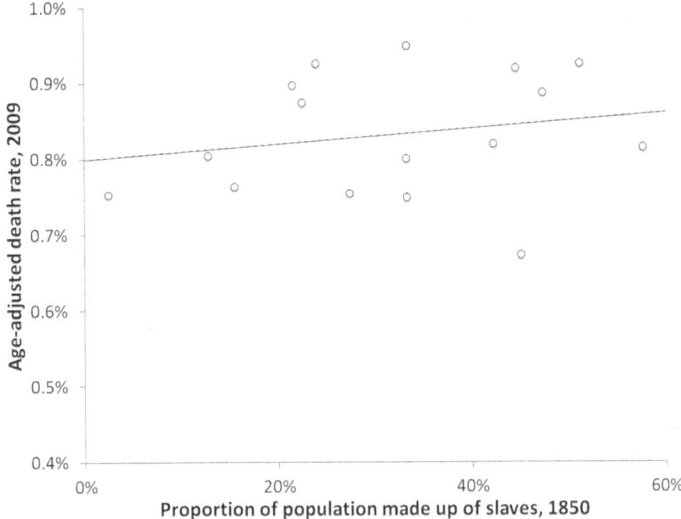

Figure 4.12   Scatterplot of relationship between 1850 proportion of population
              that is slave and 2009 death rate

Conclusion

The art and science of displaying social data on maps is one that has evolved
slowly over the past four centuries and will undoubtedly continue to evolve.
Some of the excellent maps in the most recent U.S. presidential elections
demonstrate how to place two variables on the map at the same time by color
coding one (as was done in this chapter) while distorting the size of the states to
represent the other (typically population). This method gives a great picture of
the overall relationship between the two variables, but some sense of geography
is lost in the distortion. Our method, employing separate maps and a scatterplot,
preserves the geography but requires the viewer to tie together multiple reference
points. Ultimately, there is no one correct choice and the best option is the one
that best tells your story.

Chapter 5

Innovation and Inertia in Atmospheric and Census Cartography in Nineteenth- and Twentieth-Century America

Mark Monmonier

In exploring the development of statistical cartography since 1800, this chapter focuses on two significant application domains: weather science and census geography. Mapping practices that developed in these arenas reflect advances in cartographic technique as well as advances in the understanding of their respective geographic phenomena. Even so, the word *advances* must be used cautiously lest the reader misinterpret this essay as uncritical support for the problematic notion of *linear progress*, whereby maps are assumed to become progressively more precise, reliable, and effective. Indeed, as many of the examples examined here illustrate, change in the quality of maps serving these two endeavors—like the quality of maps examined in other aspects of the history of cartography—has been neither consistently steady over time nor progressive in outcome. Because the map is a robust instrument of communication in the sense that a poorly designed map (but not a deliberately deceptive one) seldom obscures a prominent pattern, dysfunctional practices persist and new ones emerge. Indeed, an inefficient design is of little consequence when a map user familiar with the phenomenon is motivated to understand the data and its geographic message, or when a map is presented to a less savvy public, either to create the impression that the map author uses scientific tools and is therefore knowledgeable, or to proclaim that a bureaucracy is aware of a problem and eagerly pursuing a solution. In situations such as these, a widely used substandard design, even one that is woefully dysfunctional, rarely invites criticism.

This coordinated examination of atmospheric cartography and census mapping recognizes that both activities fall within the scope of *thematic mapping*, one of nine modes of mapping used by Matthew Edney to frame the study of map history during the Renaissance and the European Enlightenment.[1]

---

[1]   Edney's modes underlie the conceptual framework for Vol. 6 of the *History of Cartography* founded by the late David Woodward and the late J.B. Harley, and published by the University of Chicago Press. Edited by the author of this chapter, Vol. 6 (2015) covers the twentieth century; it is based on a system of "hierarchically integrated conceptual clusters" that include Edney's nine original modes (boundary surveying, celestial mapping, geodetic

As defined by Edney, a *mode* is a collection of cultural, social and technological relations that characterize a particular mapping practice. Within this conceptual framework, thematic mapping developed as an outgrowth of mathematical cosmography in the late 1700s to serve a centralized military state eager for rational, encyclopedic knowledge acquired through systematic measurement and quantification—a specialized form of mapping exemplified by map symbols like isolines and graphically logical, easily decoded series of progressively darker shades of gray. In general, thematic maps are medium- to very-small-scale representations of physical or social phenomena. In the physical realm, thematic maps can be useful in coping with hazards or managing a natural resource, whereas in the social realm, they can help a political or financial elite establish or maintain control. In this latter realm, for instance, maps help campaign managers and marketing executives see at a glance the locations and preferences of voters and consumers.

Differences between Application Domains

A meaningful comparison of atmospheric cartography and census mapping requires recognition of salient differences rooted in strategies for collecting and processing data. Census data, for instance, are collected for households, the characteristics of which are recorded on a census questionnaire, filled out originally by a census-taker traveling from house to house but more recently gathered through a self-reporting instrument delivered by the postal service. Because many of the questions posed by the twentieth-century census might be considered invasive, public cooperation demanded a strategy for shrouding the identity of individual persons and households, typically through aggregation to demographic categories like age-sex cohorts or geographic units like blocks, census tracts, counties, or states. Nondisclosure rules that limit the demographic detail in published census data recognize that appropriate anonymity is difficult if not impossible for areal units with few inhabitants; for this reason cross tabulations by income and occupation provided at the state level would be entirely too revealing at the block or census-tract level. For the Census of Agriculture, the fundamental reporting unit is the farm and the smallest geographic unit in published reports is the county, except where an exceptionally small number of farms precludes disclosure of county totals. The Census of Manufacturing and the other economic censuses are similarly, if not more severely, constrained.

---

surveying, geographical mapping, land surveying/property mapping, marine charting, thematic mapping, topographical surveying, and urban mapping), two new modes (dynamic cartography and overhead imaging) that emerged during the twentieth century, Edney's three original multimode endeavors (administrative mapping, map collecting, and map publishing), and one new endeavor (academic cartography) that emerged as a significant endeavor in the twentieth century.

In the meteorological realm by contrast, measurements of temperature, barometric pressure, precipitation, and winds are taken at discrete points, which constitute a sparse observation network. In the United States, the National Weather Service's roughly 900 sites with an Automatic Surface Observing System (ASOS) provide a first-order data network for capturing a variety of frequent, specialized measurements, whereas the more than 9,000 volunteers in the NWS's Cooperative Observer (COOP) program form a second-order network, which is more geographically dense but samples less frequently and is less rigorously supervised (Ryerson and Ramsay). In addition, the roughly 15,000 participants in the Community Collaborative Rain, Hail, and Snow Network (CoCoRaHS) constitute a third-order network of weather enthusiasts with limited training but nonetheless useful in providing even denser coverage of important precipitation events. CoCoRaHS volunteers typically measure only once a day, if at all, between 6 and 9 a.m., using rudimentary instruments like a simple rain gauge, a "hail pad" for capturing the size and density of hailstones, and a two-foot-square snowboard and ruler for measuring snowfall.[2] These networks assure a continual and abundant flow of data, much of it conveniently summarized on maps.

Fundamental differences in the geometry of data collection produce prominent differences among map symbols. Although weather and climate data collected by networks of irregularly spaced observation points could be portrayed cartographically with discrete point symbols, maps of atmospheric phenomena have relied largely on isolines (isobars, isohyets, isotherms) interpolated from irregularly spaced data points by a human analyst or computer algorithm. Representation as a continuous three-dimensional surface implies that additional measurements taken anywhere within the region are unlikely to reveal marked discontinuities. Less common is the fine-grid raster display constructed by computer interpolation or based on numerical modeling; these comparatively recent maps rely on gray-scale or color area symbols akin to those on most statistical maps of census data, which are seldom displayed as contoured surfaces. Although a rasterized surface could reflect spatial discontinuities, interpolation usually imposes at least minimal smoothing.

Sampling rate is the next most distinctive difference insofar as census enumerations are conducted only once every 10 years for the Census of Population and Housing, and once every 5 years for the agriculture and manufacturing censuses. By contrast, hourly observation is common for the first-order weather network used for meteorological prediction, and once-daily observation is typical for the second- and third-order networks, used mostly by climate scientists. The result is a vast number of maps, some displayed only on demand by an interactive system or summarized as a series of probability maps. Moreover, the huge number

---

[2]   Robert Cifelli et al. provide a concise summary of the history and objectives of CoCoRaHS. Membership is surely higher than the 11,000 estimated several years ago by Christopher Fiebrich.

of replicated measurements allows a preliminary sorting of the data into monthly or seasonal datasets, or into longer temporal epochs for studies of climate change. Census cartography, with a slower, more intermittent data stream, yields far fewer thematic maps.

Origins

Data that could have been used to make atmospheric and census maps remained unmapped for decades before anyone recognized that a map might reveal meaningful patterns. The United States has conducted a decennial census since 1790, principally to satisfy a constitutional mandate to reapportion the House of Representatives according to population. For the first census, a mere six questions sought the name of the head of household and the numbers of free white males 16 and older, free white males under 16, free white females, other free persons, and slaves; later censuses added questions about literacy, occupation, and real-estate ownership, among others. According to historian Susan Schulten, no effort was made to map federal census data until secession of the slave-holding states was imminent in early 1861, when military strategists in Washington used a county-level choropleth map of Virginia's slave population produced by the Coast Survey from advance tabulations from the 1860 Census, to identify counties in the western part of the state, where relatively low rates of slave-ownership reflected minimal support for the Confederacy ("How," *Mapping* 119–55; see Figure 5.1). Three days after the Virginia legislature voted to secede, Union troops crossed the Ohio River and occupied these western counties, which then seceded from Virginia and eventually formed their own state, admitted to the Union as West Virginia two years later.

However valuable to military strategists, the map was also a persuasive political argument for dividing the state. Numbers representing slaves as a percentage of county population were enhanced by graytone symbols showing a marked contrast between the eastern half of the state, where slaves constituted more than half the population of many counties, and western Virginia—shown in white—where slaves accounted for less than 5 percent of the county population. Although the map lacks a key relating each graytone to an interval of percentages, a table lists the white and slave populations of each county as well as slaves as a percentage. Lithographic copies with a publication date of June 13, 1861, note that the map was "Sold for the benefit of the sick and wounded of the U.S. Army."

Choropleth maps were a relatively recent tool, first used in 1826 in France, to map geographic variation by département in the number of inhabitants for each male schoolchild (Wallis and Robinson 18). Notable examples that followed include Joseph Fletcher's 1849 map of population density in England and Wales, compiled from the 1841 British census (see Figure 5.2, as well as Chapter 4 of this volume). Seven area symbols produced using diagonal and cross-hatched lining show each county's position relative to the overall average, and numbers

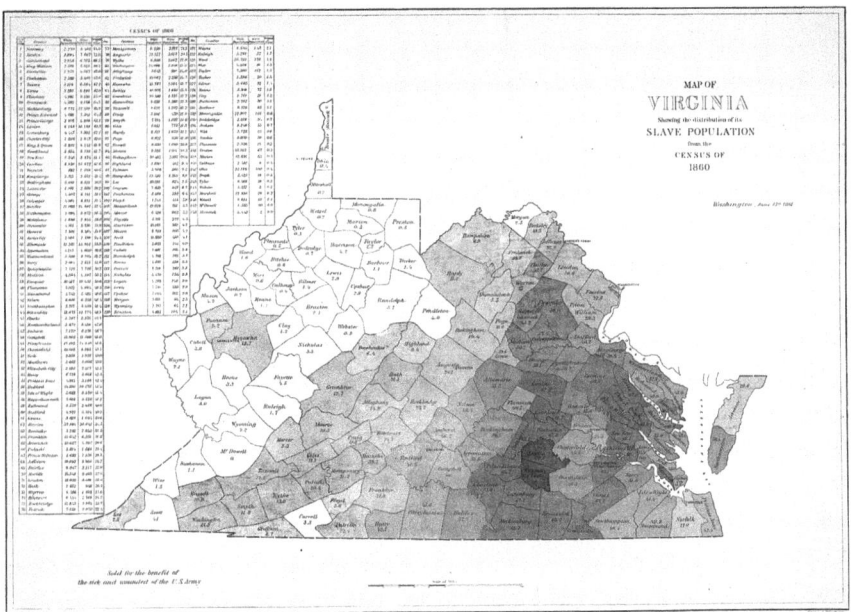

Figure 5.1   Lithographed copy of the "Map of Virginia, showing the distribution of its slave population from the census of 1860," published in Washington, DC, in 1861 by Henry S. Graham

ranging from 1 to 44 indicate each area's rank, from rural County Westmoreland, with the lowest density, to Middlesex, which includes the City of London, with the highest. Class intervals mostly reflect round-number percentage deviations from the mean (−50 percent and more, −30 to 50, −10 to 30, "− and + less than 10," +10 to 30, +30 to 70), except for the three most-dense areas, identified as outliers by a range (179 and more) well above the second highest interval (+30 to 70). Fletcher, who was Britain's Inspector of Schools as well as Honorary Secretary of the Statistical Society of London, used this and similar maps to illustrate links between maps of "moral statistics" and maps specifically related to education. "In all the Maps," he wrote, "it will be observed that the *darker* tints and the *lower* numbers are appropriated to the *unfavourable* end of the scale, whether of influences or results" (Fletcher, "Moral"). His curious omission of the countrywide average on the map as well as in the accompanying article reflects a strong sense of numerical relativism.

As the technique evolved, the typical choropleth map divided an assemblage of areal units, such as states, counties, or census tracts, into four to seven categories based on a single measurement or ratio—usually a percentage or density—and represented each category by an area symbol in a progressive series running from light (for lower values) to dark (for higher values). Numbers indicating each areal unit's percentage or rank were replaced by a map key that listed the range of values

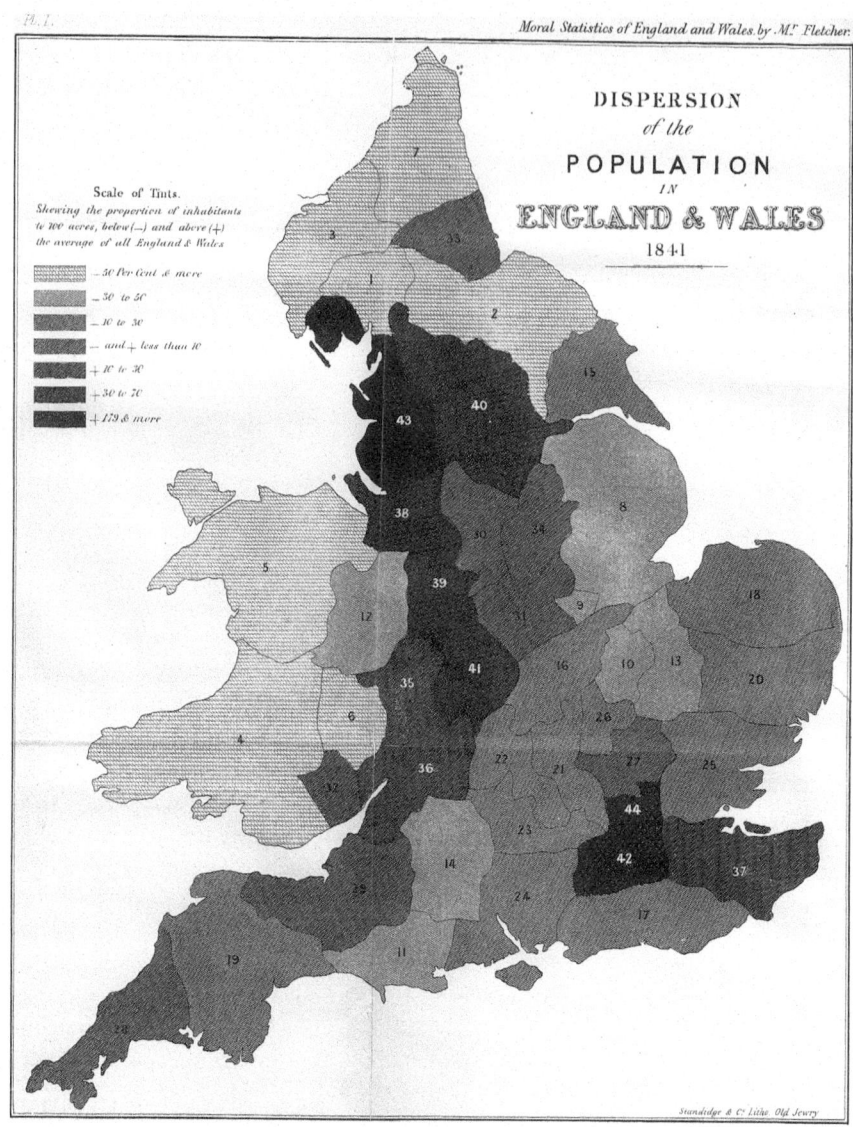

Figure 5.2   Joseph Fletcher's 1849 choropleth map of the "Dispersion of
Population in England & Wales" was one of several choropleth
maps accompanying his article "Moral and Educational Statistics
of England and Wales"

represented by each symbol, and round numbers commonly provided the cut-points
between categories. Occasionally the map maker violated logical principles that
called for using intensity data, such as rates and percentages, rather than magnitude

data, such as counts, and for an unambiguous light-to-dark progression of area symbols. This latter rule was easily compromised by an unsystematic use of hues that sacrificed aesthetic appeal for reliable decoding (Monmonier, *How* 139–73).

Some nineteenth-century statistical maps based on census enumerations were *dasymetric* maps. As a close cousin of the choropleth map, a dasymetric map also divided data values for area units into a small number of categories represented by a series of progressively darker area symbols. Unlike choropleth maps, which always reflected the defined boundaries of enumeration areas, dasymetric maps were not constrained by the boundaries of states, counties, or other areas for which the data had been collected. As exemplified in Figure 5.3, from the map of white male illiteracy in the *Statistical Atlas of the United States* edited by Francis A. Walker, superintendent of the 1870 Census, and published in 1874, a dasymetric map of a population trait considers overall population density as a limiting variable and need not acknowledge state or county boundaries. This map partitions the eastern half of the country into six categories, divided at cut-points of 5, 12, 20, 40, and 60 according to the percentage of white males 21 and older who "cannot write." Only the three lowest categories are apparent in this excerpt, centered on one of the more literate parts of the country. Note that the central Adirondacks in upstate New York is excluded because the population density here is less than two people per square mile. An overlay of shaded lines and numbers (in blue on the original) delineates zones in which population density is either 18–45 or 45+ persons per square mile. Note that western Massachusetts (in category II) is less literate than adjoining areas in New York (in category I), central Massachusetts (also in category I), and the upper Connecticut Valley and adjoining parts of Vermont (in category 0). Smooth lines between area symbols reflect the compiler's attempt to refine the less reliably homogeneous zones on a choropleth map of the same data. More difficult to compile and inherently more subjective, dasymetric maps were uncommon in the nineteenth century and rarely used in the twentieth century.

Although census cartography had largely abandoned the dasymetric technique by the end of the nineteenth century, the 1874 *Statistical Atlas* contributed a notable innovation found in all subsequent graphic summaries of the decennial census. On page 6 a small centrographic map (see Figure 5.4) developed by U.S. Coast Survey cartographer Julius Hilgard described the progressive westward movement of the country's population. For each census year the nation's center of mass was represented by a single symbol with coordinates calculated as the arithmetic means of latitude and longitude weighted by population. The summary dot symbol generally followed the 39th parallel, with the greatest movement between 1850 and 1860. Updated versions reveal a distinctly southwestward shift since 1940.

Weather science was similarly slow to recognize the map as an efficient tool for pattern recognition. The first weather map was invented around 1816 by Heinrich Wilhelm Brandes, a physicist who taught mathematics at the University of Breslau. Brandes used data collected more than three decades

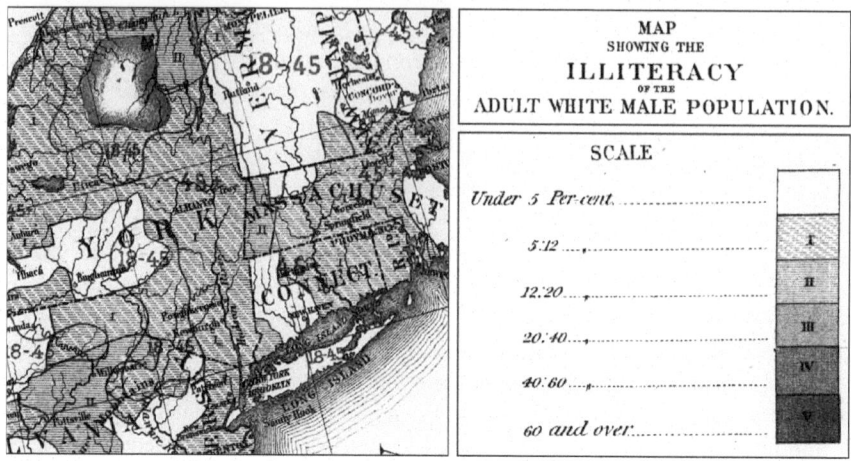

Figure 5.3   Excerpt and key from Francis A. Walker's 1874 "Map showing
                  the illiteracy of the adult white male population, compiled
                  from returns of population at the Ninth Census of the United
                  States, 1870"

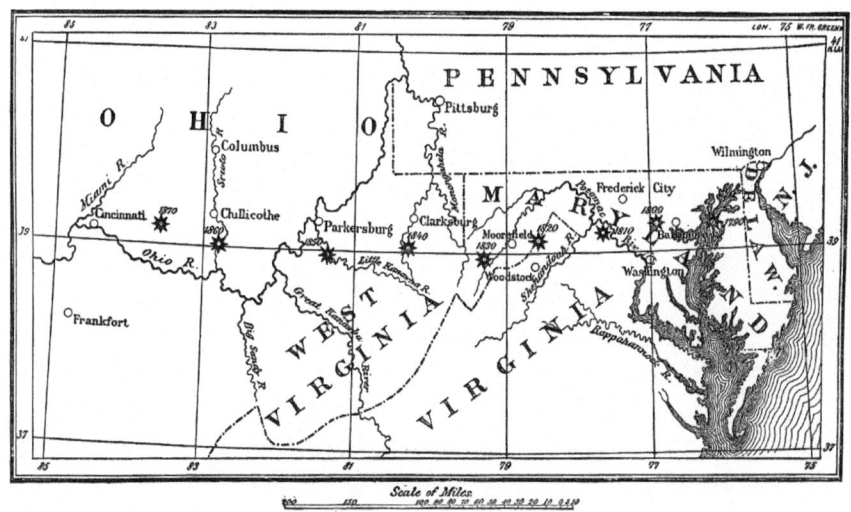

Figure 5.4   Walker's *Statistical Atlas* included this centrographic map, devised
                  by James Hilgard to describe the westward movement of the center
                  of U.S. population

earlier by members of the Meteorological Society of the Palatinate, headquartered
in Mannheim. The Palatinate network relied on postal communications to
collect and share measurements of temperature, pressure, winds, and weather

conditions—information deemed useful in exploring weather's presumed links to celestial events as well as its local impacts on insects and health. Before Brandes speculated on the spatial relationship between pressure, winds, and precipitation, apparently no one considered the data worth mapping, and no one dreamed that, thanks to the electric telegraph, surface weather maps would eventually become an operational tool for forecasting weather. I've never found a portrait of Brandes, who died in 1834. He described his first weather map in an article published in 1819 in *Annalen der Physik*, but no map was included with the short article and no artifact has survived. Even so, the data he used were preserved in print, which allowed Wilhelm Trabert to reconstruct a replica of the first weather map for his 1907 textbook (see Figure 5.5). As Brandes had described, isobars and wind arrows revealed a counterclockwise circulation around a low-pressure cell centered over the English Channel on March 6, 1783. Storms formed as air rushed into this depression.

Brandes's pioneering use of isobars (lines of equal pressure) followed by two years the publication of the premier map of isotherms (lines of equal temperature) by German naturalist Alexander von Humboldt (1817), who might have been inspired by the isogons (lines of equal magnetic declination) employed more than a hundred years earlier by Edmond Halley (Robinson, "Genealogy").[3] Lines of equal value became a mainstay of the synchronous weather map, a snapshot of the atmosphere based on observations taken at the same time. The synchronous weather map became a forecasting tool after the emergence of an extensive telegraphic network in the mid-nineteenth century encouraged meteorologists to map and predict the movements of pressure systems (Monmonier, *Air* 39–54). In a similar vein, climatologists, who studied the impacts of average weather and its variations, used maps with isotherms and isohyets (lines of equal precipitation) to describe annual and seasonal patterns, as exemplified by the map of winter precipitation in Figure 5.6, from the *Meteorological Register, for Twelve Years, from 1843 to 1854*, issued by the U.S. Surgeon-General's Office in 1855. The Surgeon-General of the Army had initiated collection of weather data at military posts in 1819, but the network was sparse and its observations inconsistent. Note that the isohyets here are merely boundaries between zones labeled directly with their average precipitation. A zone running from northern Maine through central Illinois has two averages: 7 and 7½ inches. Although zone boundaries are highly generalized, the map shows the importance of proximity to water, particularly the warm waters of the Gulf of Mexico.

Intended principally as a forecasting tool, weather maps became a complex assemblage of diverse symbols, confusing to most adults but readily intelligible to a committed cadre of meteorologists and other trained users. The maps'

---

[3]   Humboldt was concerned with climate, not weather, and his highly generalized map of average annual temperatures used smooth isotherms to describe warmer conditions in western Europe than along the east coast of North America—an effect of the Gulf Stream (see Monmonier, *Air* 25–6).

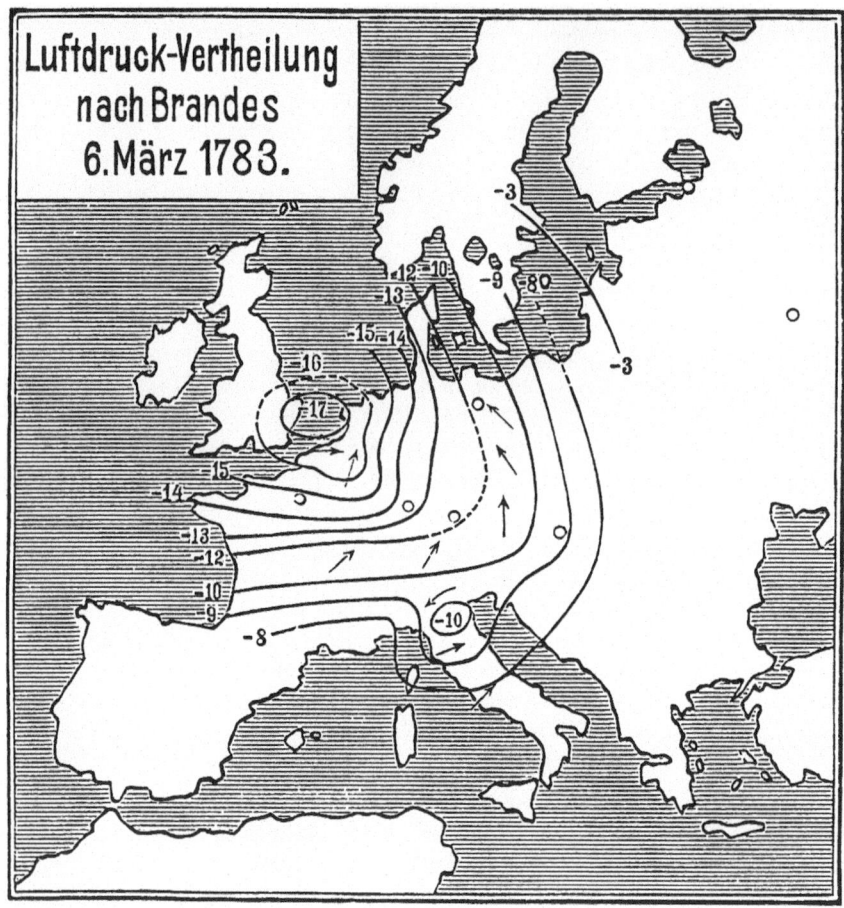

Figure 5.5   Wilhelm Trabert's 1905 reconstruction of the weather map
               described by Brandes in 1819

complexity was most readily apparent in point symbols that rely on arcane graphic codes to represent for each weather station a variety of atmospheric measurements, including wind speed and direction, the type and height of clouds, visibility, and recent trends in barometric pressure. As shown in Figure 5.7, graphic codes intended for a specialized audience are a form of graphic shorthand with meaningful positions on the station symbol. The graphic codes shown here are specific instances of a wider range of codes, which could be consulted in meteorology textbooks or forecasting guides. Several visual variables are apparent: direction (for wind flow or to distinguish upward and downward trends), shape (for cloud type), graytone value (for cloud cover), and numerousness (as with the barbs added to the wind direction symbol to reflect wind speed). For most stations the map included only a few of the various graphic

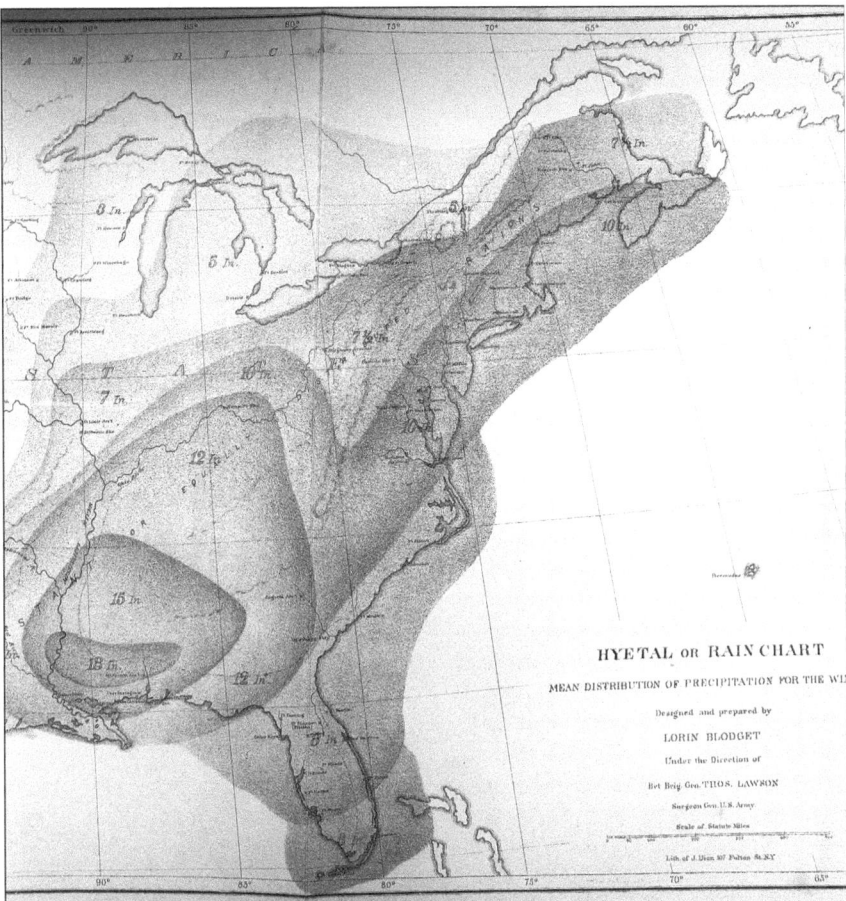

Figure 5.6   U.S. Surgeon-General's Office, eastern half of the "Hyetal or Rain
Chart: Mean Distribution of Precipitation for the Winter"

symbols and numbers shown in this specimen. Symbols and numbers were given
standardized positions presumably intended to minimize misinterpretation.

To promote its work among the wider public, the Weather Bureau prepared a
simpler version of the surface weather map for newspapers, which supplemented
the current forecast with a cartographic snapshot of very recent weather (Heiskell;
Ward). Lines representing surface fronts (cold, warm, occluded, and stationary)
appeared on U.S. weather maps in the early 1940s, after the Weather Bureau
recognized the value of air mass theory, developed in Norway during the First
World War, and by the end of the decade fronts had displaced isotherms on the
small weather maps that had become a regular feature in many daily newspapers
(Monmonier, *Air* esp. 153–96). Forecast maps for newspapers emerged in the
1950s and had largely displaced maps of yesterday's weather by the 1980s, when

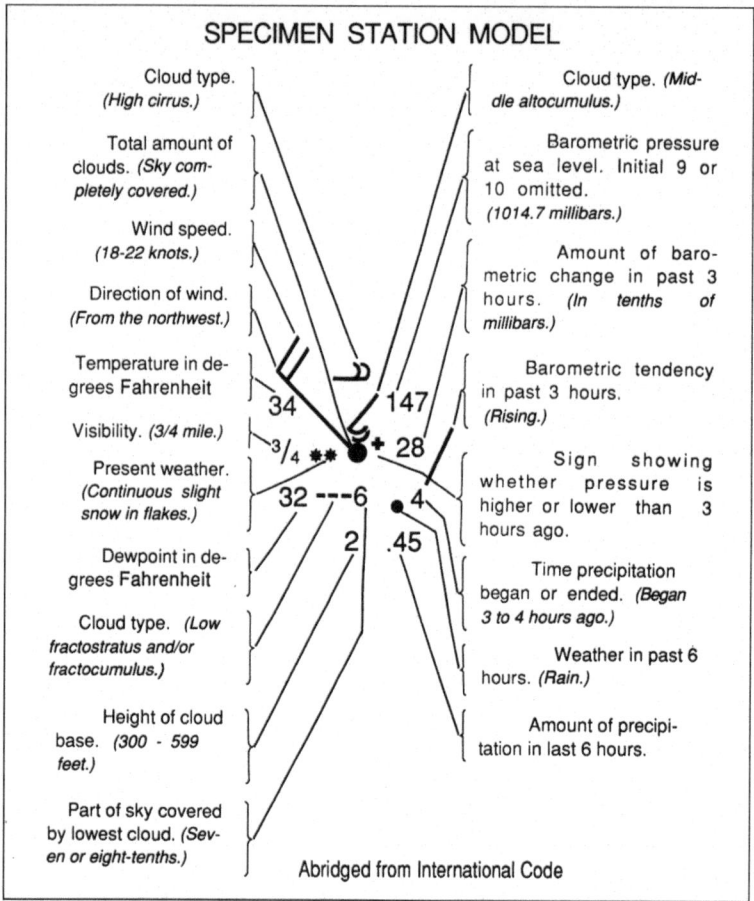

SPECIMEN STATION MODEL

Cloud type. *(High cirrus.)*

Total amount of clouds. *(Sky completely covered.)*

Wind speed. *(18-22 knots.)*

Direction of wind. *(From the northwest.)*

Temperature in degrees Fahrenheit

Visibility. *(3/4 mile.)*

Present weather. *(Continuous slight snow in flakes.)*

Dewpoint in degrees Fahrenheit

Cloud type. *(Low fractostratus and/or fractocumulus.)*

Height of cloud base. *(300 - 599 feet.)*

Part of sky covered by lowest cloud. *(Seven or eight-tenths.)*

Cloud type. *(Middle altocumulus.)*

Barometric pressure at sea level. Initial 9 or 10 omitted. *(1014.7 millibars.)*

Amount of barometric change in past 3 hours. *(In tenths of millibars.)*

Barometric tendency in past 3 hours. *(Rising.)*

Sign showing whether pressure is higher or lower than 3 hours ago.

Time precipitation began or ended. *(Began 3 to 4 hours ago.)*

Weather in past 6 hours. *(Rain.)*

Amount of precipitation in last 6 hours.

Abridged from International Code

Figure 5.7   Specimen station model included with the daily weather maps issued by the U.S. Weather Bureau, later the National Weather Service

newspaper firms recognized the importance of richly informative full-color weather packages as a strategy for competing with each other as well as with television.

Innovations and Inertia

A marked contrast between atmospheric cartography and census mapping developed in the latter part of the twentieth century, as meteorology and climatology became even more deeply map-immersed while census cartographers generally relied on traditional graphic displays, occasionally updated to recognize the increased use of color reproduction or to exploit the greater spatial detail of small units of areal aggregation, namely, census tracts, block groups, and blocks. Small-area data gained increased prominence in the wake of the 1990 Census, when political cartographers in various states sought to gain an electoral advantage by using electronic software to concentrate voters registered with the opposing party in relatively few congressional districts—perhaps the only way (aside from the use of small-area data for marketing) in which the census was a moderately reliable predictive tool (Monmonier, *Bushmanders*). For the most part, census data played largely descriptive and explanatory roles in the media and the social sciences, and the graphics used—and misused—were little different conceptually from their nineteenth-century counterparts.

Even so, the U.S. Bureau of the Census, which had pioneered development of geographic data structures for urban areas in the late 1960s and 1970s, made conceptual advances in cartographic design by involving academic researchers experienced in color theory and subject testing in the development of new products such as two-variable *cross maps* and the atlas *Mapping Census 2000: The Geography of U.S. Diversity* (Brewer and Suchan). Cross maps became prominent in the 1970s, when the Bureau issued large, thin statistical atlases based on tract data for each of the 65 largest Standard Metropolitan Statistical Areas (Meyer, Broome, and Schweitzer 109–16). Cross maps were also used for nationwide county-unit maps describing relationships like the co-variation of mean family income and the percentage of the adult population with a high-school education. Geographic cartographer Judy Olson demonstrated the inherent complexity of these maps for uninitiated users and the consequent need for carefully organized keys and categories as well as a clear explanation of the maps' goals and how to read them. This caveat explains why the *Diversity* atlas, developed by Cynthia Brewer, a student of Olson, and Trudy Suchan, a student of Brewer's, contains only one cross map: a two-variable display relating population density in 1990 to the percentage rate of population change from 1990 to 2000. In addition to providing carefully researched sequences of color symbols, juxtaposing state and county-level maps on the same page, and using inherently meaningful national means as cut-points for choropleth maps, the atlas was innovative in its portrayal of the percentages of county and state populations identifying with two or more racial groups. These maps—and indeed the atlas itself—reflect the 2000 Census's pioneering questionnaire that allowed respondents of mixed parentage to report more than one race.

What's innovative is not always efficient or effective. Except in one instance, the *Diversity* atlas inappropriately portrays count data with choropleth maps. According to the theory of visual variables, count (or magnitude) data such as the total number of persons and the number of African-Americans are more logically mapped using circles, squares, or other magnitude symbols that vary in size, in accord with the principle "bigger means more, smaller means less" (Bertin, *Semiology*; Monmonier, *How* 19–24). By contrast, the atlas appropriately uses choropleth maps for intensity data such as population density or the African-American percentage of the population because intensity data are most reliably decoded when represented using intensity symbols in accord with the principle "darker means greater intensity, lighter means lower intensity." The only appropriate maps of count data in the atlas are a pair of state and county maps of total population—"Number of people by state" at the top of the page and "Number of people by county" below—on which a symbol's area is proportional to the count it represents. As shown in Figure 5.8, the county-level map of graduated squares clearly contrasts the huge populations along the Boston–Washington corridor and in several vast metropolitan regions with the relatively sparsely inhabited areas in between. Perhaps the only excuse for the choropleth map of the same data in Figure 5.9 is that the latter's area symbols are easier to read in congested areas, where the graduated squares in Figure 5.8 overlap. Although arbitrary round-number cut-points (10,000/25,000/50,000/100,000/500,000) cannot obscure the highly robust pattern of metropolitan concentrations, a choropleth map's broad categories can easily obscure or exaggerate subtle differences between neighboring non-metropolitan counties. These subtle differences suffer accordingly on the choropleth maps of counts for the atlas's various one-race and two-or-more-race portrayals—distributions denied a supplementary map with graduated squares.

Why the atlas did not include maps like Figure 5.8 for each series of count data accorded a choropleth map is puzzling. The single pair of maps with graduated squares confirmed that available mapping software could easily create the images, but each additional map would add to the cost of reproduction. Was the Census Bureau overly concerned about printing cost, or did it prefer maps more amenable to a fuller use of color? Except for its clever "Nighttime Map" of the U.S. population—with white dots on a dark blue background—the Bureau has never used dot-distribution maps for demographic data despite using automated software to create hundreds of small-format county-level dot maps of crop and livestock data since the early 1970s for graphic summaries of the quinquennial Census of Agriculture (U.S. Bureau of the Census 14–15).

The marginal cost of additional maps and related displays became moot with the advent of interactive systems for exploratory data analysis (EDA), whereby a user can freely explore the visual efficacy of different cartographic symbolization schemes as well as different cut-points and different color series for choropleth maps. Though currently available only to users with appropriate software, interactive mapping promotes richer but often-complex analyses based on multiple linked windows like the experimental scatterplot matrix and map in Figure 5.10. This example shows a movable "brush," currently in the cell in the second row and left-hand column; data points inside the brush are highlighted here, corresponding data points are highlighted in the scatterplot array's other cells, and corresponding areas are highlighted on the map (Monmonier, "Geographic"). Users can explore a large dataset by swapping in other variables, electing a larger (4 × 4, 5 × 5, etc.) scatterplot matrix, or changing the year or time period under consideration. In principle, preprogrammed graphic sequences called *graphic scripts* could be used to introduce an analyst to a new data set and a *user profile* could tailor a script to

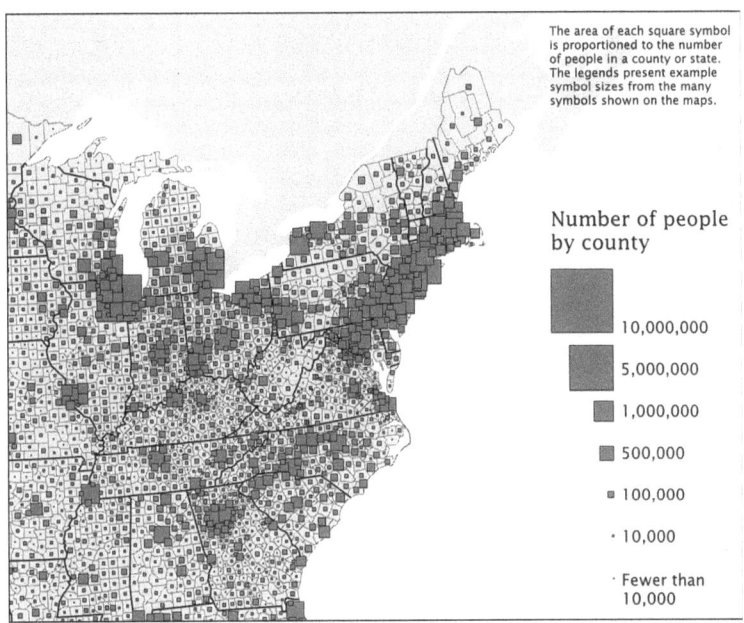

The area of each square symbol is proportioned to the number of people in a county or state. The legends present example symbol sizes from the many symbols shown on the maps.

Number of people by county

10,000,000

5,000,000

1,000,000

500,000

100,000

· 10,000

· Fewer than 10,000

Figure 5.8   Black-and-white image of Brewer and Suchan's magnitude map of county populations, appropriately portrayed with graduated squares

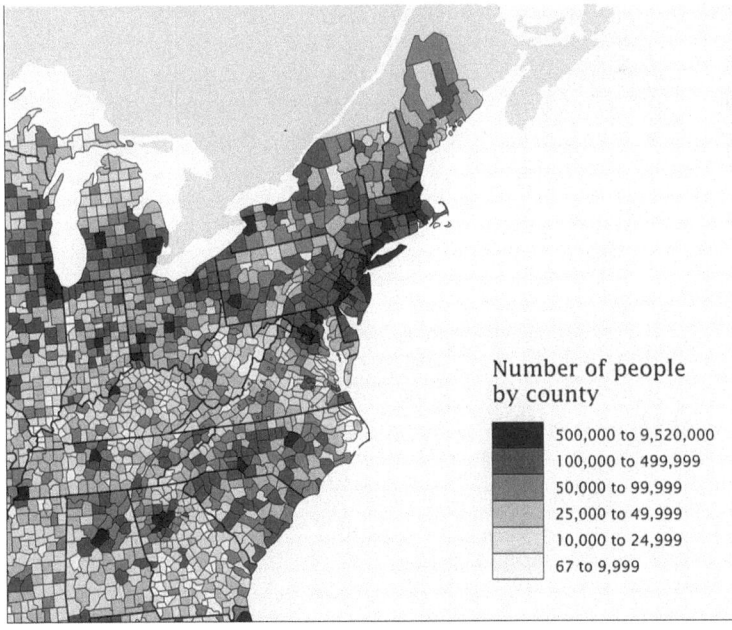

Number of people by county

500,000 to 9,520,000
100,000 to 499,999
50,000 to 99,999
25,000 to 49,999
10,000 to 24,999
67 to 9,999

Figure 5.9   Black-and-white image of Brewer and Suchan's choropleth map of county populations

the analyst's residential history, interests, level of experience, and other factors (Monmonier, "Authoring"). Though brushing and scatterplot matrices can be found in interactive statistical graphics software, graphic scripts and user profiles remain largely hypothetical.

By contrast, atmospheric cartography became not only graphically richer than census mapping but more heavily dependent upon electronic computing as an everyday tool of the "operational meteorologists" who predict weather and issue storm warnings. Weather forecasting had always been immersed in maps, but new observation and modeling techniques available since the Second World War led to numerous cartographic strategies for visualizing atmospheric data. By the end of the twentieth century computational models initialized with current weather conditions reached forward in time to provide ever longer-range predictions, satellite imagery of clouds and water vapor offered insightful glimpses to the formation and movement of storm systems, and an integrated network of Doppler radars provided detailed "nowcasting" of precipitation and severe winds. Television made these maps readily available in real time to millions of viewers. Although graphically complex and potentially challenging for lay viewers, carefully selected displays were typically presented in a user-friendly viewing environment by a knowledgeable and personable weathercaster who interpreted their patterns (Carter, "Map" and "Weather").

Radar maps, on which echo intensity (reflectance) is measured in decibels, employ many more categories than conventional choropleth maps, thereby violating warnings against illogical multiple hues. Even so, radar maps work exceptionally well because meteorologists receive intensive training in interpreting reflectance displays. Doppler radar can operate in *severe-storm mode* to show variation in the intensity of inbound and outbound winds, the close juxtaposition of which might indicate a tornado, or in *precipitation mode* to show the size and density of particles of precipitable moisture. Doppler precipitation displays with nearly two-dozen distinct colors, typically running from light blue through green, yellow, orange, light red, and red, can reflect qualitative differences among drizzle, heavy rain, hail, and snow as well as quantitative differences in intensity. Operational meteorologists work with these products constantly and can customize the assignment of colors to their individual needs, experience, and intuition. Television weathercasts and online weather websites typically offer a standard set of colors, in which the greens represent rain, the blues represent snow, pink represents mixed rain and snow, and more intense shades of each represent more intense precipitation. A short temporal sequence of successive maps uses motion as a visual variable, to show the movement of frontal systems, thunderstorm cells, and bands of lake-effect snow. Most animations are *loops* that move forward in short steps, pause at the most recent map, and then repeat the sequence from the beginning. Some animation systems let the viewer *rock* the display by moving forward and then backward between the oldest and most recent maps. Interactive systems allow the person controlling the display to zoom in or out, or to toggle on or toggle off reference features like principal boundaries, roads, and place names.

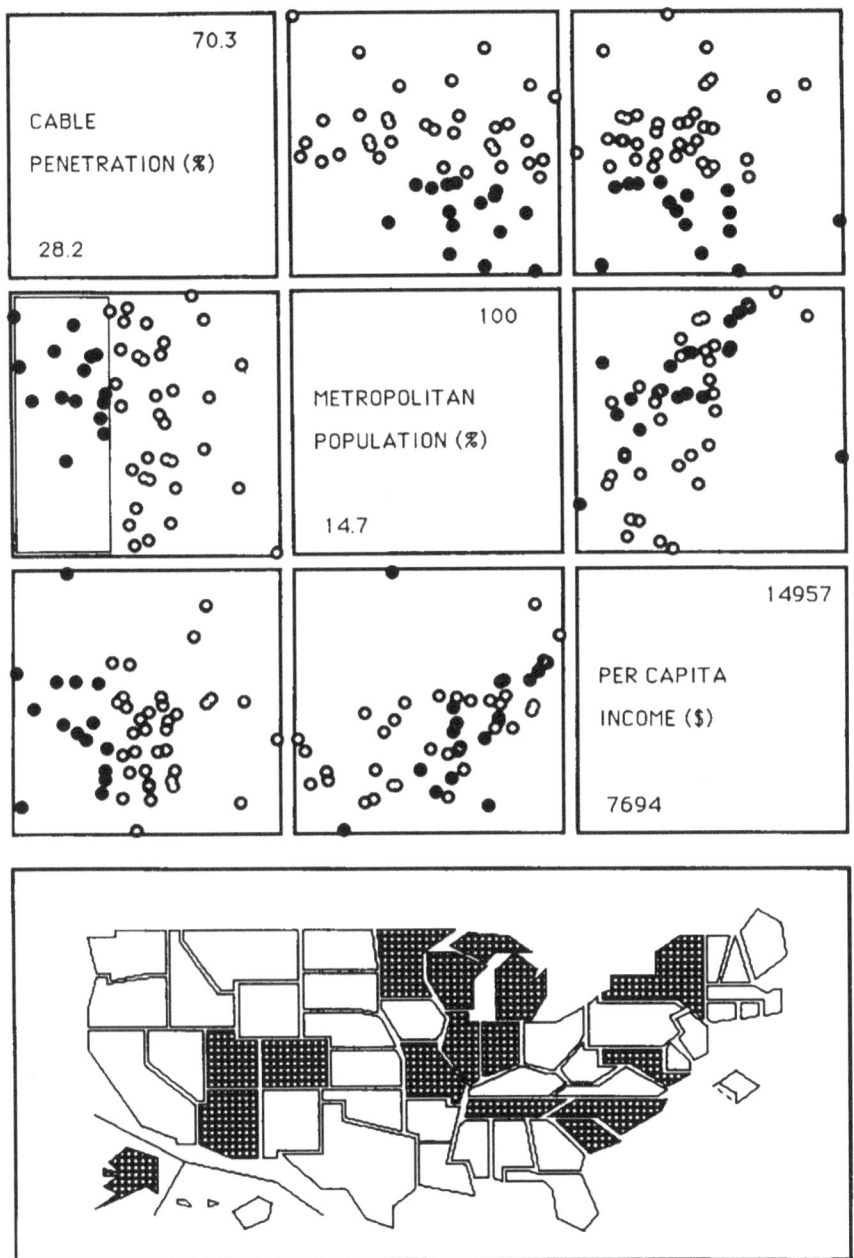

Figure 5.10　Scatterplot matrix for exploring relationships among three
variables includes a map and a movable brush for selecting areas

Some interactive systems offer arrays of maps as both a menu and a tool for comparative visual analysis—what Edward Tufte has touted as "small multiples" (*Visual Display* 170–75). A case in point is the array of nine snow-related maps in Figure 5.11, from NOAA's National Snow Analyses website. The viewer can choose the entire conterminous United States or a specific region, select a particular day of the year, and compare at a glance themes like snow water equivalent, average snowpack temperature, and snow melt. By clicking on one of the nine cells, a viewer can then call up an animation with a larger map that focuses on a specific theme and shows a day-by-day sequence for an entire season or year, an hour-by-hour sequence for a two-week period, or an hour-by-hour loop for a single day. What's more, animations can be paused and the map further enlarged.

As Figure 5.12 illustrates, not all maps based on complex measurements and modeling are graphically complex. Simple lines and letters added to a sparse outline map provide a probability-based summary of an ensemble of atmospheric circulation models that have been run forward in time much longer than 36 or 72 hours. The ensemble includes similar models with different initial conditions, the probabilities reflect the extent to which the models agree on temperatures or precipitation amounts significantly above or below the climatic average for the particular time period, and the maps show parts of the country where abnormally high or low outcomes are predicted. Isolines on these maps show probabilities for above-average (A), below-average (B), and normal (N) conditions, the probabilities of which sum to 100 percent. The example in Figure 5.12 is the three-month temperature outlook map released in mid-June 2012 for the following July, August, and September. In this example, the likelihood of above-average temperatures is better than 33 percent for a large part of the country, except for fringe areas in New England, along the Pacific Coast, and in the Upper Midwest near the Canadian border, all of which have equal chances (EC) of above-normal, below-normal, or normal temperatures. The dark areas inside contours labeled "50" indicate a likelihood better than 50 percent for above-average temperatures in large areas centered on Missouri and Utah. On the temperature maps, areas likely to experience above-average temperatures are shaded in orange and areas likely to have below-average temperatures are shaded in blue. On the corresponding precipitation maps, above- and below-average areas are shaded, respectively, in green and a parched-looking tan.

Although consistent, temporally stable measurements, analyses, and graphic standards are essential for reliable examinations of climate trends, graphic conservatism is not always beneficial. I attribute this graphic conservatism to bureaucratic inertia, that is, a bureaucracy's tendency, like an object moving through space, to keep doing what it has been doing simply because there's little incentive for nonelected officials to abandon well-established, socially accepted practices that seem to work and have not encountered opposition. And even though atmospheric cartography seems less encumbered by bureaucratic inertia than census cartography, one example of the persistence of an ill-conceived design stands out: rectangular projections with straight-line parallels that entail

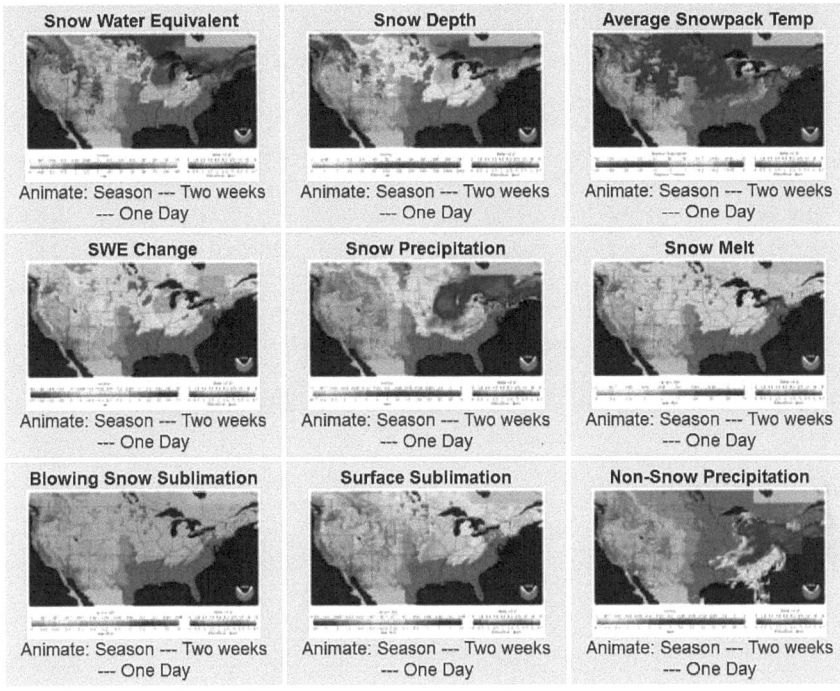

Figure 5.11   An array of maps on NOAA's National Snow Analyses website
serves as both a menu and a device for comparative visual analysis

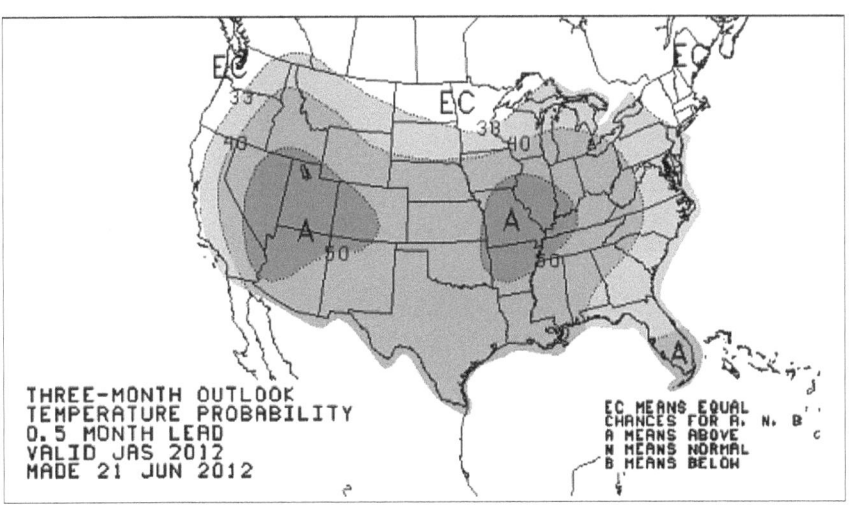

Figure 5.12   Three-month outlook map exemplifies the graphic simplicity
of longer-range forecast maps (National Oceanic and
Atmospheric Agency)

pronounced east-west stretching on national maps (Figure 5.11) and severely distort distances and angles. This distortion is even more apparent on regional maps, particularly for more northerly regions like the Upper Midwest, where a degree of longitude is notably shorter on the ground than a degree of latitude and straight-line parallels are severely stretched when the projection is not re-centered within the region. Rectangular projections for a variety of atmospheric maps are a legacy of the 1960s and 1970s, when computer maps were plotted on line printers and other comparatively coarse raster displays that rendered curved parallels of latitude as jagged lines but worked well with straight-line parallels. Although the high-resolution monitors and laser printers that replaced them were adept at drawing the smooth curves of less-distorted map projections, the rectangular frameworks were not updated.

It is difficult to say whether distorted distances and angles adversely affect a weather scientist's interpretation of the map: the map, after all, is a remarkably robust graphic device, particularly for well-trained, experienced users. Even so, the persistence of rectangular projections for atmospheric maps of a mid-latitude country like the United States is ample evidence of bureaucratic inertia and a puzzling tolerance of substandard graphics. I say "puzzling" insofar as the International Meteorological Committee, which met at Salzburg, Austria in 1937, dealt decisively with standards for map projections, and the U.S. Weather Bureau adopted its guidelines the following year (Gregg and Tannehill; Griggs). But several decades later digital cartography undermined the Salzburg guidelines with a more efficient, if not always effective or aesthetic, way of making maps. Although the persistence of rectangular projections might not matter to trained map users, it contradicts the notion of linear progress.

Chapter 6

Mountains of Wealth, Rivers of Commerce: Michael G. Mulhall's Graphics and the Imperial Gaze

Miles A. Kimball

Michael George Mulhall (1836–1900), the author of *Mulhall's Dictionary of Statistics* and a number of other illustrated statistical texts, was a ubiquitous feature of late-nineteenth-century statistics—often quoted, often corrected, sometimes ridiculed, but always at hand when one wished to make an argument backed by numbers. In the development of statistical graphics, he is best known for applying pictograms of various sizes to show relative quantity. For an example, see Plate 21 (Plate IV of the 1899 fourth edition of the *Dictionary of Statistics*), which uses orange steers to depict the "Production of meat, lbs. yearly per inhabitant" at the top of the chromolithographed sheet, and green steers at the bottom to show consumption in the same terms.

Of course, today it's easy to sneer at such a graphical production. For instance, does meat include only beef? Why are the steers orange for production and green for consumption?[1] Do cattle really vary that much in size? Perhaps Australia is so successful in the production of meat because they raise giant steers. (And pity Italy with its tiny ones!) Edward Tufte suggests that such graphics are particularly susceptible to "the lie factor," in part because it's difficult to compare the area of such an irregular object with a similar object of a different size (see "The Shrinking Family Doctor," Tufte, *Visual Display* 69). Willard C. Brinton critiqued this kind of graphic in his 1914 *Graphic Methods for Presenting Facts*, and H. Gray Funkhouser notes that graphics in Mulhall's style—which he calls "picture statistics"—have significant limitations:

> A serious defect in this type of comparison by figures diminishing in size is that misleading impressions are often conveyed by them. The reader cannot always determine whether comparisons are to be made in terms of one, two or three dimensions. In addition, the eye is usually not sufficiently trained to make accurate comparisons by areas or volumes. The authors of these diagrams have been at fault at times in not recognizing the geometric principle that areas are

---

[1] At this point in the history of printing, one could obtain pretty much any color of lithographic ink one desired. See Kimball, "London."

to each other as the squares of corresponding linear dimensions. (Funkhouser, "Historical" 349)

Without reading the data labels, how are viewers to know from the size of a steer that Australia produces precisely 1.5 times as much meat per inhabitant as Argentina? In Mulhall's graphic, we have no way of knowing if the size of the steer is determined by height, length, or area. (In fact, it appears that the steers may be indexed by the area of their bodies—and given that their bodies are vaguely rectangular, I suppose that's a step in the right direction.) Also, while the graphic may make it possible to compare generally one country's production or consumption of meat with another's, it makes it more difficult to compare national dynamics—production versus consumption within a single country—which might give a sense of the balance of production and exports or imports. For example, it takes a minute of scanning to see that Great Britain's production of meat is small compared to its consumption; evidently even then good English beef came from Argentina.

Yet despite Mulhall's checkered performance as a producer of statistical graphics, there are some good reasons for looking at his work more carefully. Since Edward Tufte's use of eighteenth- and nineteenth-century examples in *The Visual Display of Quantitative Information* drew popular and scholarly attention to the history of statistical graphics, we have tended to focus on the superlatives of *first* and *best*. We were encouraged to do so especially by Tufte's often-quoted championship of Charles Joseph Minard's *carte figurative* of Napoleon's march on Moscow as "the best statistical graphic ever drawn" (Tufte, *Visual Display* 40). This focus is a natural and necessary stage in reclaiming such a rich, forgotten trove of graphic expressions of data, and it helps us find historical examples to encourage excellent (and discourage foolish) graphics today. But it also tends toward a progressive, teleological story of the development of statistical graphics that runs swiftly by many interesting tributaries, oxbows, and stagnant pools. To avoid a Whiggish history of statistical graphics, we must look seriously at some of these dead ends—not just as mistakes to be avoided but as historical data because they can help us understand aspects of the history of statistical graphics that might be overshadowed by the brilliance of more innovative or exemplary practitioners.

In addition, because the historiography of statistical graphics tends to read as a string of innovations and improvements, our attention has centered on many men (and one notable woman) who were for the most part socially prominent, highly educated professionals—scientists (such as Alexander von Humboldt, John Herschel, and many others), health professionals (such as Florence Nightingale and John Snow), cartographers (such as Heinrich Berghaus and A.K. Johnston), and professional statisticians (such as Minard, Charles Dupin, Adolphe Quetelet, and the bureaucratic authors of the *Albums de Statistique Graphique* and the 1874 *Statistical Atlas of the United States*). We should recognize that there was also a considerable stock of what Alfred Meitzen and Roland P. Falkner called "private statistics," collected or presented by statistical amateurs rather than by experts or by state statistical bureaus (166). Perhaps the most obvious example is William Playfair, who came out of nowhere with no formal training to apply visual methods to statistics with remarkable polish and style, though with occasional missteps (Friendly, *Gallery*). Other obviously brilliant statistical amateurs include André-Michel Guerry and Francis Galton.

As a retired journalist, Mulhall may perhaps have filled out the bottom of the ranks of amateur statisticians, but his books were so popular, so commonly available, and so widely used that his significance rises due to his broad impact. We can glimpse this impact—for good or ill, in his age and after—from several pieces of evidence. Mulhall's peers recognized his contributions to statistics by naming him a Fellow of the Statistical Society of London (later the Royal Statistical Society) in 1880. References and reviews in contemporary publications repeatedly referred to Mulhall as "the eminent statistician"—a sobriquet he may have encouraged, but that others gladly took up. And Mulhall was cited in official and unofficial works not only through his own productive period, but throughout the twentieth century and into the twenty-first. As recently as 2002, in an article (published by Cambridge University Press, no less) on the history of the silk industry in Lyons and London, Alain Cottereau refers readers of a table on the national production of silks to the 1909 fourth edition of *Mulhall's Dictionary of Statistics* (151). To his credit, Cottereau hedges a bit: "[Mulhall's] approximations are risky," he says, but then goes on to claim that they "seem to be the least unreliable and fit in more or less with the most trustworthy international data" (151). Other modern-day scholars are more circumspect, but they still mention Mulhall. Irving B. Kravis, for example, provides some of Mulhall's figures in discussing studies of national incomes and prices, but notes that "Mulhall ... reports his methods within a paragraph only in later editions, and his sources not at all" (20). Discussing gold standards, Marc Flandreau similarly refers to Mulhall's "wealth estimates for 1895," but points out that "Mulhall's estimates had severe limitations" (42). Nonetheless, the appearance of Mulhall in such recent works of scholarship testifies to his long reach.

Finally, just as J.B. Harley proposed for maps, I think that in Mulhall's graphics, "[i]nstead of just the transparency of clarity we can discover the pregnancy of the opaque" (8). Mulhall's faults in statistical graphics ironically make his works a particularly clear lens through which to see one of the great motivators behind the development of statistical science and its graphic presentation: nationalism. We like to think of the twenty-first century as the age of globalization, but the eighteenth and nineteenth centuries, driven by sail and then by steam, saw the first real explosion of global trade and all of the international competition and imperial conquest that came along with it. With competition comes a desire for comparison, and statistics and by extension statistical graphics were developed in part to provide the means of comparing the nations of Europe and their current and former colonies. As Johan van der Zande and many others have pointed out, statistics arose in part from what in the eighteenth century Gottfried Achenwall called *statistik*, or a collection of facts about nation-states (411). It's only natural then that some of the first modern statistical graphics appeared in William Playfair's 1786 *Commercial and Political Atlas*, which visually compared the balance of trade between Great Britain and the other nations of Europe and Asia.

In this regard, Mulhall's graphics reveal a perspective I will call an imperial gaze. Mulhall gives us examples of how graphics were developed and used in part to compare the national (and thus by nineteenth-century terms, inevitably racial) qualities of different countries. If statistical graphics tell a story, then the visual rhetoric of Mulhall's graphics tells the story of international competition, British cultural supremacy, and cautionary tales about the possibility of an oncoming decline.

To see this aspect of the graphics and Mulhall's works, we will also need to trace the antecedent visual culture that influenced Mulhall's idiosyncratic use and arguably abuse of the pictogram. A second strand of my argument will therefore address an implicit desire in these graphics to bring statistical data into synoptic view. Examples from atlases comparing the principal mountains and rivers of the world show a striking resemblance to Mulhall's graphics on international commerce and wealth. This resemblance encourages the realization that as Western culture exerted power in ways that could (visually) move mountains and place them all together in one view extends as well to the power of statistical graphics to visualize the invisible. While Mulhall may have known just enough about statistics to be dangerous, his graphic works contributed a visual rhetoric that saw nation pitted against nation in a race for dominance of the entire globe.

While I do focus here on a single person, I do not intend to contribute to the mostly biographical scholarship that has characterized work on the history of data display to this point. As I said above, this scholarship has reclaimed important historical figures, but it tends to limit itself to finding the "first" and "best," mostly overlooking deeper structures of visual rhetoric and social history that surround these historical figures and their innovations. So in this chapter, I offer what one could perhaps call a counter-biography, or a deep biography. By this I mean a close analysis of the work of a graphical practitioner in the heyday of development of visible numbers, but focusing on someone who was certainly not first and may well have been worst. The aim is less to record his life than to find what it reveals of attitudes toward nationalism and statistical thinking, as well as the more profound questions of how data displays have participated in a socially constructed visual rhetoric.

Mulhall: Journalist, Emigrant, and Immigrant

Michael George Mulhall was born in Ireland in 1836 and educated at the Irish College in Rome. At 24, he followed his brother Edward Mulhall to Buenos Aires, where in 1861 they began publishing the *Buenos Ayres Standard*, the first English-language newspaper in South America (Howat). The brothers collaborated on the authorship, printing, and publishing of *The River Plate Hand-Book, Guide, Directory, and Almanac* (1863), which appeared in several editions and served as a *vade mecum* for British immigrants to Argentina. Separately, Michael Mulhall wrote three similar volumes of descriptive statistics while in Argentina—*The Cotton Fields of Paraguay and Corrientes* (1864), *Rio Grande Do Sul and Its German Colonies* (1873), and *The Handbook of Brazil* (1877).

With these works of *statistik* to his credit, Mulhall returned to England in 1878 as a successful author (Murray). Some years later, Mulhall included an evocative passage while discussing immigration to Great Britain in *Fifty Years of National Progress, 1837–1887*:

> At the same time there has been a constant tide of *immigration*, the average in the last 10 years reaching 132,000 per annum, most of them returned colonists. Some were disappointed emigrants, others men who returned home after years of industry, to enjoy leisure and fortune in Great Britain. (Mulhall, *Fifty* 13)

This passage could suggest how Mulhall saw himself as a returning emigrant—not as a failure, but as a successful contributor to the Empire who "returned home after years of industry." Another indication of this viewpoint is that he seems not to have returned to Ireland, but settled in Sussex.[2] This choice could suggest that despite his Irish Catholic heritage, and although he maintained his interest in and connections to Argentina and Ireland, he saw himself primarily and proudly as British.

Mulhall's Statistical Works

After his return to England, Mulhall began to publish statistical books in earnest. The best known and most successful of these works were his illustrated dictionaries of statistics. Mulhall claimed his to be the world's first dictionary of statistics:

> Hitherto every science has had its Dictionary except Statistics, one of the reasons, perhaps, why so many philosophers deny this branch of study to be a science. This is the first Statistical Dictionary ever published in any language. (*Mulhall's* i)

The dictionaries appeared in four editions published by Routledge in London and New York between 1884 and 1899:

- *Mulhall's Dictionary of Statistics* (1884).
- *Mulhall's Dictionary of Statistics*, 2nd edition (1886).

---

[2]   Mulhall's dedication to *The Progress of the World* (1880) is signed from "Grasslands Crawley." Crawley is a town in Sussex, some 30 miles south of the City of London.

- *The Dictionary of Statistics*, 3rd edition (1892).
- *The Dictionary of Statistics*, 4th edition (revised to November 1898).

The fourth edition included a second part with updated statistics between 1890 and 1898; it was also reissued in 1903 and 1909, with no appreciable changes. In addition, after Mulhall's death, Augustus D. Webb's *The New Dictionary of Statistics, a Complement to the Fourth Edition of Mulhall's Dictionary of Statistics* appeared in 1911, following the same general approach as Mulhall's other editions, but including no graphics.

Each of these editions followed the same pattern: an alphabetical listing of topics (some predictable, some obscure) with mostly tabular data and a paragraph or two of discussion. For example, in the fourth edition readers could find long entries on commerce, cotton, and crime; marriage, meteorology, and money; and prices, railways, shipping, and wealth. These are the kind of entries most likely to receive graphical illustration: the fourth edition included diagrams on agriculture, banking, commerce, education, food supply, iron and coal, mining, population, steam-power, and wealth. But one could also find shorter and more exotic entries. For example, Mulhall offers these statistics on fasting: "1684. Four men taken alive out of a mine in England, after 24 days without food"; "On December 14, 1810, a pig was buried alive by fall of a cliff at Dover, and on May 23, 1811, it was dug out alive, after 160 days." And this on kangaroos: "Kangaroos can jump a fence 11 feet high." And this on Koumiss: "Extracted by the tartars from mares' milk, a gallon of milk giving three ounces of Koumiss brandy."

Mulhall's dictionaries were wildly successful, at least economically. Advertisements for the third edition (1892) claim a print run of 6,000 copies. In Victorian terms, this would have been a runaway hit for a reference book, particularly one this expensive. For example, the 1891 second "enlarged edition" was produced in "Super-royal Octavo" with 640 pages, priced at £1 11s. 6d (Advertisement 66). Comparing historic prices to values today is inherently difficult, but this price in 2010 British pounds would be somewhere between £133 using a retail price index approach and £1,630 compared as a share of Gross Domestic Product ("Measuring Worth").

In addition to his dictionaries, Mulhall was an avid contributor to the reviews and journals of his day, including those published in England and the United States, where his statistics complimenting the marvelous growth of that former colony were quite welcome. Mulhall also published a string of books focusing on national (and especially British) progress. These works included the following:

- *The Progress of the World in Arts, Agriculture, Commerce, Manufactures, Instruction, Railways, and Public Wealth since the Beginning of the Nineteenth Century* (1880); no diagrams.
- *Balance-Sheet of the World for Ten Years, 1870–1880* (1881); 11 diagrams.
- *History of Prices since the Year 1850* (1885); 32 diagrams.

- *Fifty Years of National Progress, 1837–1887* (1887); 1 diagram.
- *The United States and the Future of the Anglo Saxon Race, by Rev. Josiah Strong; and the Growth of American Industries and Wealth, by Michael G. Mulhall* (1889); no diagrams.
- *Industries and Wealth of Nations* (1896); 32 diagrams.
- *National Progress in the Queen's Reign, 1837–1897* (1897a); no diagrams.

Mulhall's Graphic Practice

While not all of these works included diagrams, most included a number of lithographic or chromolithographic plates holding statistical graphics. The graphics were consistently provided on the recto side of a tipped-in glossy (sized) sheet, with the verso remaining blank. When in color, they were overprinted lithographically with a small range of pale colors, including pink, blue, green, orange, yellow, and buff. This color was frequently applied in the manner of hand-coloring copper or steel plates, such as one can see in earlier works such as Playfair's. For example, rather than covering an entire slice of a pie chart, the color will be applied only to the line dividing one slice from another.

Of the graphic genre for which he is best remembered—the pictographic chart—Mulhall produced many instances. For example, in his 1881 *Balance-Sheet of the World for Ten Years, 1870–1880*, he produced the following pictographic charts:

- Cannons to show "The Art of War," or military expenditures (Plate K).
- Bags of grain to show food supplies (Plate L).
- Bales of cotton to show cotton consumption (Plate C).

However, not all of his graphic productions were of this type, and many were of genres which we still use today:

- Column and bar charts; for example, "Education," Plate IV of *The Dictionary of Statistics*, 4th ed.
- Stacked bar charts; for example, "Stocks of Gold and Silver" in *The History of Prices since the Year 1850*.
- Pie charts; for example, "Mining," Plate VII of *The Dictionary of Statistics*, 4th ed.
- Mosaic charts; for example, "Occupations of Mankind," Plate III of *Industries and Wealth of Nations*.
- Small multiples, such as rows of squares or circles to represent different quantities; for example, "Carrying Trade of the World by Railway," Plate G of the *Balance-Sheet of the World for Ten Years, 1870–1880*.

- Line graphs; for example, "Price-Level of Great Britain, 1840–84," in *History of Prices since the Year 1850.*[3]

Despite the ease with which we can apply ridicule to Mulhall's pictographic charts, in many of these more conventional cases, Mulhall's practices would not be out of line with best practices today. They typically include clearly labeled axes, they don't clutter up the display with unnecessary gridlines, and they avoid obvious common errors like elevated zero points.

However, Mulhall also produced a number of curiosities, to which I will return anon.

Mulhall's Reviewers—and Critics

Mulhall's reputation was undeniably blemished—increasingly so as his career continued and more of the weaknesses of his work came to light. But in part that's where his work should engage our interest. For if we acknowledge that his work is flawed, we also must acknowledge that many people continued to use and applaud it for over a century. What does this tell us about the history of statistics and statistical graphics?

At the very least, Mulhall's success testifies to the hunger and enthusiasm for statistics of all kinds. It is far more common to find compliments on the "eminent statistician" than criticisms. Mulhall capitalized on these compliments in the second and subsequent editions of the *Dictionary* by including puffs from a variety of readers on the title page or elsewhere in the front matter:

'This admirable dictionary.'—Emile de Laveleye.

'The quintessence of statistics.'—Leroy Beaulieu.

'We want an edition in French.'— Yves Guyot.

'His statistics are most reliable.'—Baron Malortie.

'Mulhall's history of prices is accurate.'—Neumann Spallart.

'*His figures are remarkably correct.*'—Report of the U.S. Secretary of State.

'A German edition of the Dictionary would be useful.'—Tech. Blatt. Berlin.

'As useful as the Census report.'—*Graphic.*

---

[3]   This line graph, which shows the imports and exports of Great Britain, would not have been out of place in Playfair's *Commercial & Political Atlas* (1786).

'No books of reference have higher claims.'—*Glebe.*

'Display a vast amount of research.'— *The Times.*

'Remarkably well arranged and clear.'— *The Economist.*

'Inexhaustible treasury of facts.'—*Economiste Français.*

'This wonderful work stands alone.'—*Boston Beacon.*

'An unrivalled arrangement of statistics.'—*Academy.*

'Books of reference as trustworthy as they are unique.'—*Scotsman.*

'A boon to the student or public writer.'—*Irish Times.*

'The result of laborious and skilled research.'—*Contemporary Review.*

'Compiled in a convenient and intelligible form.'—*Spectator.*

'Written with great care and intelligence.'—*N.Y. Nation.*

'His works are well known to our readers.'—*Revue des Deux Mondes.*

'Clear, accurate, and comprehensive.'— *Toronto Globe.*

'They are a mine of facts.'— *Weekly Register.*

'No library should be without them.'—*Colonial Register.*

'Bring a vast number of facts within small compass.'—*Daily News.*

'The model of a statistical work.'—*Mark Lane Gazette.*

While a number of these testimonials have proven difficult to track down, several are accurate, including those from *The Times* of London and *The Economist.* These testimonials, Mulhall's sales figures, and his claims for his dictionary's primacy provide evidence of Mulhall's success as a tireless self-promoter whose books were obviously his bread-and-butter. But they also show that a good number of people found Mulhall's dictionaries worthwhile. Many of the reviews listed above are complimentary, along the lines of this 1880 example: "Mr. Mulhall has done his work with great care and intelligence, and he has found room for a number of curiosities of statistics, vital and other" (quoted in Mulhall, "Progress of the World" 332). These reviews may have been written by journalists rather than experts, but even some experts, like the economists Emile de Laveleye, Paul Leroy-Beaulieu, and Yves Guyot, the statistician Franz Xavier von Neumann-Spallart, and prominent figures such as Karl, Baron Malortie, at least appeared to approve of Mulhall's work.

But Mulhall had his critics, as well, which suggests that as the enthusiasm for statistics grew, so did the sophistication of the people who consumed them. As one might suspect, criticisms arose mostly from people who knew somewhat more about statistics than the typical periodical reviewer. Some complained of the loose bagginess of Mulhall's all-but-the-kitchen-sink approach to gathering statistics; a reviewer for *The Athenæum* said of Mulhall's *Dictionary*, "Far better would the author have done had he called his volume 'A Statistical Scrap-Book, or Numerical Curiosities in Art and Science'" (*Mulhall's* 22) A more common charge against Mulhall was that he did not provide the sources for his numbers or explain his methods for combining them. For example, Richmond Mayo-Smith criticized Mulhall's *Industry and Wealth of Nations* (1896) in the *Political Science Quarterly* as follows:

> This work needs but little notice. It is by the author of the well-known *Dictionary of Statistics*, and has the same faults as that work, namely, uncritical use of statistical data, unwarranted assumptions in order to fill up gaps and lack of adequate references to authorities by which the figures can be verified. The book is dedicated to his 'Fellow-workers in the Field of Statistical Research'; but I doubt if in any other science a compendium would be regarded as a work of research, or a mass of unverified or unverifiable data and conclusions would be welcomed by fellow-workers. (351)

Similarly, E.G. Ravenstein's review in *The Academy* takes Mulhall to task regarding his *The Progress of the World in Arts, Agriculture, Commerce, Manufactures, Instruction, Railways, and Public Wealth since the Beginning of the Nineteenth Century* (1880):

> This work contains a vast amount of statistical information on the principle States of the world. It bristles with figures; but so striking are some of the features presented, so interesting some of the subjects dealt with, that even general readers may derive from its perusal a considerable amount of pleasure. The author has ransacked a vast array of official documents and unofficial compilations, and had his qualifications as a scientific statistician been at all commensurate with his industry as a collector he might have produced a work of high authority and great usefulness. Unfortunately for his readers, many of the figures presented are altogether misleading … Many of the numerical statements made are mere guesses … It will be apparent from what we have said that the numerical statements presented by the author, and the conclusions based upon them, must be received with caution. (74)

Regardless of his sloppy documentation and data collection, Mulhall was able to cite authorities when pressed, as in the epistolary debate in *The Times* between Sir Francis Bell and "X.Y.Z," who disagreed with some of the statistics Bell had quoted from Mulhall. After claiming to be "unwillingly drawn into a controversy," Mulhall did provide sources for the figures in question ("Australasian Statistics" 139). Unfortunately in doing so he also had to admit to having transposed two numbers. Even so, his critic continued, rightly, to claim that "Mr. Mulhall, like ordinary folks, should furnish his proofs when called upon" ("Australasian Statistics" 144). This controversy over Australasian statistics characterizes the troubled relationship Mulhall seems to have had with the RSS, of which he was a fellow.

"Un Statisticien Fantaisiste"

For a further example, the *Journal of the Royal Statistical Society* published in December 1887 a translation under the title "The Abuse of Statistics" of an article by Alfred de Foville which originally appeared in the *Journal de la Societe de Statistique de Paris*, there titled "Un Statisticien Fantaisiste" ["A Fanciful Statistician"]. De Foville takes Mulhall to task, starting with the graphics in the 1885 *History of Prices*, which he belittles mercilessly: he describes the "little vessels sailing over the water" which Mulhall used to show the tonnage of merchant shipping; armies represented by "quite a battery of little guns delicately coloured in pale pink and green"; and "at the word *Houses* quite a street of little buildings" (Foville 704–5). De Foville poked fun repeatedly at Mulhall's use of color, referring to it in glib terms: "His *History of Prices* has to delight the eye eight diagrams coloured in blue and buff. In one portion of this work we find weights prettily presented in buff colour which are intended to show the proportion of taxes in various States to earnings" (704). De Foville also suggests that "[t]he master of graphic statistics might possibly take some exception to these diagrams, and would certainly find many anomalies in them" (704–5). For example, de Foville notes that "sailing vessels are represented by steam ships, and the 78 million of tons attributed to the British flag occupy more space than the 152 million of tons which is estimated to be the burthen of all the existing mercantile navies, English and foreign" (705). Putting the nail in the coffin, de Foville remarks of Mulhall's graphics,

> These variegated landscapes certainly have the effect of considerably enlivening a volume of statistics, but if we carried the principal [sic] a little farther we might represent all statistics by pictures, even those relating to demography. For example, the births might be represented by cradles, the marriages by bouquets and wraths of orange blossom, and the deaths by coffins. (705)[4]

---

[4]   This criticism seems to presage the controversy surrounding Sam Dragga and Dan Voss' article "Cruel Pies," in which they faulted statistical graphics for being inhumane

Clearly, de Foville found Mulhall's graphic practice laughable, and somewhat pitiable.

The center of de Foville's criticism, however, is the inaccuracy he sees in Mulhall's statistical work, claiming that "an enormous number" of figures are "absolutely incorrect," and "in a vast number of cases the numbers given appear to be purely imaginary" (705, 706). To support this charge he cites problems in a wide variety of Mulhall's figures, including the consumption of salt and the number of dogs in France. But de Foville saves particular opprobrium for the mischief Mulhall's incorrect figures have created. He cites a case in which newspapers cited Mulhall's figure for Irish pauperism as being 3,668,000 people, a number so exaggerated that the Secretary of State for Ireland protested in the House of Commons. As it turns out, de Foville reports, Mulhall started with government returns for the number of evicted people, but treated this figure as the numbers of families, which he then multiplied with the arbitrary "assump[tion] that the Irish family consists of seven persons" (707). Although Mulhall was Irish, this does indeed seem to be a rather off-handed assumption.

Statistics and the Imperial Gaze

Given Mulhall's many foibles and failures, why should we pay attention to his work?

De Foville's article also introduces as a witty aside the primary themes of Mulhall's approach to statistics, and indeed, significant themes of all nineteenth-century statistics: nationalism and imperialism. Foville notes that in Mulhall's graphic on shipping, "the 78 million of tons attributed to the British flag occupy more space than the 152 million of tons which is estimated to be the burthen of all the existing mercantile navies, English and foreign. But the sea has its mirages, and patriotism also perhaps" (705). Mulhall's work centers on the comparison of nations, presented as competitors in a global, imperialistic war against each other and against themselves. In so doing, Mulhall engages in what is perhaps the nineteenth century's greatest project: the building up of one nation above another through nationalism and imperialism, which as we shall see was a rhetorical as well as a physical struggle. Here lies the significance of Mulhall and his graphic practice: as a barometer of visual culture and nationalist rhetoric, as well as a case study of visible numbers and their participation in a social rhetoric of progress and power.

Although *The Progress of the World*, Mulhall's first book upon returning to England, includes no diagrams, its illustrated cover strikingly conveys the tone of the nationalism and imperialism of his statistical graphics (see Figure

---

and suggested adding pictographic or even cartoon drawings to re-insert reminders of humanity into graphics about accident statistics. For example, they suggest adding crosses into Minard's graphic on Napoleon's march to Russia to emphasize the many deaths that occurred.

6.1). On a rich blue cloth, a gilt sun rises—or sets—behind billowing clouds, its rays labeled with the grounds upon which the book measures progress: Arts, Architecture, Steam, Manufactures, Railways, Electricity, Commerce, Public Wealth, Instruction, and Press. Of this cover de Foville said, "It is bound in blue, and judging from the exterior, it is a very attractive volume, bearing on the cover the representation of a sun—*Solem quia dicere audeat*? [Who shall dare to say the sun is wrong?]" (704).

Perhaps to justify his entry into statistics, Mulhall, the retired journalist and recent inductee to the Statistical Society, dedicated his book "To the press of Great Britain, which so zealously promotes the moral and material progress of the age" (*Progress of the World*). With no other preamble, Mulhall's *Progress of the World* launches into statistics divided into three parts: "The World," "The British Empire," and "Foreign Countries," which includes chapters on the European countries, as well as those of Asia, South America, and finally, the United States. In each part (or in part three, in subdivided chapters for each country), Mulhall outlines progress in the areas articulated by the cover, as well as a few more: population, agriculture, banking, food, and charity.

What's interesting about this otherwise un-illustrated book is its revelation of Mulhall's ethnocentric—even jingoistic—viewpoint about the progress of the British Empire. This viewpoint extends to most of his work throughout his career, including his graphic productions. In fact, Mulhall published a number of works with "progress" in the title, including *Fifty Years of National Progress, 1837–1887; National Progress in the Queen's Reign, 1837–1897*; a series of five articles in the *North American Review* entitled "Progress of the United States," which related statistics about the New England, middle, southern, Prairie, and Pacific states; and finally, in the year before he died, "Five Years of American Progress." This persistent focus on progress suggests the importance of the idea to Mulhall's view of the world and the relationships of nations within it. Arguably, economic competition is at the base of all imperialism. During Mulhall's lifetime, however, the focus of imperialism definitely shifted to this focus, in that imperial success grew to be measured in wealth, rather than in land. The great European land-grab for colonies had slowed, having almost already reached its full development, with only parts of Africa yet to be divvied up. At this point the focus of imperialism shifted from annexation of lands to warfare between imperial powers (the Franco-Prussian War, the first and second Boer Wars, Napoleon III's intervention in Mexico, the Spanish-American War, etc.) and, more significant to Mulhall, toward economic competition.

Mulhall also was most productive during a period in which the details of economic competition had become significantly more visible to governments and the public. This visibility arose because nations began (in the eighteenth century but especially in the early nineteenth century) to make serious, concerted, and long-standing efforts to gather data about international trade, including imports, exports, and transportation, as well as national industry, such as manufacturing and agricultural production. And for our purposes, this visibility became real,

Figure 6.1   Cover of Mulhall's *The Progress of the World*, 1880

rather than metaphorical, with the development of visual media for displaying statistical information. Beginning with Playfair's graphics about exports and imports from and to Great Britain, many of the innovative movements toward a graphic statistics arose out of the desire to show the wealth and health of imperial nation-states—not just in isolation, but in competition with other imperial powers. This visualization of nations ties back to the earliest impulses of *statistic*: to measure and compare the relative strengths and weaknesses of national competitors.

Despite its focus on data and quantification, however, this effort to measure national power was not unbiased. For example, it's worth noting that Playfair's *Commercial and Political Atlas* is for the most part a collection of ribbon graphs showing the balance of imports and exports "In favor of Great Britain" and "Against Great Britain." Great Britain, in other words, is the viewpoint at which Playfair sets his viewer; his charts show the relative power of nations in economic competition from the perspective of one nation.

Like Playfair's, Mulhall's work shows that he saw the British Empire at the center of the world, just as it is at the center of his first statistical book. Mulhall makes it his mission, much as he ascribes to the press in his dedication, to encourage the progress of the Empire by laying out its growth in comparison to other countries—both to magnify British accomplishments or to warn against its possible eclipse by other nations. The foundation of this biased visualization of international competition is not only bragging on one's own nation's accomplishment, but the nagging worry that other nations might be accomplishing more.

As the cover of *The Progress of the World* suggests, while the sun may never set over the British Empire, it might be rising for someone else—most notably, the United States. Mulhall's positioning of the United States as the last nation considered in this book implies a somewhat passive-aggressive attitude towards Great Britain's former colony and growing competitor. In the book, the Ottoman Empire, all of the South American countries, and even Sarawak (now part of Malaysia) come before the United States, as if Mulhall were saving the biggest threat for last. Mulhall is not openly critical of the United States; on the contrary, he particularly admires the country's natural resources, commenting for example that "for internal industry or foreign commerce, the United States are singularly favored. They can put boundless territories under tillage and their coastline is equal to half the Earth's circumference" (Mulhall, *Progress* 492). He also repeatedly compares the United States to its European competitors: "The area of lands taken up by settlers in the last forty-five years is equal to the aggregate extent of France, Spain, and Portugal, or five times that of the United Kingdom" (Mulhall, *Progress* 494). The last paragraph of the book includes this passage:

> American industry and population increase much faster than in Europe, and so does the wealth of the nation. Every day that the sun rises upon the American people it sees an addition of £500,000 to the accumulated wealth of the Republic,

which is equal to one third of the daily accumulation of mankind. (Mulhall, *Progress* 525)

These comparisons make it clear that Mulhall sees the United States as a juggernaut on the horizon, taking aim not only at Great Britain but at all of the European powers. This reading is underscored by his series of articles on the progress of the United States. At the beginning of an article in the *North American Review*, Mulhall speaks in open admiration: "The progress of the United States in the last fifty years so far surpasses that of any other nation in ancient or modern times, whether viewed in regard to population, or to industry and wealth, that the subject is one of extraordinary interest, not only to Americans, but to European spectators" (Mulhall, "Progress" 566). Contrast this with the opening of a book Mulhall published the same year, *National Progress in the Queen's Reign, 1837–1897*: "There is no country *in Europe* which has made such material progress in the last 60 years in the United Kingdom. No European statesman or economist will be found to dispute this fact" (*National* 1, emphasis added). Mulhall may be able to trumpet the European preeminence of Great Britain, but on a global stage he argues that his own country and all of its European competitors have been eclipsed by the United States.

The tension in the parallel language in these two passages suggests that Mulhall is deeply invested in the idea of not only human, but national progress. He wishes his own country to excel, but he is willing not merely to acknowledge but to applaud progress in other countries. Moreover, doing so provides an implicit criticism of the British Empire as it falls further behind its former colony.

The Imperial Gaze: Mountains and Rivers

We can see this visual argument in Mulhall's statistical graphics, as well—and particularly in two idiosyncratic graphic forms that show data as overlapping isosceles triangles of different heights, or wavy lines of different lengths (see figures 6.2 and 6.3). Both of these examples appear in the fourth edition of the *Dictionary of Statistics*, but Mulhall used these graphic forms several times previously. For example, in the 1881 *Balance Sheet of the World for Ten Years, 1870–1880* he used triangles for Plate A, "Industries of All Nations" and for Plate H, "Accumulated Wealth of Nations," and he used wavy lines for Plate I, "Income of Earnings of Nations." As statistical graphic forms, these have little to recommend them objectively, and they offer little apparent advantage over more traditional bar or column graphs. The wavy lines are relatively innocuous, and we could simply describe them as stylized bars, though the need for such stylization is not immediately evident. It is possible that the penchant for mimetic or representative graphics we see in Mulhall's pictographic charts comes into play with the wavy lines in representing commerce in terms of river or ocean transport. But that does not explain why he used the same wavy lines to show the "Income of Earnings of Nations," which does not necessarily have a direct relationship with water.

The triangles, however, are significantly flawed as a means of representing statistical graphics visually. Although as usual Mulhall does not provide his sources or full data, the triangles appear to be indexed solely by height, as represented on the y-axis. However, as similarly shaped triangles differ in height, the ratio of their areas differs as the square of the ratio of their sides. In other words, a triangle twice as high as a smaller similarly shaped triangle will have an area considerably greater than twice that of the smaller triangle, making the larger triangle seem visually more significant. This problem is compounded in Figure 6.3 in that the triangles in the center of the display have a smaller vertex angle (they're taller and thinner) than the triangles on the left and right (which are shorter and broader)—so they're not even similarly shaped. In addition, the triangles overlap as they extend out from the center, inconsistently concealing part of the area of the triangles they occlude. These variations make it very difficult for a viewer to make accurate comparisons between the triangles.

Beyond simple incompetence, why would Mulhall use such ineffective graphics?

To understand the significance of these graphic forms, we must look to their antecedents: the popular synoptic views of the relative heights of mountains and lengths of rivers—and sometimes of lakes, islands, and waterfalls. These graphics appeared in many atlases beginning in the late eighteenth century and extending through the twentieth century, but they reached perhaps their zenith in the mid-nineteenth century. A web archive of images of such graphics appears on the Donald Rumsey Map Collection web site. Examples appear in figures 6.4, 6.5, and 6.6.

These graphics, of which there are scores of other examples, tend to show the mountains of the world stacked up in one view as if they made up a single mountain range. Typically, but not always, the graphics include a labeled y-axis against which the heights of mountains could be read. Some instances split the view into two displays, one for the western and one for the eastern hemisphere (Mitchell, *Western*); others split up the mountains by continent (Johnson, *Johnson's*). Most often the mountains were arranged so as to show the tallest mountains in the middle, with shorter ones laid out to each side (Mitchell, *Heights*). In other examples, the mountains were arranged tallest left or tallest right, creating an effect very much like a bar graph of data points. Finally, some examples were arranged artistically, as if the plates were intended to be seen as a picturesque view of a mountain range (see again Figure 6.5).

Rivers tended to be a later addition, appearing in the blank spaces around the mountains. For example, Rumsey includes William Darton and W.R. Gardner's 1823 *Comparative Heights of the Principal Mountains and Lengths of the Principal Rivers*, which has the mountains arranged along the bottom from shortest on the left to tallest on the right, and the rivers hanging from the top, with the longest on the left and the shortest on the right. In effect this created two nesting right-angle triangles, which would fit nicely on a single sheet. Or

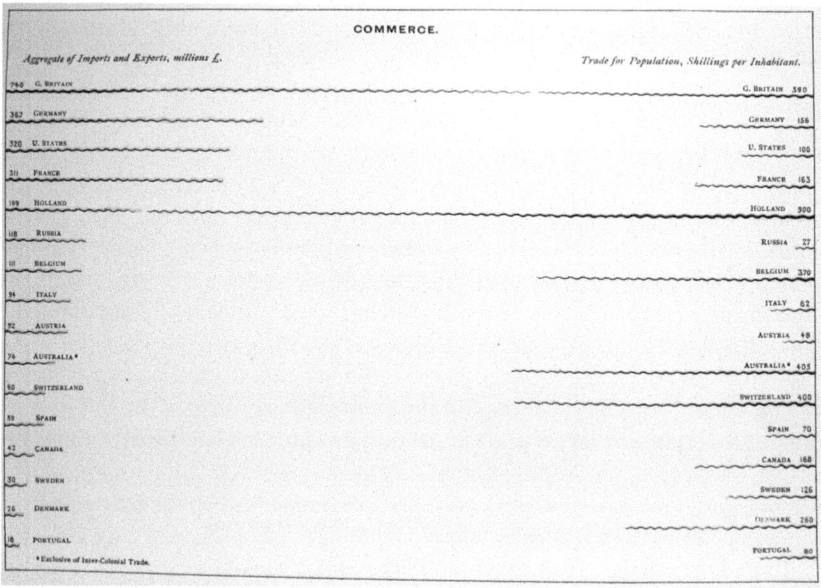

Figure 6.2 "Commerce," Plate III of *The Dictionary of Statistics* (4th ed., 1899)

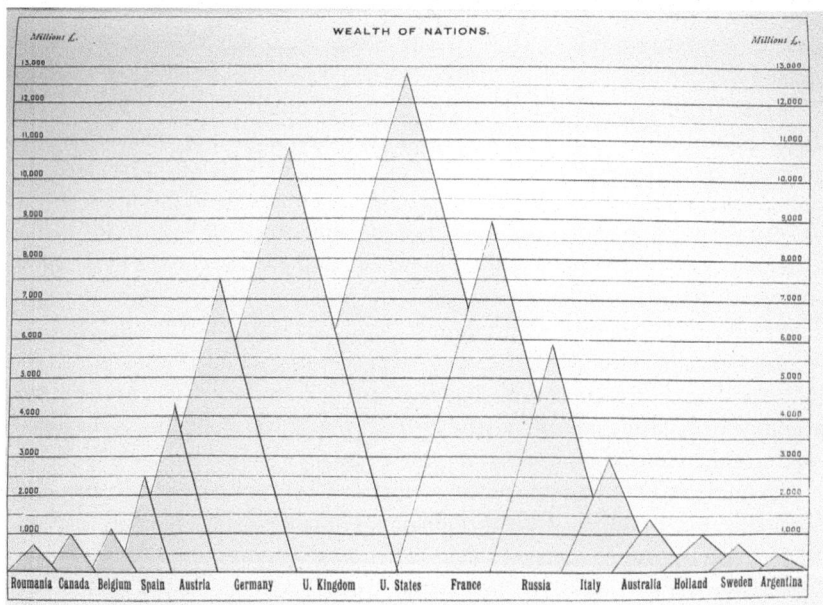

Figure 6.3 "Wealth of Nations," Plate X of *The Dictionary of Statistics* (4th ed., 1899)

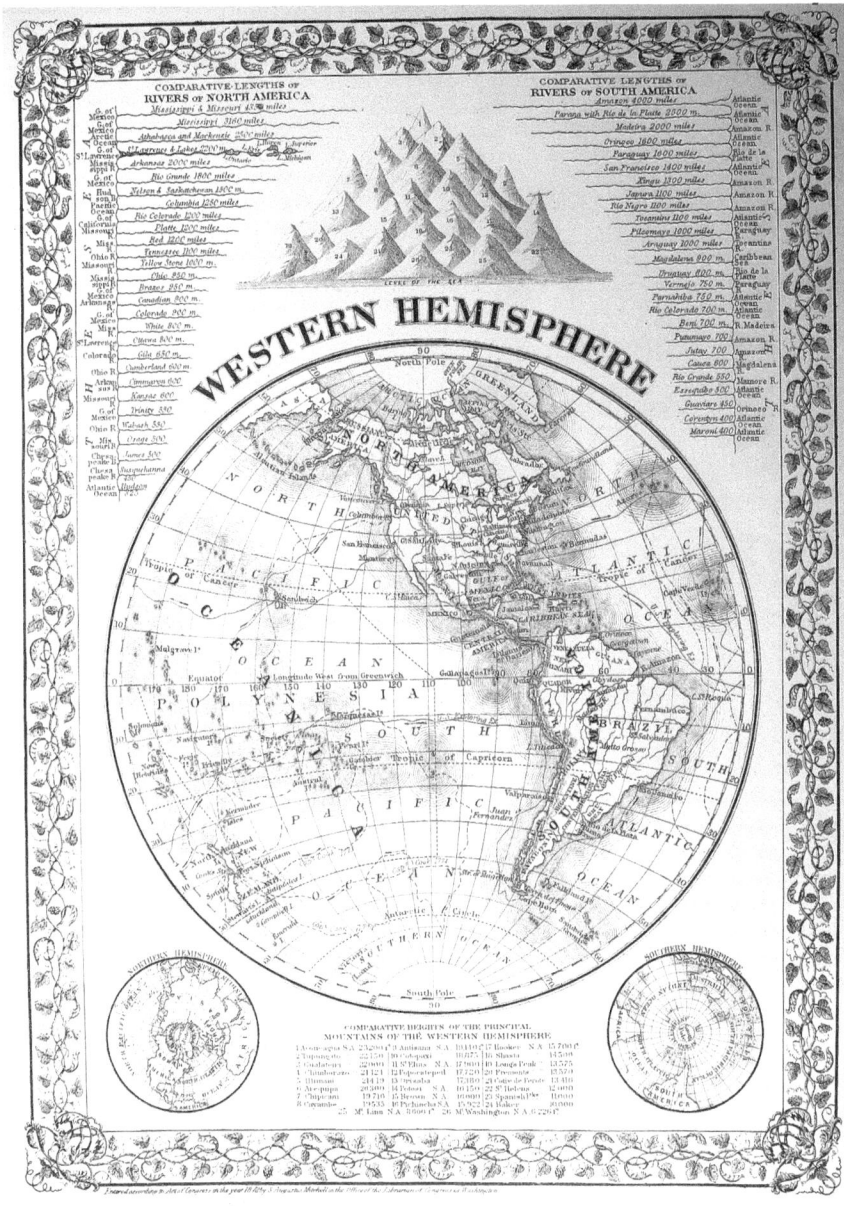

Figure 6.4   Western Hemisphere, engraved and published by S. Augustus Mitchell (1872)

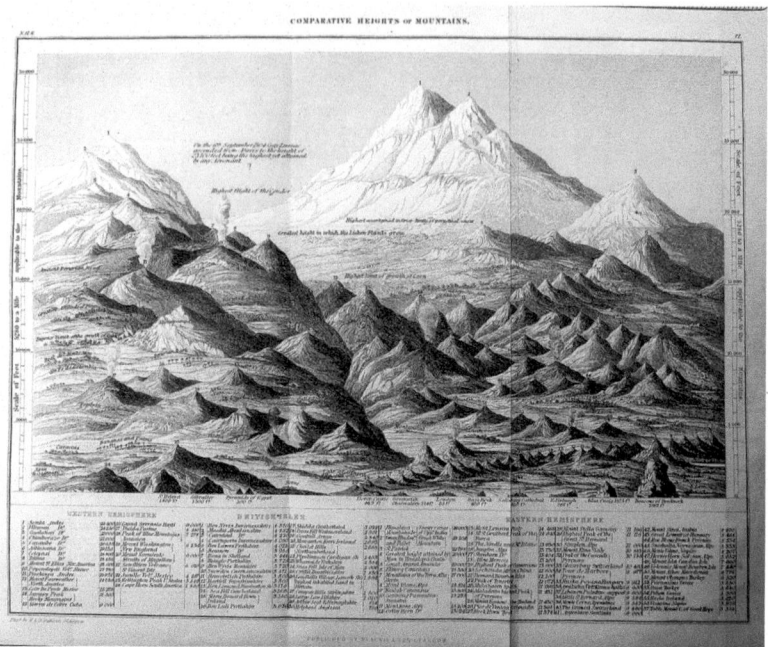

Figure 6.5   *Comparative Heights of Mountains*, engraved by W. & D. Duncan
and published by Blackie and Son, Glasgow (1853)

with the displays that centered the tallest mountains, the rivers could be draped
almost like a curtain from the top to the left and right of the centered mountain
range, as in Figure 6.6.

There were also some interesting combinations and modifications. In some
atlases the rivers and mountains were inset into traditional geographic maps, or
on the edges of geographic displays of the hemispheres. Other notable geographic
features were sometimes included for comparison, as well, including especially
waterfalls, rivers, and islands. Few of these displays presented geographic
information beyond the continent or hemisphere in which the mountains appeared.
However, some did represent the rivers as starting at a shore and extending into
a generic land mass. For example, John Thomson and William Home Lizar's
1823 *Comparative View of the Lengths of the Principal Rivers of Scotland* shows
the rivers extending from a horizontal coast up into (fancifully high) mountains,
and the Society for the Diffusion of Useful Knowledge (SDUK) produced a very
complex display showing rivers and tributaries extending from a central sea out
radially to the edges of the sheet.

These displays—and particularly those that arrange the mountains with the
tallest in the middle, as in figures 6.4 and 6.6—bear a striking resemblance to
Mulhall's triangles, to the point that Mulhall's work might be considered an
abstraction of the same genre. Furthermore, in the *Dictionary of Statistics* Plate

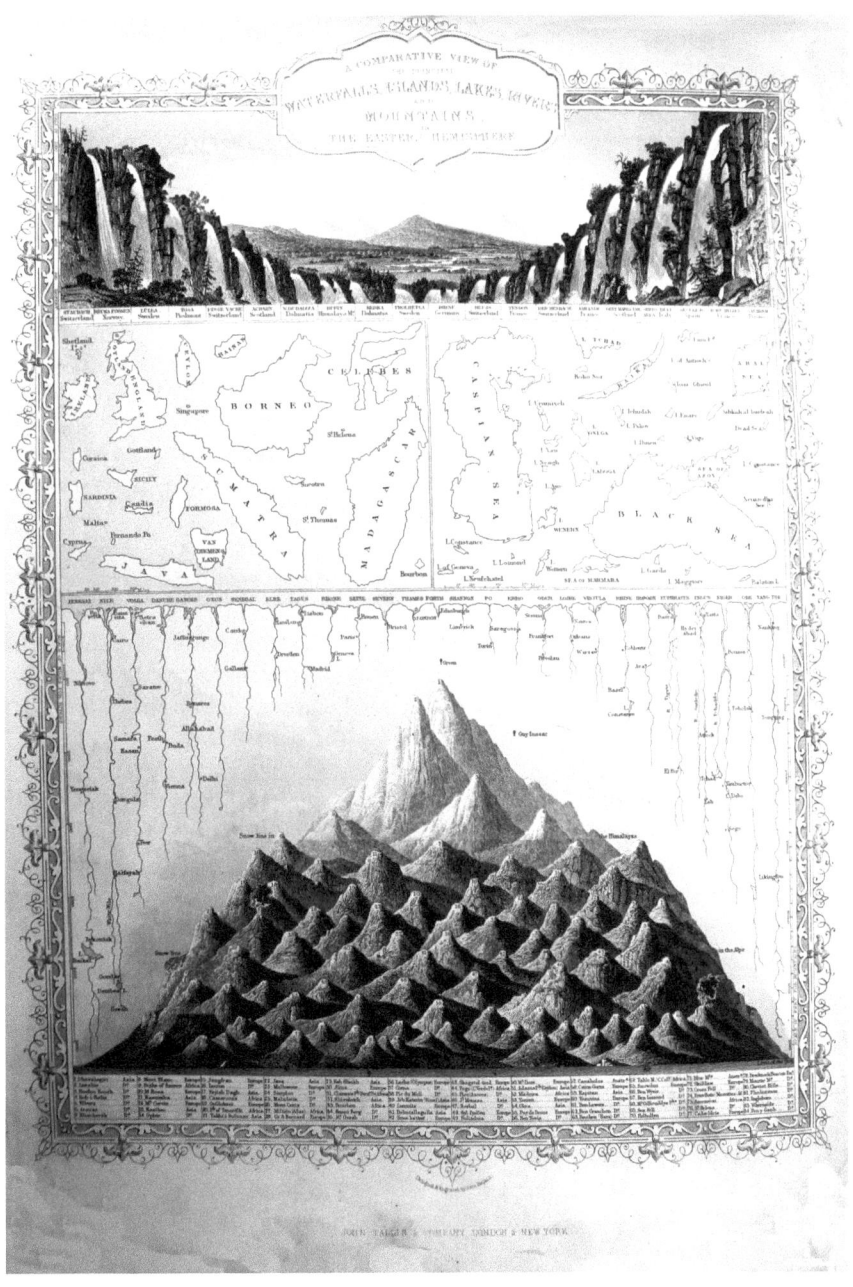

Figure 6.6   *A Comparative View of the Principal Waterfalls, Islands, Lakes,*
             Rivers and Mountains, in the Eastern Hemisphere, engraved by
             John Rapkin, published by John Tallis & Company, London and
             New York (1851)

X (Figure 6.3), the right-hand sides of the triangles are marked with thicker lines than left-hand sides, providing a sense of shadow with the lighting source to the left. This shading is consistent with the vast majority of the atlas displays of principal mountains of the world, which almost always give the sense of depth by providing shadows to the right and the lighting source to the left of the mountain range.

What's going on here? Why would Mulhall choose to follow this particular graphic fad or convention, when his graphics are not displaying mountains or rivers, but national wealth and commerce? I would argue that both Mulhall's displays using triangles and wavy lines to show national wealth and commerce and the synoptic displays of mountains and rivers share a similar viewpoint: the imperial gaze. In cultural terms, the synoptic views express not only geographic data (which could be shown as easily with more abstract forms, such as bar charts), but a vision of the world arranged for a single imperial viewer. The popularity of these graphics testifies to the ubiquity of this viewpoint. In essence, through the technology of graphic display, these images metaphorically allowed viewers to wrest the very mountains from their roots to present them for convenient perusal in an atlas or the wall of a drawing-room. This is a powerfully hubristic claim, entirely consonant with the hubris that led the imperial nations of Europe to ransack the globe for profit. In similar terms, Mulhall invokes this synoptic model to present the entire world's commerce and wealth to one view.

And that one view is unmistakably imperial. For example, consider Figure 6.7, a "Comparative View of the Principal Rivers in the Four Quarters of the World," published in 1832 by the Society for Promoting Christian Knowledge. As an example of the genre of comparative views, this graphic has little remarkable about it, showing the rivers grouped into Europe, Asia, America, and Africa. But the accompanying table of data on the rivers depicted in the chart has an interesting first column, labeled simply "ratio of length" (see Figure 6.8). The ratio of length to what? The answer to this question is not explicitly stated, but a quick glance down the column past the entry for the Forth (1/2), the Tay (6/7), the Trent (9/10), and the Shannon (12/13) reveals the Thames, at a ratio of 1:1. In other words, all the rivers of the world are measured by the standard of the Thames. This approach might have a good rationale; comparing the unfamiliar to the known is an acceptable way to show difference of scale. But it also reveals the ethnocentrism of the imperial gaze, which measures the entire world through the framework of British values.

Kevin Brown rightly points out that these synoptic mountain displays were influenced by Alexander von Humboldt's famous 1805 profile view of mountain vegetation and geology in the Andes, although some synoptic displays predate von Humboldt. But the synoptic displays more closely resemble many well-known nineteenth-century paintings of mountains, such as J.M.W. Turner's (1775–1851) *Llanberis* (1800), or Caspar David Friedrich's (1774–1840) *The Watzmann* (1824–25), or even work as late as Albert Bierstadt's (1830–1902) *Sunrise on the Matterhorn* (1875). The Romanticism of the synoptic views—the presentation

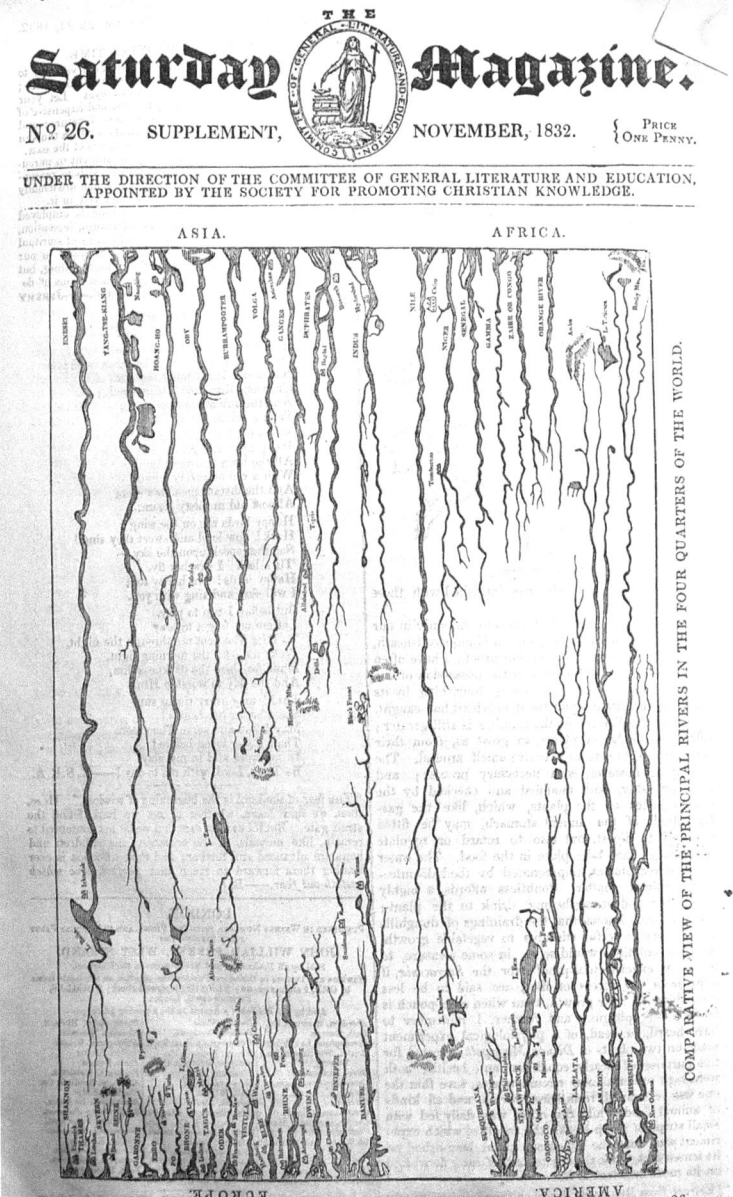

Figure 6.7   *Comparative View of the Principal Rivers in the Four Quarters of the World*, published by the Society for Promoting Christian Knowledge (1832)

Ratio of Length.	Names.	Locality.	Rise.	Places passed by.	Discharge.	Length in Eng.Miles
	RIVERS OF EUROPE.					
⅓	Forth	Scotland	BenLomond Mntns.	Glengyle, Stirling, Edinburgh	North Sea, near Dunbar.	110
½	Tay	Scotland	Grampian Hills	Lochs Dochart and Tay; Dunkeld, Perth, Dundee.	North Sea, by the Firth of Tay.	165
⁷⁄₁₀	Trent	England	Norton, Staffordsh.	Stone, Rudgeley, Burton, Farndon, Gainsborough,	North Sea, by the Humber.	200
¹¹⁄₁₂	Shannon	Ireland	Mountains of Leitrim	Loughs Allen, Rea, Dergeart; Killaloe, Limerick, Tarbert.	Western Ocean.	205
1	THAMES	England	Cotswold Hills	Reading, Henley, Windsor, Staines, Brentford, London, Gravesend,	North Sea, at Sheppy Isle. 26—2	215

Figure 6.8   Table accompanying *Comparative View of the Principal Rivers in the Four Quarters of the World*, published by the Society for Promoting Christian Knowledge (1832)

of many mountains as one sweeping mountain range, centered on the viewpoint of a single graphic connoisseur who had the means and desire to buy such a thing—is worth thinking about. This viewpoint posits a powerful, self-reflexive viewer, much like the one Friedrich portrayed in his most famous painting, *The Wanderer above the Sea of Fog* (1818), which shows the back of a man looking out on a mist-swathed mountain range. In fact, it's worth noting Julius Schrader's (1815–1900) portrait of von Humboldt (1859) in the New York Metropolitan Museum of Art, which shows the aged explorer with a Humboldtian Chimborazo in the background, almost as a hybrid of the artistic and statistical versions of these images; all it needs is a legend and a y-axis.

Although the relationship of Romanticism to nature has been debated at length for decades—see for example Jonathan Bate's *Romantic Ecology* or Stephen Hunt's *Green Romanticism*—the consensus seems to be that Europeans of the Romantic period were not merely fascinated with nature in its own right, but in nature as a reflection of the individual, whose identity was framed by the impression of nature on the soul. Thus we have Wordsworth's passage on climbing Mt Snowdon in *The Prelude*, which is much less about a mountain than it is about Wordsworth's self-apotheosis as a sublime poet. And as Edmund Burke pointed out, the sublime itself was an internal, aesthetic sensation gained by viewing vast and terrible things—like mountains.

More broadly, this sublime experience was not only about artistic or personal identity, but national identity. And that national identity, when the nation is an empire, is a complex web of pride, tension, and contradiction. As an empire, a nation loses its central identity precisely as it gains power over the people it subjugates. Those people become part of the empire, inevitably changing it from within. Thus some of the most commonplace markers of traditional nineteenth-century "British" culture—tea from China; sugar from Barbados; curries, cashmere shawls and marmalade from India—are in fact imports from subjugated peoples. The interface between the imperial self and the colonial other is thus as contested as that between British men's macassar hair oil from Indonesia and the antimacassars, or doilies, British ladies crocheted to keep the oil off their furniture.

Much recent scholarship on Romanticism, including most prominently Edward Said's *Orientalism*, has emphasized the interconnections between Romantic art, identity, and empire. Saree Makdisi has commented on these interconnections this way: "In other words, the energy of British imperial culture was directed not only outward, to other places and peoples, but also inward, toward the very heart of England itself and the forms of otherness suddenly discovered to be lurking within it" (429).

In this regard we can understand Mulhall's employment of these graphic forms not only as a celebration of imperial power and international competition, but as a commentary on his own place in that milieu. Thus Mulhall, the emigrant from England's first colony, Ireland, returns to immigrate to England to be an alien, an interloper, and yet a part of the empire, a celebrant of it and apologist for it, as well as a voice warning of its eclipse. He has seen the juggernaut on the horizon. He has lived in the Americas and seen their amazing natural wealth. And for all Mulhall's British patriotism, his graphics show the British Empire as a foothill to the United States, at least in terms of its economic wealth and promise.

Statistical Graphics and Identity

I began this chapter with the claim that we can sometimes learn more from history's detours than from its highways, which themselves seem straighter in retrospect than they were in actuality. In Mulhall, we see a case in which the graphic statistician returned to a form from the past—the synoptic view of mountains and rivers—to invoke and extend a rhetoric of imperial power. The invocation was a link to the European project of conquering the world, whether physically or commercially; the extension was to the possibility of triumph or decline, as nation competed with nation for supremacy.

Looking at Mulhall reminds us that statistics are not disinterested, and statistical graphics are even less so. Our interests are what help us choose what to quantify statistically. Some subset of that data is of such great interest that we feel the need to see or display it more clearly. So we create graphics for the data we care most about—either for ourselves (our nation, our wealth) or for our audiences, whom we wish to convince of something. This is particularly true of statistical graphics in the eighteenth and nineteenth century, which were labor-intensive and expensive to produce.

What we find interesting or compelling in a graphic is in part what expectations and preoccupations we bring to it. For example, consider Minard's famous 1869 graphic of Napoleon's March to Moscow. Many of us find gripping and evocative this depiction of human misery because it is a cautionary tale warning against the tragedies of war and the follies of pride. But consider how this graphic would look to someone else—say, for example, Pyotr Ilyich Tchaikovsky, whose *1812 Overture* appeared in 1882, just 13 years after Minard's graphic. Tchaikovsky wrote the *1812 Overture* to celebrate the Russian victory at the Battle of Borodino, often counted as the turning point of the war before the disastrous (to the French) siege of Moscow. To a patriotic Russian like Tchaikovsky, Minard's graphic could possibly evoke pity for fallen foes, but it would more likely resemble a chronicle of triumph for Russia than a cautionary tale for France. From this perspective, it is a visual record of how Russia (both the state and the brutal Russian winter) beat back an imperial aggressor and utterly destroyed his army. With no other alteration, Minard's graphic reads equally well from both of these radically different viewpoints.

Looking at Mulhall's graphics brings this dynamic into high relief. The contradictions of an immigrant celebrating the empire he has joined—even while remaining marginalized by it and while warning of its demise—is something we can see especially clearly because Mulhall's graphic practices are so dissonant with what we think of now as the march of progress from the first innovations of statistical graphics to the best practices of today.

Chapter 7

"A scheme of cross-roads, orderly and mad": British Trench Maps of the First World War

Marguerite Helmers[1]

The First World War calls to mind many images, none as indelible as those of the trenches. In particular, the muddied sloping walls of the British trenches on the Western Front have become iconic, a shorthand, visual argument for the miseries of army life, for what Wilfred Owen termed "the pity of war" ("Strange" 65) The first trenches were dug in 1914 and as the war progressed, over 25,000 miles of trench lines extended through Belgium and France, created by German, French, and British forces. In 1914, the British Ordnance Survey was called to serve the war efforts of the British Expeditionary Force. While maps produced by that agency during the First World War presented topographical data, they also opened conceptual spaces for the military commanders: sites of possible advance and areas for caution. As Denis Wood contends, the "point of the map" is "to present us not with the world we can *see*, but to point *toward* a world we might *know*" (12). Maps suggest structure, a way to understand topography, events, and narrative. Amy Propen argues a similar point, writing that "Maps not only *reflect* renderings of the world, but they also *create* these renderings. This creation process relies on particular cartographic conventions that shape how the map makes meaning ... that is, the map can help paint a picture about where things are and how the landscape ought to be perceived" (237).

The extent of fighting from August 1914 to November 1918 was vast in terms of territory, personnel, and politics. The progress of the engagements

---

[1]   I would like to thank the map librarians at the British Library map room for their assistance in preparing this article. Librarians at the American Geographical Society Collection at the University of Wisconsin Milwaukee also enabled me to compare British and American Expeditionary Forces maps. At the Imperial War Museum, thanks go to the librarians who helped me locate appropriate photographs of desolated areas. The interlibrary loan department at Polk Library, UW Oshkosh is, without fail, most prompt and responsive, even with the most obscure request. James Power and Somme Battlefield Tours provided new and old trench maps and made arrangements for me to visit the areas described in this essay. This project was funded by a generous grant from the University of Wisconsin Oshkosh Faculty Development Board. Nick Dvoracek in the UW Oshkosh IDEA Lab edited the maps for publication. Finally, my family deserves credit for their patience: Emily, for her insights about art and design; Caitlin, who helped edit the manuscript; and William, who walked the trenches with me. As usual, they have gone over the top with their assistance.

that occurred during these five years is displayed on official maps published after the war, which offer a digest of territorial gains and losses. The data on official, recapitulative maps is clean and simple, with bold arrows and shading demonstrating tactical advances across miles of ground. In this chapter, rather than analyze the summative presentation of data on the official maps, I turn to British trench maps used in the field during the war. The trench maps are more of a work in progress in terms of data acquisition and presentation; rather than representing what was known after the war ended in 1918, the maps show what was known during the actual campaigns, a knowledge that was often perilously incomplete and available only to a small percentage of those on the front lines. Peter Chasseaud notes that most front line troops never saw the secret editions of the maps that covered their area (*Topography* 7). The purpose of trench maps was to identify the German trenches and gun placements. Because they offered a way for commanders to understand topography, they were instrumental in directing the tactical course of events of the Western Front.

The story of cartography in the First World War reveals the important shift in attitude about the primacy of maps to military operations that occurred from the war's onset to the war's end. From the moment early in the war when the British Expeditionary Force believed it did not need the services of cartographers to the production of over 3 million maps by war's end, military command apprehended that maps could predict action, showing not only what had been achieved, but what operations were possible. Maps such as "planning maps," "harassment maps," "situation maps," and "barrage maps" offered the coordinates necessary for neutralizing the guns and defenses of the opposition. The proliferation of the maps reflected the shift in tactics from a war of movement to the breaking of the stalemate. The need for accuracy that demanded the expertise of the Ordnance Survey also advanced development of three mapmaking techniques. With sound ranging, the coordinates of the enemy guns were located by an array of microphones that enabled observers to measure the differences in time between the initial sound of a gun firing, followed by the sound of the shell moving through the air, and culminating in the impact of the shell. Flash spotting was less precise, relying on visual information to site the position of a gun. Aerial observation initially relied on oral reports and sketches, but by 1915 reconnaissance officers were using cameras to assist in mapping the front. The need for accurate representations of the counter positions even led to the refinement of cameras. Wood's insights into the quest for greater accuracy in maps is central to my argument: the more it was argued that maps had become scientific windows on the world, the less people were concerned with the fact that a map is a representation that presents a partial truth:

> All you have to do is ignore the way the window isolates this view at the expense of another, is open at only this or that time of day, takes in only so much terrain, obligates us to see it under this light ... or that. This is the sleight of hand: if you're paying attention to the glass, you're not paying attention to what you're seeing through the window. Not that accuracy is not worth achieving, but it

was never really the issue, only the cover. It is not precision that is at stake but precision with respect to what? What is the significance of getting the area of a state to a square millimeter when we can't count its population? (21)

Wood asserts that the cartographer "repressed the magnitude and significance of his intervention in what passes in the map for a transcription of nature," thus pointing toward the rhetorical nature of all cartography: maps represent what the cartographer is interested in, silencing the rest (77). Turning to the significant contributions of cartographer J.B. Harley to the contemporary theories of understanding maps, Wood quotes Harley's insistence that:

> Maps are never value-free images ... Both in the selectivity of their content and in their signs and styles of representation, maps are a way of conceiving, articulating and structuring the human world which is biased towards, promoted by, and exerts influence upon particular sets of social relations. (quoted in Wood 78)

Therefore, although accuracy is particularly important in military operations, we must acknowledge that even military maps are artifacts created by a selection of symbols and the display of data. Mark Monmonier succinctly presents this idea, writing, "maps, like speeches and paintings, are authored collections of information" ("How" 2). Propen also points out that the "map also creates meaning through its *selectivity*, or the inclusion and exclusion of information" (238). Ben Barton and Marthalee Barton amplify the relationship between choice and power in their study of maps, commenting, "Rules of inclusion determine whether something is mapped, what aspects of a thing are mapped, and what representational strategies and devices are used to map those aspects. These rules amount to either explicit or implicit, overt or covert, claims to power" (55).

   In fact, most trench maps contained an important disclaimer that reminds readers of their constructed nature:

> The fact that an obstacle is not represented on the map does not necessarily mean that there is none there. It is often impossible to distinguish obstacles or to identify their character.

Such is the official warning to readers that the trench maps encode aporia, spaces of absence and hesitation. Rhetorically, an aporia causes the reader to doubt or offers an ambiguity. Terry Eagleton called the aporia one of the key devices of deconstructive theory, the aim of which "is to show how terms come to embarrass their own ruling systems of logic; and deconstruction shows this by fastening on the 'symptomatic' points, the *aporia* or impasses of meaning, where texts get into trouble, come unstuck, offer to contradict themselves" (116). Both Propen and Monmonier fix upon the grid and the technique of plane tabling as examples of the ways that maps lie, presenting a version of reality that is actually a fiction (see

"How"). Propen comments that "the grid functions as an ideologically charted representational device that distorts while its goal is to convey accurate models of the terrain by positioning space along equal lines of latitude and longitude" (236). Planimetric distance, that which is measured on a plane, "compresses the three-dimensional sheet by projecting each point perpendicularly onto a horizontal plane," thereby underestimating "overland distance across the land surface" (Monmonier, "How" 32). Thus, in trench maps, the generalized topography makes trenches into straight lines and No Man's Land into a plane, despite the terrain being a series of pits, valleys, shell holes, rutted tracks, and barbed wire. Therefore, "Not only is it easy to lie with maps, it's essential. To portray meaningful relationships for a complex, three-dimensional world on a flat sheet of paper or a video screen, a map must distort reality" (Monmonier, "How" 1).

By reminding combatants to read beyond the printed data, the cartographers of the Great War were perhaps unwittingly prescient about the outcome of many battles. Certainly in the Battle of the Somme in 1916, that which was "not represented on the map" was as instrumental to the outcome as any obstacles that were recorded. As events of the First World War transpired often contrary to the data on the map, the potential grand narrative of a just war against an invader of neutral territory became, in poet Siegfried Sassoon's words, "a war of aggression and conquest."

This chapter is not designed to be a comprehensive study of trench maps on the Western Front. Instead, I use a selection of maps from the 1916 Somme engagement to tell a story about the involvement of the Ordnance Survey in military cartography, to demonstrate the considerable detail of the grid system that attempted to make the bombardments along the Western Front even more precise, and to show that the data from the maps hides the data from the diaries of the men who fought in the battles. The title of this chapter is taken from a line in Antonia S. Byatt's poem "Trench Names," which remarks on the monstrous human work of digging tunnels in the ancient woods and fields of France to establish a new landscape of war:

> The sunken roads were numbered at the start.
> A chequer board. But men are poets, and names
> Are Adam's heritage, and English men
> Imposed a ghostly English map on French ...

I begin with the larger history, the background of the Battle of the Somme and history of the Ordnance Survey as official cartographers to the military. I then narrow the study to a few map sheets from the Somme engagement to show the magnitude of the grid system. Following the presentation of the data, I discuss the outcomes of the Somme assaults, the effects of battle on the combatants, and the ways that the battle changed the landscape and the maps across a 12-month period. In conclusion, I return to the trench names and reflect on the power of the maps to recreate topography and create a sense of place. Thus, the essay moves through topography, events, and narrative, with cartographic data as its focus.

July 1, 1916: The Somme

In contemporary memory of the Great War, the Battle of the Somme has come to signify prolonged agony and almost certain death. Among the more recognizable names of combatants are poets, artists, and musicians, including J.R.R. Tolkien, Edmund Blunden, Siegfried Sassoon, George Butterworth, and Ivor Gurney, making the battle additionally significant for its haunting influence on literature and the arts.

The Germans had occupied the area around the River Somme in Picardy since 1914. Due to a lack of military action, the area was regarded as a "quiet front," which had given the Germans ample time to dig their trenches deep, 30 feet into the chalky soil, and establish a system of machine gun posts covering approaches from the west, north, and south (Keegan 289). General Sir Douglas Haig planned an attack along the 30-kilometer line of the front with the French General Joffre; the plan was for the British to attack north of the River Somme and the French to cover the southern sector. Due to disastrous losses in the previous years of the war, Haig had assembled a new fighting force made up of untested recruits from towns across England. These were composed of families, friends, and coworkers who were grouped together to fight in a recruiting scheme called "Pals battalions." The Pals were joined by four Territorial and four regular divisions who had seen service earlier in the war. In all, 20 divisions were grouped together as the British Fourth Army to the west of the German strongholds in the towns of Beaumont-Hamel, Thiepval, Ovillers, and La Boiselle. They faced six German divisions who occupied the towns, the hills, and the roads.

Haig planned a bombardment of over 1 million shells to precede the attack on July 1, 1916. In theory, the field guns would cut the wire in front of the enemy's trenches, destroy the artillery, and damage the defensive lines of the Germans. Prior to the battle, the British dumped 3 million shells along their front "to feed 1,000 field guns, 180 heavy guns and 245 heavy howitzers" (Keegan 291). For one week, the British fired on the German lines. On July 1, they detonated 17 mines under German positions along the front and at 7:30 a.m., the whistles were blown to send the first wave of British troops toward the German encampments under what was called a "creeping barrage" of advance shell fire designed to keep the opposing forces in their trenches, should any have survived the first week's heavy shelling. Haig's objective was the town of Bapaume, seven miles beyond the front line, uphill, and through a countryside already pitted by shell holes, trenches, and barbed wire. The British army would not achieve this objective on July 1. Fighting continued along the front until November, with several intermittent attempts at advance by the British. Over 19,000 British soldiers were killed on the first day of fighting alone and over 1 million were killed from all sides throughout the length of the Somme engagement. Synonymous with pain and suffering, the Somme is remembered for its loss of life on a massive scale:

> The simple truth of 1914–18 trench warfare is that the massing of large numbers of soldiers unprotected by anything but cloth uniforms, however they were trained, however equipped, against large masses of other soldiers, protected by earthworks and barbed wire and provided with rapid-fire weapons, was bound to result in very heavy casualties among the attackers … [T]he conditions of warfare between 1914 and 1918 predisposed towards slaughter and that only an entirely different technology, one not available until a generation later, could have averted such an outcome. (Keegan 293–4)

In this chapter, while not providing a military history of the Somme engagement, I have found that some background to the scale of the fighting along this one sector of the front line is necessary to understand the dichotomy between planning and execution. Trench maps of the Somme region present two related and possible narratives: that of the tactics and the casualty figures and that of the human experience and the human cost. The same maps that offered sites for guns and the disposition of troops also marked the places where human life was deployed into terrible conditions and extinguished at tremendous emotional cost to survivors and to nations. Many writers have explored the relationship between the battles and the mythology that emerged in the writing, especially those pieces published in the official press and by the war poets Owen and Sassoon. Looking for something larger than the unbearable truth that men were wasted at the will of commanders, writers who study the landscape of the Western Front have translated this site of military engagement and subsequent massive loss of life into a sacred and symbolic space. Thirty-two million maps were produced during the First World War; 8 million servicemen were sent to war from Britain; 451 cemeteries on the Somme alone are maintained today by the Commonwealth War Graves Commission. From each of these statistics, a different story may be told.

In his work of military geography *The Old Front Line* (1917), the British author (and later poet laureate) John Masefield described the topography around which the British were situated. Masefield was sent to the front to write an official history, but was denied access to official papers. Thus, he concentrated his work on topography, making *The Old Front Line* an unusual work of dystopian pastoral. A particular sadness of expression infuses Masefield's prose throughout the work. But while Masefield marks the passing of time in the greater narrative of the war, he also takes a moment to note the trench names, remarking on them as "paths to glory":

> This description of the old front line, as it was when the Battle of the Somme began, may some day be of use. All wars end; even this war will some day end,

and the ruins will be rebuilt and the field full of death will grow food, and all this frontier of trouble will be forgotten. When the trenches are filled in, and the plough has gone over them, the ground will not long keep the look of war. One summer with its flowers will cover most of the ruin that man can make, and then these places, from which the driving back of the enemy[2] began, will be hard indeed to trace, even with maps. It is said that even now in some places the wire has been removed, the explosive salved, the trenches filled, and the ground ploughed with tractors. In a few years' time, when this war is a romance in memory, the soldier looking for his battlefield will find his marks gone. Centre Way, Peel Trench, Munster Alley, and these other paths to glory will be deep under the corn, and gleaners will sing at Dead Mule Corner. (1)

As Masefield predicted, today, the landscape surrounding the Somme is returned to farming. Contrary to his predictions, the "frontier of trouble" would not be forgotten; it would become hallowed ground to be visited by thousands annually since 1919. The trenches are in fact still visible, with many, such as those at Delville Wood, preserved as sacred groves. Thus, far from "not long keep[ing] the look of war," the ground has been indelibly marked by the events of 1916. And, "with maps," the events are retraced by tourists and families. The towns with such famous names instrumental to the Somme engagement—Fricourt, Thiepval, Beaumont-Hamel—are situated amidst gently rolling hills. To travel the roads around Albert is to experience a duality of vision, seeing the fields as they are now but populating them with the divisions of the Third Army as they were in 1916. To travel to the places of the past, the trench maps are important in order to visualize and make meaning from the ordinary farm landscape. Furthermore, they enable narratives, bringing "the old front line" and its stories to life.

Mapping the Western Front

The story of the design of the trench maps emerges from three directions: the established design of Ordnance Survey maps; the overlaying of existing maps from France and Belgium; and the necessities of war resulting in the need for developing a large corps of surveyors and mapmakers in France. Even so, it wasn't until late in the war that maps were standardized. In August 1914, the British army did not realize the scale or type of war in which they would be engaged. Envisioning a rapid war of movement that would rely on cavalry advance, surveyors were told that their services would not be needed and members of the

---

[2]   Throughout this chapter, I have refrained as much as possible from using the term "enemy" to describe the forces opposing the British Expeditionary Forces, opting instead for tactical terms such as "objective" or "counter objective." Should the word remain in the text, it is restricted to quoted material.

OS were sent to the front as fighting men, not as mapmakers. At the beginning of the war, only 10 people worked on mapping the Western Front, and these were primarily intelligence officers. However, as Harold Winterbotham pointed out, "After the opening battles in Flanders the line became more or less stationary, and the advent of 'Granny,' the first 9.2 inch howitzer, quickly followed by others of her kind, gave us our first ideas as to what tasks would fall to our [the mapmakers'] lot" (Winterbotham 13). Battery commanders could not work with the enlargement of the French 1:80,000 map, new trenches on the German and allied sides emerged daily, and enemy gun positions, dugouts, railways, new roads, and ammunition dumps all had to be mapped (Winterbotham 13). By 1918, over 4,000 people were employed by the army and the Ordnance Survey in creating maps to deal with the different situations of the war (Winterbotham 13). The chief need of the army, Winterbotham noted, was in "locating the German guns":

> As time went on these were more and more carefully concealed. Woods, houses, and villages all hid a certain proportion; and the professional camouflage expert did his best for the remainder. (13)

British mapmakers, needing to compensate for the lack of accurate extant maps of the front, began a new survey of NE France, covering 12,000 square miles (Winterbotham 14). This survey began in February 1915, six months after the start of the war. By July 1915, each Army had its own topographical section (*Report on Survey* 8).

The Ordnance Survey mapmakers faced many challenges. For one, "the rolling downs of Picardy" has "few conspicuous features to fix," Winterbotham wrote,"and the churches lie mostly in the hollows" (15). "[D]ust from constant traffic" of infantry on foot, cavalry, bicycles, and motorized supply and medical transport obscured measurements (Winterbotham 14). The topography that required mapping was constantly changing due to bombardments that laid towns to waste, putting new demands on the mapmakers. Winterbotham complained, "What is the highest and most conspicuous point of a building to-day will probably not be so to-morrow … both artilleries are busy destroying all points of vantage" (hyphens in original, Winterbotham 15). Consequently, marking trig points was difficult, as "many of the churches whose positions we knew had been knocked down" (Winterbotham 14). Nonetheless, it was important that "for the proper placing of batteries every hedge and field must be shown" (Winterbotham 17). In the maps of the town of Thiepval, at times, the highest point was often the remains of a hedge.

Around 1916, Field Survey Battalions were given two printing machines, which printed maps at a size of 35 by 22.5 inches. The presses ran "round the clock" (Winterbotham 18), producing 700 colored maps per hour. By 1917, to keep up with the demand for maps, Winterbotham brought his own printing press to the front to print daily situation maps (Chasseaud, *Topography* 11). Duplicates of the large scale base maps were overprinted with specialized intelligence. In addition to trench maps, battery positions were the most important information

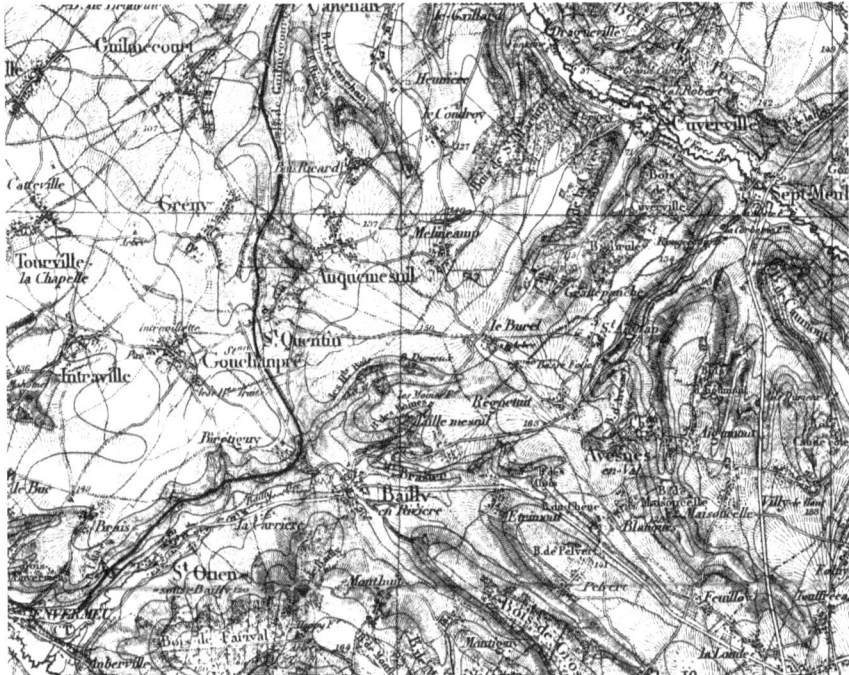

Figure 7.1   The British Expeditionary Force initially used extant French maps
before recognizing that their scale was inappropriate to the type of
military operations they needed to employ
Source: © William Ready Division of Archives and Research Collections, Mills Memorial
Library, McMaster University. 204ww1map, box 3, envelope 94. Detail.

for the British commanders. Small numbers of secret editions of the maps showed
British and German trenches, including the trench names. "Maps cars" made daily
runs to Corps and Divisional Headquarters, traveling 40 to 100 miles along roads
often badly damaged by shell fire. Even so, access to the maps was limited, and
thus "the troops in the trenches made their own diagrams and sketches, which
were freely distributed" (Winterbotham 17).

As Masefield indicated, the names of the trenches, although the most ephemeral
in terms of landmarks, were some of the most enduring features of the Western
Front. Names were given to the British and German trenches by the soldiers.
Chasseaud records that trench names had a "considerable history" (*Rats* 10):

> The name originally bestowed by a unit in the line might first be painted on
> a board, and entered onto a rough sketch map by that unit. It might then be
> transferred by a draughtsman of an RE Field Company, under the Divisional
> CRE, responsible for trench construction and maintenance, onto that
> Company's master manuscript map, using the topographical or trench map

produced by the Army Topographical Section or, from February 1916, Field Survey Company (FSC) as a base map. Such drawings would then be used by the Topographical Section or FSC to compile the secret edition trench map showing the British trenches. (*Rats* 10)

Trench names appeared as early as September 1914, and as Chasseaud reports, quickly became part of British popular culture. A popular newspaper, *The Illustrated London News* "carried several drawings by an officer in the Ploegsteert Wood sector [in 1915], showing various British military constructions and dugouts dating from the winter of 1914–15, with name board proclaiming *Somerset House, Hotel de Lockhart, Plugstreet Hall* and *Scawby*" (Chasseaud, *Rats* 10). Trench names were a vernacular rhetoric of the front, often juxtaposing fond memories of places at home with the miserable and featureless landscape that surrounded the soldiers. While the irony of many of the names is evident, the potential for the vernacular names to impose a new meaning on the events of the landscape was significant.

Maps of the Somme

Two main offensives were launched by the BEF against the German positions along the Somme: July 1, 1916, as part of the major action along the Somme front; and, because the July 1 action had failed to gain ground, September 26, 1916, as part of the Big Push. Lyn MacDonald points out the difficulty of the British offensive, noting that, "On July 1 the British had been in position for a bare ten months. But the enemy had been there for almost two years" (*Somme* 6). The Germans were well entrenched in Thiepval, as Gerald Gliddon reports:

> Thiepval itself had been transformed into a fortress and much work had been done by the Germans in order to make it impregnable. Redoubts, blockhouses, and concrete vaulted shelters had been built on the surrounding ground and a continuous line of trenches went around the village. (380)

The town of Thiepval is located at the top of a slight, "deceptively gentle" elevation (MacDonald, *Somme* 6), which provided the Germans with a view to the west and south. West of the town, along either side of the River Ancre, is marshland. Thiepval Wood covered the area between the river and the town. From the river to the town of Thiepval, the ground rose from 70 feet above sea level to 140 feet. The forward positions of British firing trenches were situated 20 to 30 feet lower than the town along the eastern and northern edges of the wood. The British heavy guns were located to the southwest. MacDonald quips, "The British Military had taken over Thiepval Wood as surely as they had taken over Aldershot":

For the first year of the war, a tottering signpost, drunkenly askew, had kept up a pretence that it was '*Propriété privé. Entrée Interdite.*' But it had long ago given up the ghost. Now battered trenchboards nailed to the trees bore directions in uncompromising English: 'To Johnson's Post.' 'To Iniskilling Avenue.' 'To Hamilton Avenue, Campbell Avenue, Elgin Avenue.' 'To Belfast City.' 'To Paisley Dump.' (*Somme* 66)

Of 12 known OS maps produced of the Beaumont-Hamel region, 3 maps in particular offer a way to understand the topography of the area around the town of Thiepval, the events that transpired from July 1 to November 18, and some of the stories of "the old front line." All of these maps are at the revised and now standard scale for trench maps and barrage maps, 1: 10,000:

Sheet 57dSE1&2 2A. (February 7, 1916)
Sheet 57dSE1&2 2D. (August 1916)
Sheet 57dSE2 4A(S). (February 1917)

The main grid system for the active combat area in Belgium and France was plotted on a 1:40,000 map. Each grid square was assigned a letter from A to X (reading left to right), further divided into numbered squares, and refined to a secondary grid system designated with lower case alphabetic coordinates. This 1:40,000 flat sheet was then subdivided into four 1:20,000 sections, marked as NE, NW, SE and SW. The 1:20,000 grid was further subdivided into four 1:10,000 maps, which are the maps most commonly used for trench and barrage maps and which are the maps most typically used for research today.

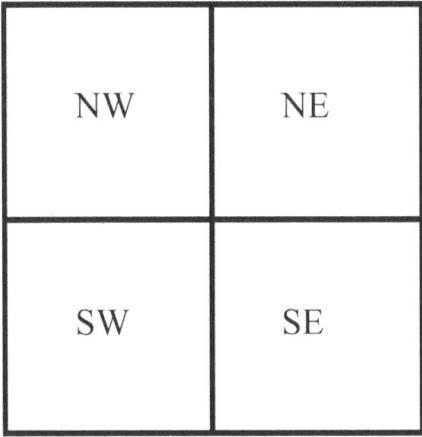

Figure 7.2   Sample grid referencing system for 1:40,000 flat sheet that was subdivided into four 1:20,000 sections, marked as NE, NW, SE and SW. Each of these 1:20,000 sections became its own map

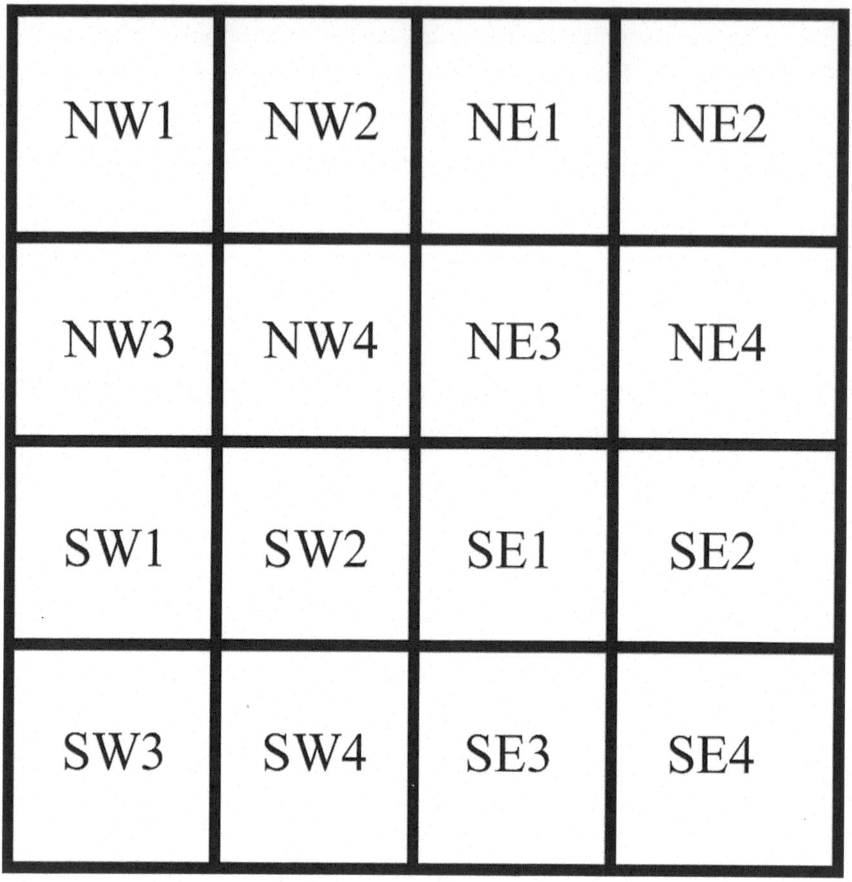

NW1	NW2	NE1	NE2
NW3	NW4	NE3	NE4
SW1	SW2	SE1	SE2
SW3	SW4	SE3	SE4

Figure 7.3   1:20,000 trench grid referencing system. The 1:20,000 grid was
         further subdivided into four 1:10,000 maps

The sheets for 57D covered the area around the farming village of Hebuterne in northwest France. The map marked 4A(S) is a secret edition (S), released in limited number and which showed British defenses. All maps were produced by the 4th Field Survey Company of the Fourth Army, commanded by Lieutenant General Sir Henry Rawlinson. The base maps resulted from air photography, survey, and cadastrals; the maps were corrected every two months from the new outline of December 2, 1915, to October 14, 1916, and then again in February 1917. The map designated 57dSE1&2 Edn.2A uses version 2 of the Ordnance Survey base map for Beaumont-Hamel, with corrections noted to February 6, 1916 (hence the use of A). The August 1916 map—57dSE1&2 Edn.2D—uses version 2 of the Ordnance Survey base map with the forth correction of detail (D), updating intelligence about German trenches through August 15. The map dated

February 17, 1917 4A[S] draws on the fourth revision of the base map with the first overprinting of trench information with detail from February 17, 1917 (A).

Despite the considerable detail provided by the OS and the intelligence reports, commanders made their own map notations in pencil. For example, Major General Sir Oliver Stewart Wood Nugent, commanding the 36th Ulster Division positioned at the northern edge of Thiepval Wood, used his copy of 57dSE1&2 Edn.2A to mark the objectives for the brigades under his command, the 107th, 108th, and 109th. The objectives are circled, lettered, and ironically renamed after familiar Irish sites in the province of Ulster: Kilrea, Lisburn, Cavan, Moy, Lurgan, Clones, Coleraine, Portadown, Enniskillen, Strabane, Omagh, Dungannon, Lisnaskea, Portrush, Bundoran, and Derry City. The northernmost objective for the Ulsters—the western edge of the village of Grandcourt (map reference R8d)—was marked anew as "Portrush," a seaside village along the northern coast of Ireland. In naming sites and objectives, then, there are three layers: the original French town names; the British coordinate references; and the vernacular points of reference, both those that were formalized as trench names and those that were added in pencil by the troops and commanders.

Maps distributed to the artillery and the infantry were tailored to the job they had to do; the artillery was issued a barrage map with lifts and timings indicated in minutes on the overprinting:

> The guns would operate in a series of carefully planned 'lifts.' For the last hour before the attack, the bombardment would fall with redoubled intensity on the enemy's front line—lifting at Zero [zero hour, 7:30am] to rake forward to his second line and to shell it for precisely as long as it would take the infantry to subdue the first defenses and start off for the second … It worked superbly on paper. (MacDonald 46)

Therefore, any military action was the result of a complex interplay of different maps, interlaced decisions, as MacEachern writes about mapping in general: Maps are not "communication vehicle[s]," but "one of many potential representations of phenomena in space that a user may draw upon as a source of information or an aid to decision making and behavior in space" (MacEachern 12). Added to the maps were the Summaries of Intelligence based on information gained from "captured documents, on the interrogation of prisoners, the reports of raiding parties and observations of the result of the bombardment" (MacEachern 52).

Prior to occupation in September 1914, Thiepval "had a few shops, a café or two, a sizeable church and sixty odd houses occupied by the farm workers who worked the fertile lands around" (MacDonald, *Somme* 7). The standard base map information for Thiepval as it looked in early 1916 is printed on 57dSE1&2 Edn.2D in sepia ink on linen, with German lines overprinted in red ink. On the S version, the British trench lines overprinted in blue ink (these are not visible in Figure 7.4). Both sides of the conflict would have been seeking significant trigonometric landmarks, and, as MacDonald writes, "what caught the eye from

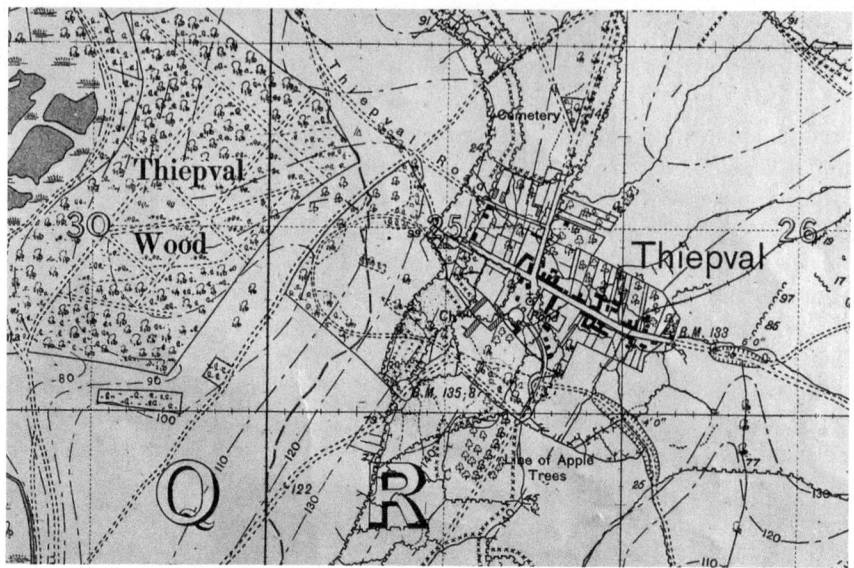

Figure 7.4   Detail of Thiepval Wood and village, showing the line of apple
                 trees, 57dSE1&2 2D
Source: © The British Library Board, Maps C.14.h-k (sheet 57.d.SE1&2). Detail.

miles away was the Chateau" (*Somme* 7). The chateau and town cemeteries were
the sites of heavy German machine gun placements. To the north of the town, on
a rise to 150 feet, the guns positioned at the area known as the Schwaben Redoubt
could aim down across No Man's Land. The easternmost point in the line, points
designated Diamond Wood, Thiepval Point South, and Maison Gris sap, were only
100 yards from the German front line trench. To the north, along the Thiepval
Road, No Man's Land extended for about 500 yards between the opposing forces.

The overprinting of 57dSE1&2 Edn.2D shows the British front line in a faint
dotted blue outline. German defenses are printed in red. German trenches keep
to the standard map reference diagram in use with the 1:10,000 sheets: Firing
trenches were represented by castellated lines; entanglements were represented by
small Xs; other trenches were represented by thick lines; and disused trenches by
dotted lines. While the sepia base map of the town depicts a country village, the
German defenses are strongly lined in red, and weave throughout the streets and
fields. From the German front line in the west, a firing trench runs up the main
road; it surrounds the chateau, encircles the pond, and leads up to the church.
Trench lines extend into the houses and barns and snake through the orchards in
the yards beyond. The road lined with apple trees is also lined with barbed wire.
Barbed wire runs from the edge of the village through the cemetery.

Just as the red lines of the German trenches ensnare the town, the British
trench lines cross and re-cross Thiepval Wood. Any wood that provided cover

for troops inevitably came under heavy shelling by the opposing forces and so it was that Thiepval Wood came to be denuded of trees just as it was scarred by earthworks and dugouts. The trench maps came to demonstrate their aporia, became unstuck, their warning that "it is often impossible to distinguish obstacles or to identify their character" revealing the ambiguity of the grid. As Wood queries, "What is the significance of getting the area of a state to a square millimeter when we can't count its population?" (21).

The events of July 1, 1916, would prove tragic. In order to conceal the movements of the 36th Division, the artillery launched a heavy barrage against the high ground that included 4-inch Stokes mortars to provide a smoke screen throughout the Ancre Valley (Gliddon 382). Moving from Thiepval Wood, the leading waves of infantry reached the German front line, "but hardly were they across however, when the German barrage fell on no man's land and on the rear companies of the first line battalions and also on the second line. As their barrage lifted, flanking machine gun fire began from the dominant position of Thiepval Cemetery" (Gliddon 382). The aporia of the maps proved fatal, and what could not be seen was instrumental in the defeat of the British: "The enemy was able to come up from the cellars in the village and to pour bullets into the backs of the 109th Bde. As well as the 108th Bde" (Gliddon 382). In the meticulous planning leading up to the assault against Thiepval, army headquarters had equipped each soldier with a small tin triangle, sewn to the backs of their uniforms. "[T]hey were intended to reveal to distant observers the progress of the infantry's advance" (MacDonald, *Somme* 53). Instead, they became the occasion for mass destruction. In Thiepval Wood, there was confusion and misery:

> At Paisley Dump all the trenches in Thiepval Wood and the tracks to Authuille and across the Ancre met. The wounded lay there in hundreds and the noise was inhuman. (Gliddon 385)

In the aftermath of July 1, as night fell, a sound arose from No Man's Land, "like nothing they had ever heard before":

> Lieutenant Hornshaw was to remember it as a sound that chilled the blood; a nerve-scraping noise like 'enormous wet fingers screeching across an enormous pane of glass.' It was coming from the wounded, lying out in No Man's Land. Some screaming, some muttering, some weeping with fear, some calling for help, shouting in delirium, groaning with pain, the sounds of their distress had synthesized into one unearthly wail. (MacDonald, *Somme* 66, 67)

Private Ernest Deighton lay wounded in a shell-hole between the first and second German trench lines. He reported later on his disorientation, his lack of sense of place:

When night came I were in a deuce of a state. I must have been fainting off and on, what with the loss of blood. You'd no idea of the passage of time. I didn't know where I were. I only knew there were Germans in front and Germans behind and I had no idea which way were the British lines ... Lights going up all the time. All this noise. Them shelling from their side and us shelling from ours, and machine-gun in between. (quoted in MacDonald, *Somme* 76)

Rescue parties and those able to drag themselves back into their own trenches recall "absolutely trampling on the wounded" (MacDonald, *Somme* 78). "You couldn't help it," recalled Private Tom Easton. "It's bad enough when you're getting bloody wounded, but it's bloody murder when they're trampling on you as well. Oh, they were crying out! I can hear them now" (quoted in MacDonald, *Somme* 78). The "precise and detailed orders" from the Army and the careful grids of the OS maps could not possibly represent the confusion of sights, sounds, and smells coming from the lines during the battle (MacDonald, *Somme* 39).

By mid-August, with the objective of Thiepval still impregnable, rains commenced. The dead in No Man's Land had not been retrieved in six weeks. They had been baked by the sun and eaten by rats, some reported to be the size of cats. When the rains came, the River Ancre flooded the valley and swamped the duckboards leading into Thiepval Wood, the water bringing with it all manner of waste and human remains. The water ran downhill from the town of Thiepval and into the wood "and turned the steep communication trenches into glissades of slime and mud, soon stirred into squelching soup by the constant passage of soldiers slithering to and from the line" (MacDonald, *Somme* 230). While the rains continued, so did the attacks, for, "[s]o long as the Germans held Thiepval, they would be able to overlook almost the whole British advance and direct their guns to crush it" (MacDonald, *Somme* 249).

Private Arthur Hales of the 2nd London Field Ambulance recalls being sent into his position to await casualties ("stretcher cases") to send to the Casualty Clearing Station. His description of the night on the Somme reveals what the maps cannot, the confusion and darkness that belied the careful naming of the trenches:

How [the officer] found his way is a marvel to me for, even if there had been any landmarks, it would have been impossible to find them on such a night. After about an hour, we struck a surfaced road which made the marching, if anything, more difficult. We were split up and forced through narrow passages between the waggons [*sic*], up to our eyes in mud ... The officer had to ask the way and then we set off again—not on the road, but across a trackless main of mud, down a slope. It was impossible to stand upright. Most of us did it on our hands and knees. (quoted in MacDonald, *Somme* 329)

Trooper Reg Lloyd of the Cheshire Yeomanry recalled the topography of the Somme into which he was sent after a mere few weeks of infantry training. Due to prolonged heavy bombardment, recognizable features of the landscape such as trees, buildings, walls, gates, and roads, were obliterated into a featureless landscape of mud and rubble:

> Going through this area it was just as if an earthquake had been there. It was all mud and I was frightened to death. Eventually we came to a notice board. That's all. Just a notice board in among a bit of rubble. And the notice board said *Pozières.*That was all there was! Just a notice board that said *Pozières* to tell us where we were. (quoted in MacDonald, *Somme* 330)

Figure 7.5   Trench sign amidst the destruction reading "This Was Forges"
Source: Photograph by William Steichen. © Art Institute of Chicago.

The last map to be produced of this area was 57dSE1&2 Edn.4A(S) on February 17, 1917. By this time, the German army was in retreat to beyond the line of defenses known as the Hindenburg Line; the regular overprinting on the map shows both British and German trenches in red. Therefore 57dSE1&2 Edn.4A(S) is a map of denouement and resolution, showing what had been as well as what is. In July and August 1916, the Mouquet Farm, about 2 kilometers east of Thiepval, had been firmly in control of the Germans, requisitioned as a large German munitions dump. By February 1917, after heavy shelling by the British Fourth Army, the German army had retired, allowing British occupation of the area—or what was left of it. Aerial and ground photographs of this area show open stretches of dusty plane with an occasional fractured tree marking the horizon.

Sheet 57dSE1&2 Edn.4A(S) describes the topography of this desolation. Roads into and out of the town are marked "Obliterated" or "Damaged by shell fire." The chateau had been reduced to rubble; its position on Sheet 57dSE1&2 Edn.4A(S) is consequently indicated "Chateau (Site of)."

Figure 7.6   View of the ruined village of Thiepval, from Thiepval Wood, September 1916
Source: © Imperial War Museum, Q1076.

Sheet 57dSE1&2 Edn.4A(S) records the destruction of the Mouquet Farm as "Ferme de Mouquet (Site of)." The most obvious revision from Edition 2D to Edition 4A is a reidentification of Thiepval, which due to heavy shelling becomes "Thiepval (Site of)." The second edition of the base map was populated with buildings and clusters of trees, what one would expect to see on the map of a small village in northern France; however, where the German defenses along the main streets were heavily lined in red, the mapmakers now use the dotted lines that represent a former or disused trench.

Where once the full expanse of woods could be marked, by February 1917 a single "tree" or "bush" noted on the map was 500 yards from the next (32a; 26d). So empty was the landscape that a four-foot post could be noted as a significant topographical feature (27b 4.9).

Conclusion

The trench maps of the First World War provided a specific type of intelligence to commanders and combatants. Their spare design, based on standard Ordnance Survey practice, combined with the various uses to which they were employed, supports the idea that maps are vehicles of communication from designer to reader.

While it is intriguing to see the maps as a way of ordering the chaos of the war environment, in the case of the trench maps this perspective would not only be limiting, but also would deny the power-knowledge claims of the trench maps and the lived experience of the combatants in the theatres of war. Because trench maps were developed from military need, they serve "the practical interests of the State machine," and the map establishes a terrain of surveillance, "representing space which facilitates its domination and control" (Crampton and Krygier 22). As various writers have pointed out, how we visualize space determines how we behave within that space: "The way we order things in planetary space is constitutive of what we see and what we conceal," writes Heriberto Cairo (1011). Such discourses and representations "inform the actions of combatants" of all levels, including at the highest, national juridical levels of government, "combatants" as nations (Cairo 1011).

Certainly, the trench maps can be seen as a psychological division between "us" and "them." In the maps that I have examined, the north–south axis divides the grid between left and right, so it is possible to read the map from the left as we do print, progressing from the "centered" British perspective to the terrain on the right, that which exists to be conquered. The two colors differentiate the nationalities of the combatants, but also constitute the Germans as the "Other," in Edward Said's sense of a person who can be defined in radical difference to the self. In fact, Central Powers fighting in the war were known as "Jerry," "the Boche," or "the Hun." In this topography of war, it is not man versus man, but nation versus nation. The boundaries constantly shift, often subtly. But these moving boundaries demonstrate that territory can be won, can change hands to be owned or possessed by one side or the other. And the boundaries are not even that accurate, for saps ran forward into the territories of the other; raiding parties crossed into No Man's Land and into opposing trenches by night; tunnellers were building mines; and airplanes flew reconnaissance overhead. The boundary line of the map was therefore somewhat hypothetical, an "as if" boundary suggesting a fortified and impermeable landscape.

Jeremy Crampton and John Krygier underscored the need for readers to see maps as active: "they actively construct knowledge, they exercise power," they "*make* reality as much as represent it" (15). Harley, Wood, and Barton and Barton established that maps are related to human experience by representing concepts and beliefs on a spatial plane. By advocating "a cartography of reality" that is phenomenological in focus, Wood changed the nature of interpreting cartographic data. More than represent a three-dimensional surface in two dimensions, the map encodes beliefs, practices, processes, and human experience.

Chasseaud's *Rats Alley: British Trench Names of the Western Front 1914–18* takes up the notion of phenomenology, exploring the meaning of names on the maps, going beyond the history of the region or the history of map production to suggest that, for the soldiers, a place name was a way of "understanding reality." He writes, "[A name] is a sign and a signifier; it designates and it identifies. It is also a code-word; it encapsulates. If it does not already have multiple meanings and associations, it soon acquires them. That icy grip around the heart and sinking feeling in the stomach on being told you are destined to hold, or attack, a certain trench happens because you know the history of that bloody place" (Chasseaud, *Rats* 21–2). Chasseaud continues by noting that naming helped "control [the] environment, to render it comprehensible, to reduce its fearful dimensions closer to the human scale" (*Rats* 22): [A name] introduced a comforting, if illusory, element of home and safety into a lethal, dirty and uncomfortable reality" (Chasseaud, *Rats* 29).

The names on the maps performed a significant psychological and cultural function for the British soldiers and the public, both at the time and in the present. "Being printed on the map next to the feature, the name creates mental links and associations between the ground, the map, the soldiers fighting over that terrain, the commanders planning operations in the rear, and the politicians, newspapers and public at home" (Chasseaud, *Rats* 27). In his work on British trench names of the Western Front, Chasseaud includes a useful table that lists two registers of discourse for trench names, one functional (the primary function of tactics) and the other social (the secondary, psychological benefit of familiar names). It is clear from this comparison that both the functional and social aspects of naming were about control. For the commanding officers, names controlled battlefields and troops. For the combatants, names controlled a confusing warren of earthworks in an unfamiliar country.

Table 7.1   Functions of trench names, adapted from Chasseaud, *Rats* 26

Trench name	
Primary function (Tactical)	*Secondary function* (Psychological/Cultural)
Important for commanders and staff	Important for front line troops
Content: Precise geographical location (pinpoint map references)	Content: Control of hostile environment
	Security, reassurance & mnemonic
	Homely associations
	Regimental or group associations
	Nostalgia
	Humor
	Irony, etc.

Depending on the context, a single name such as Jacob's Ladder in the Thiepval sector would have a different resonance: gathering point, launch point, or place of fear. In the poem "Trench Nomenclature" from his memoir *Undertones of War*, Edmund Blunden observes, "Genius named them, as I live! What but genius could compress / In a title what man's humour said to man's supreme distress?" (322). Byatt follows in the more recent poem "Trench Names":

Nonsense smiles
As shells and flares disorder tidy lines
In Walrus, Gimble, Mimsy, Borogrove—

As the irony of Blunden's poem points out, even at the time, the soldiers were aware that their lived reality was not neatly fitted to a page of orders or the coordinates of a map. In their memoirs, Siegfried Sassoon and Blunden often speak of the extent to which the troops were lost and the points at which the map made no difference to the sense of location.

Maps of the battlefields of the First World War may be understood as maps of heterotopic space, a term used by the French philosopher Michel Foucault to describe a variety of spaces that are both connected to public life and separated from it. Heterotopias are threshold spaces, marked by political or religious boundaries, circumscribed by ritual and symbolism, and ones that are isolated and "penetrable" (Foucault 26). Foucault called the heterotopia "a space that is other, another real space," even a "mirror" of the real space it mimics (27, 24).

While the battles were in progress, towns and farms were deserted by the residents, replaced by new residents, the soldiers. Horses and cows were appropriated from farm work to war work. Farm fields were dug into to create trenches. Names of familiar places back home were tacked to signboards along the trenches, giving a sense of familiarity to that which was unfamiliar. And local names were corrupted into new names: Ypres became Wipers, England was Blighty, Auchonvilliers was Ocean Villas. Thus, in heterotopias, "the real sites, all the other real sites than can be found within the culture, are simultaneously represented, contested, and inverted" (Foucault 24). Santanu Das points out that the common experience of soldiers along the Western Front was of "the viscosity of the trench mud": "mud insidiously took away human subjectivity: it rendered the living human being a Thing, formless and foundationless" (Das).

At the same time, the lived reality of the trench opened up a new unfamiliar space, that of home, which was once familiar but made strange by the experiences of war. Many soldiers wrote of the difficulty they had in returning home to family and pre-war routines when on leave. After the war ended, these tensions between the spaces of experience were even more pronounced as soldiers could not fit into their work. Those who returned severely wounded were "purgatorial shadows," shaded by their strange memories, different appearance, and altered abilities (Owen, "Mental" 76). The war went ever on and on for those who returned disabled or with shell shock. It could not be contained by the dates or

Figure 7.7   British infantry consulting a map in a trench
Source: © Imperial War Museum, Q10740.

the geographical boundary lines on a map. Eventually, many of the artists at the front transferred their impressions to fiction, poetry, painting, and song.

The heterotopia is informed by and contributes to the master narrative of a nation, while also positing a different story, one that is more ordered, clean, or precise than unbounded reality. While the maps in the history books appear precise in their presentation of data, the stories of the First World War cannot be presented so simply. Where official maps show ordered advance, solid waves of front lines, legible town names, trench maps represent the disorder of the moment, scraps of colorful and homely trench names popularized among the troops and towns that were identifiable only by rubble.

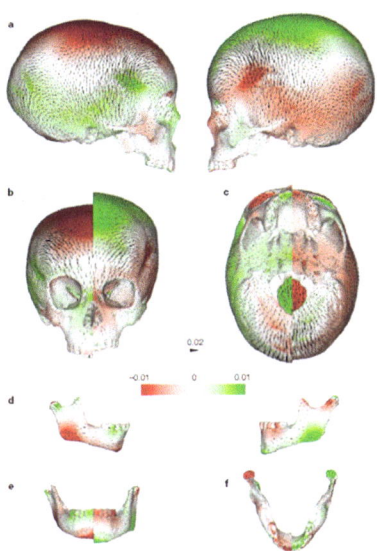

Plate 1   On the left, the Neanderthal skull; on the right, the human skull.
In the online version, red indicates shrinkage, green, expansion.
The metric is centroid size, a measure derived from geometric
morphometric analysis. The arrow above the scale shows double the
scale difference for the images above it

Source: Ponce de Leon et al. 536.

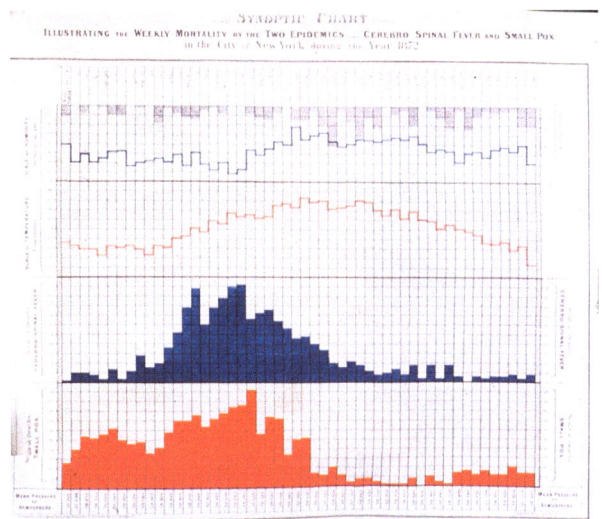

Plate 2   Synoptic Chart: Illustrating the Weekly Mortality by the Two
Epidemics—Cerebro-Spinal Fever and Small Pox in the City of
New York in the Year 1872

Source: National Institutes of Health, National Library of Medicine, History of Medicine
Division.

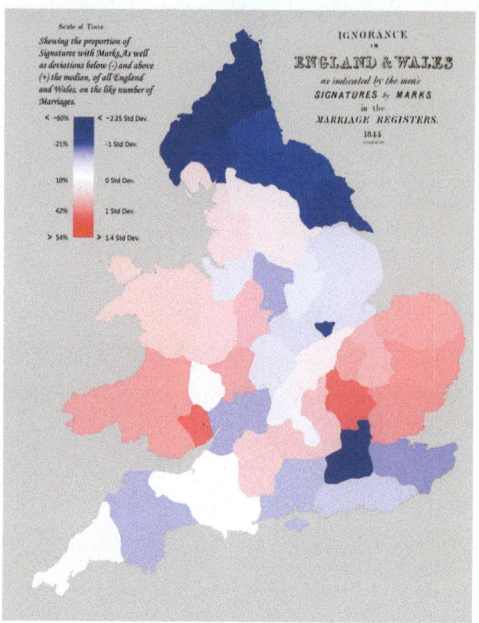

Plate 3   Fletcher's "Ignorance in England and Wales," updated with
               current techniques

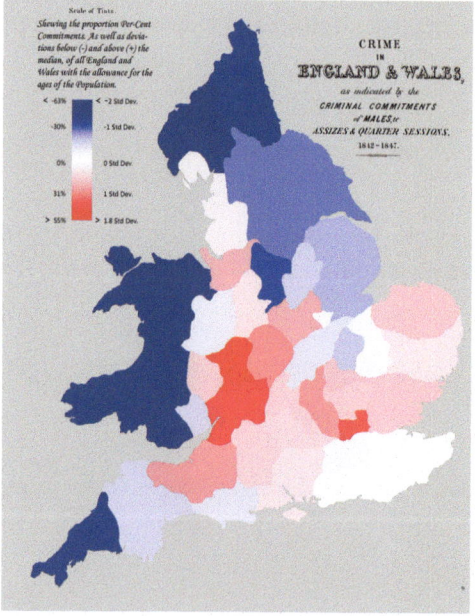

Plate 4   Fletcher's "Crime in England and Wales," updated with
               current techniques

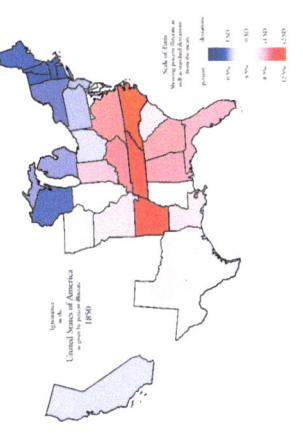

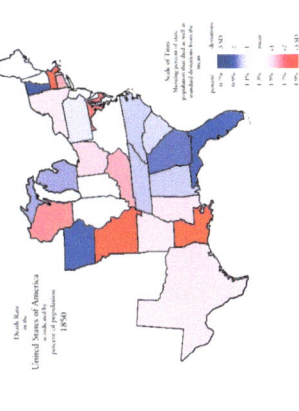

Plate 5   Slavery in the Southern United States, 1850, based on data from Helper

Plate 6   Illiteracy in the United States, 1850, based on data from Helper

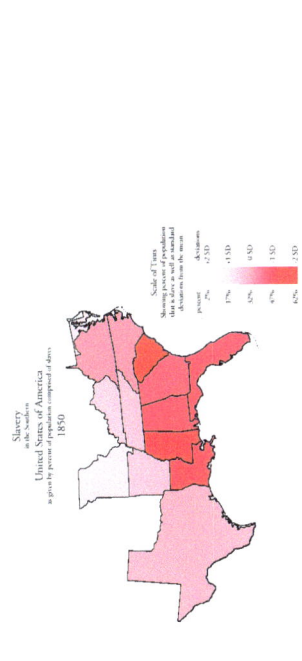

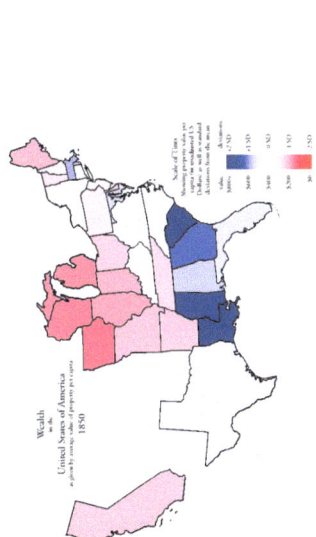

Plate 7   Wealth per capita in the United States, 1850, based on data from Helper

Plate 8   Death rate in the United States, 1850, based on data from Helper

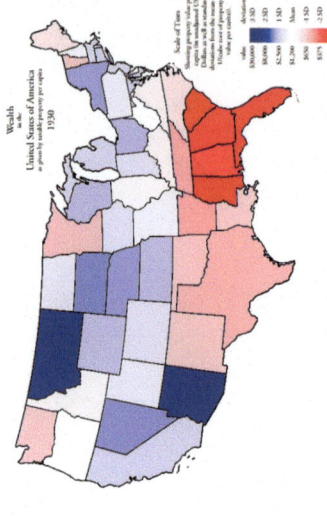

Plate 9   Illiteracy in the United States, 1930

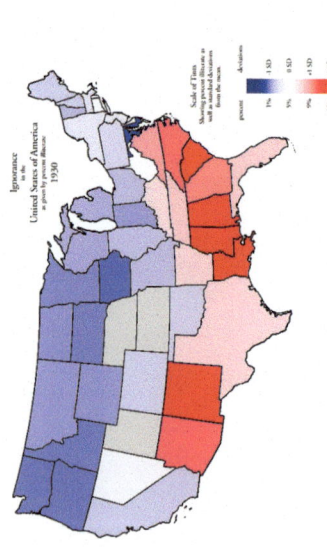

Plate 10   Wealth per capita in the United States, 1930

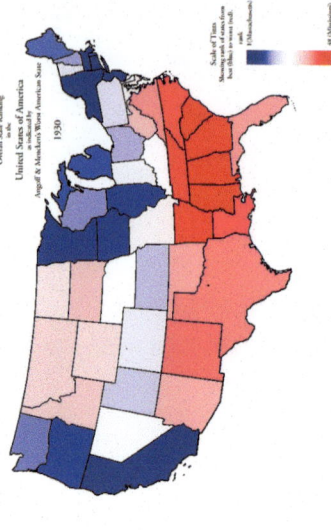

Plate 11   Death rate in the United States, 1930

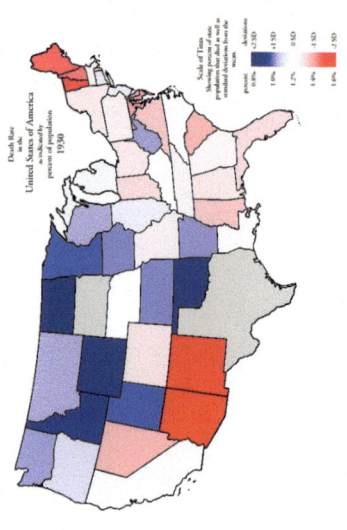

Plate 12   Overall state rank in the United States, 1930

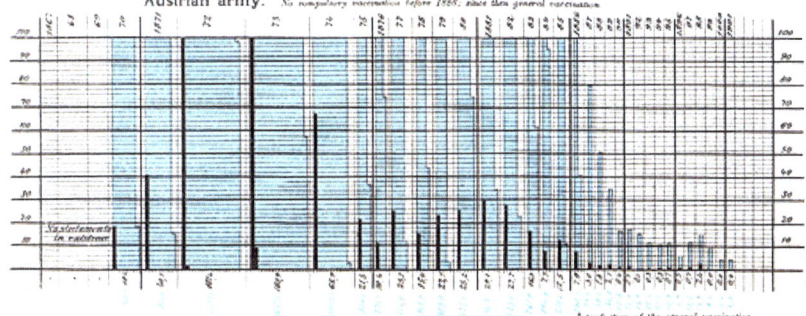

Plate 13   Incidents of disease and death from smallpox in the Prussian
army (compulsory vaccination since 1834) and Austrian army
(compulsory vaccination since 1886)

Source: Courtesy of the Historical Medical Library of The College of Physicians of
Philadelphia.

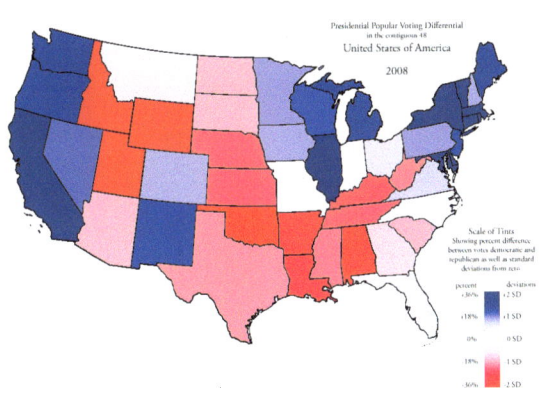

Plate 14   Percent difference between Democratic and Republican voting in the
2008 U.S. Presidential popular vote

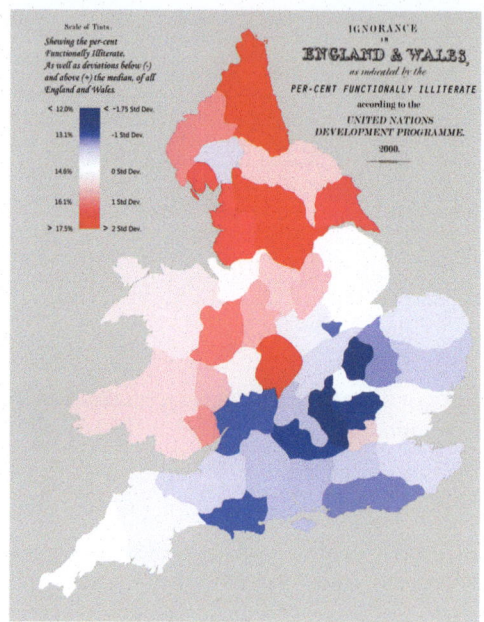

Plate 15   Ignorance in England and Wales, 2000

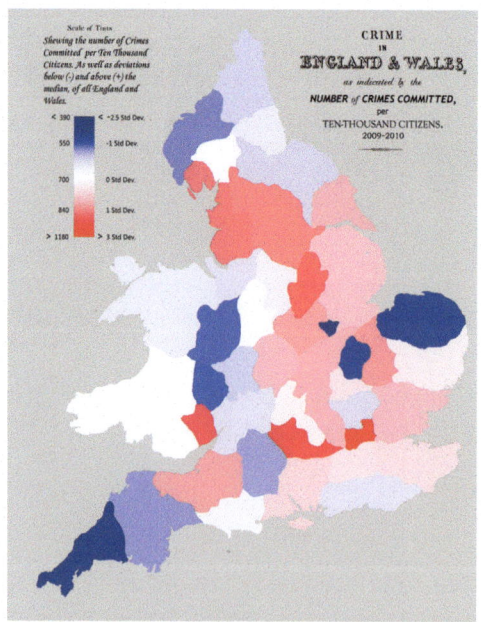

Plate 16   Crime in England and Wales, 2010

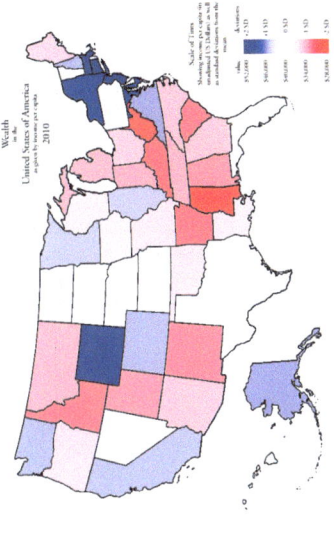

Plate 17   8th grade NAEP Reading scores in the United
States, 2009

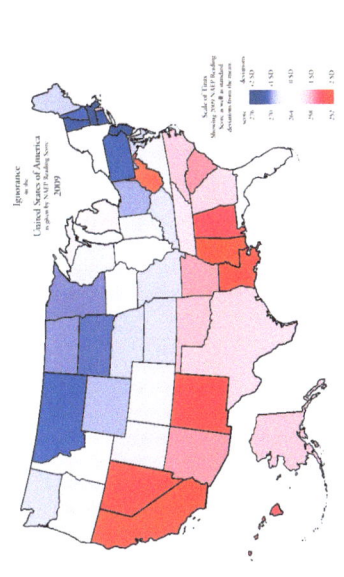

Plate 18   Per capita income in the United States, 2010

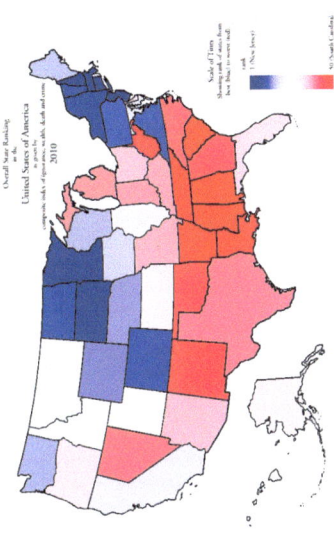

Plate 19   Death rate in the United States, 2009

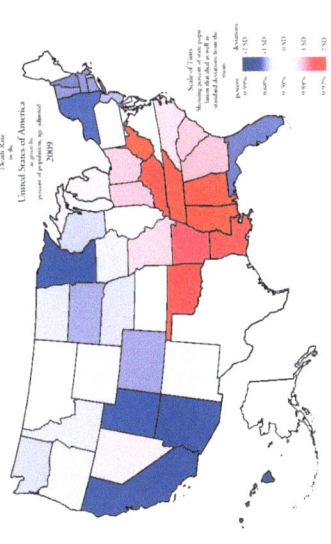

Plate 20   "Worst American State" revisited with 2010 data

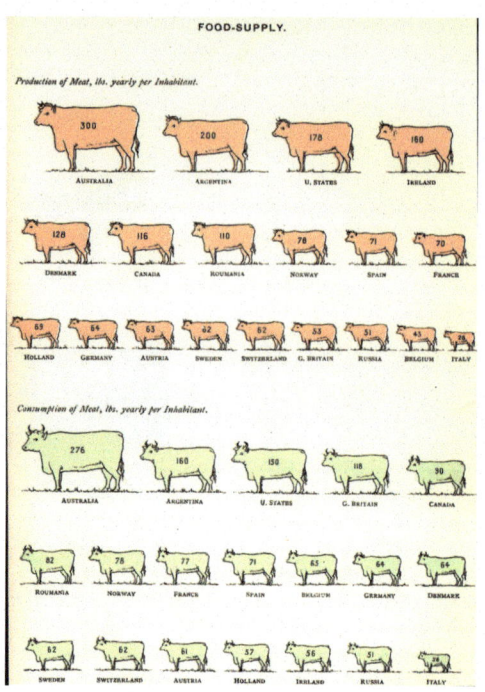

Plate 21   Food-Supply, plate V of Michael George Mulhall's *Dictionary of Statistics* 4th ed., 1899

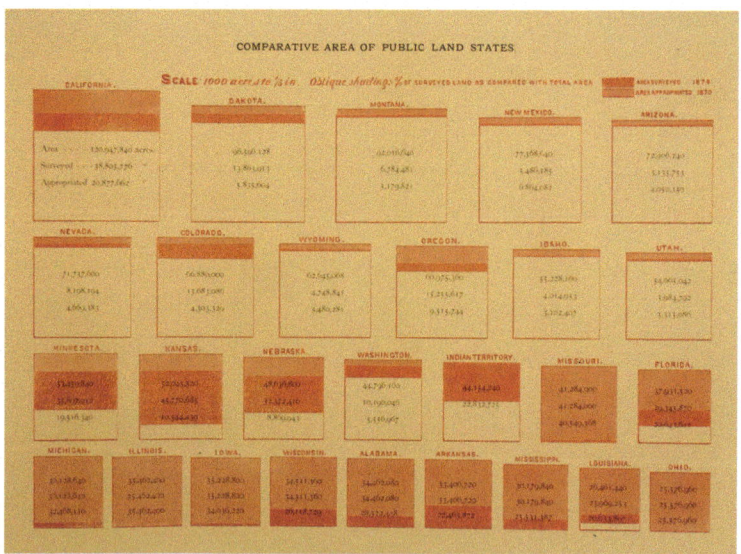

Plate 22   Area charts from the 1876 *Centennial Album of Agricultural Statistics* showing the status of U.S. land

Source: Dodge, Plate 22. Courtesy Special Collections Department/Iowa State University Library.

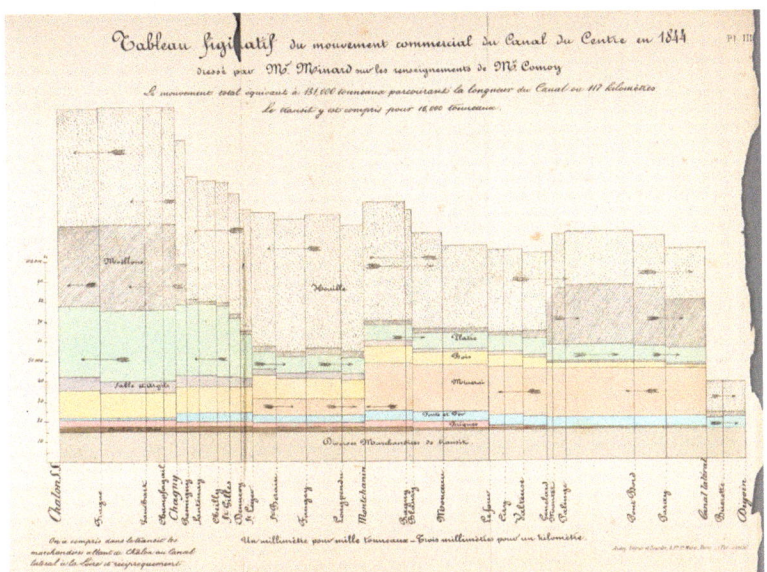

Plate 23   Area chart designed by Minard showing freight transported on a
            French canal

Source: Minard, Plate 3. Courtesy of the General Research Division, The New York Public
Library, Astor, Lenox and Tilden Foundations.

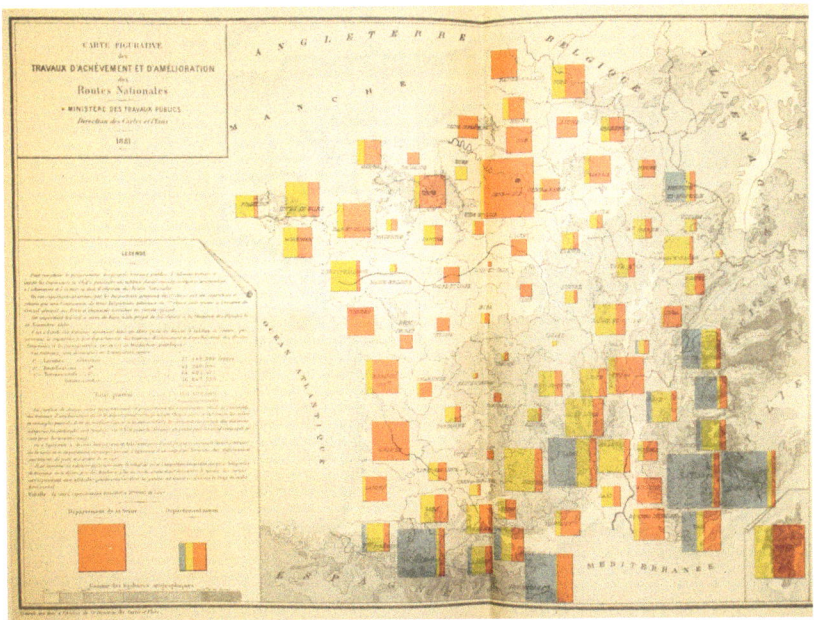

Plate 24   Area charts from the 1881 *Album de Statistique Graphique* showing
            the status of road improvements across France

Source: Ministère des Travaux Publics, Plate 20.

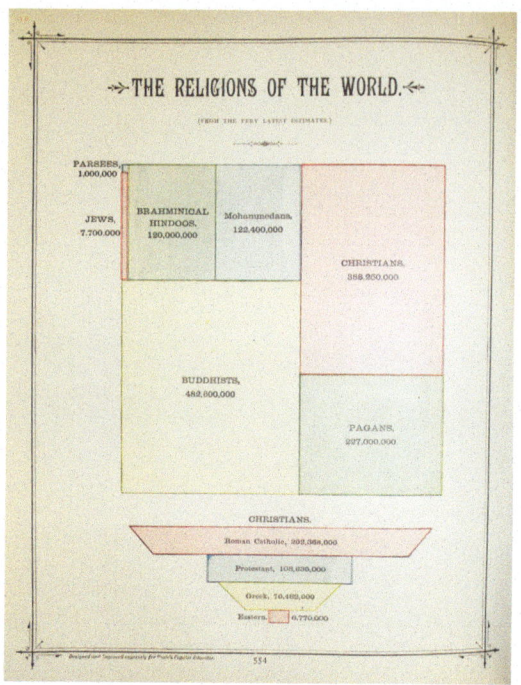

Plate 25   Area charts showing world religions in the late nineteenth century
Source: Peale 554.

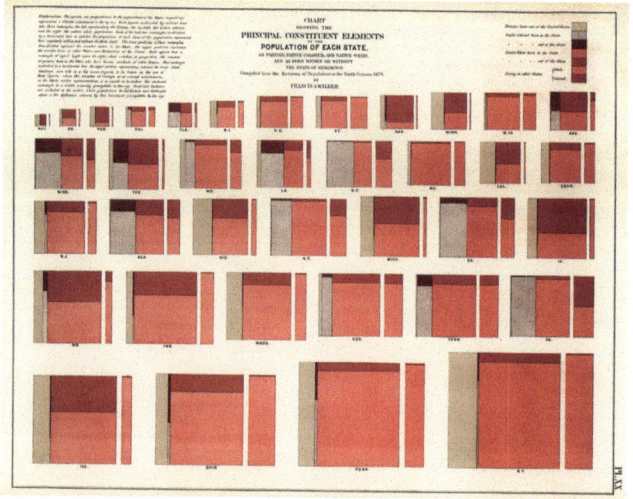

Plate 26   Mosaics from the first U.S. statistical atlas showing the population
                of states (squares on the left) and the population lost by states
                (rectangles on the right)

Source: Walker, Plate XX. Courtesy of the Library of Congress, Geography and Map
Division.

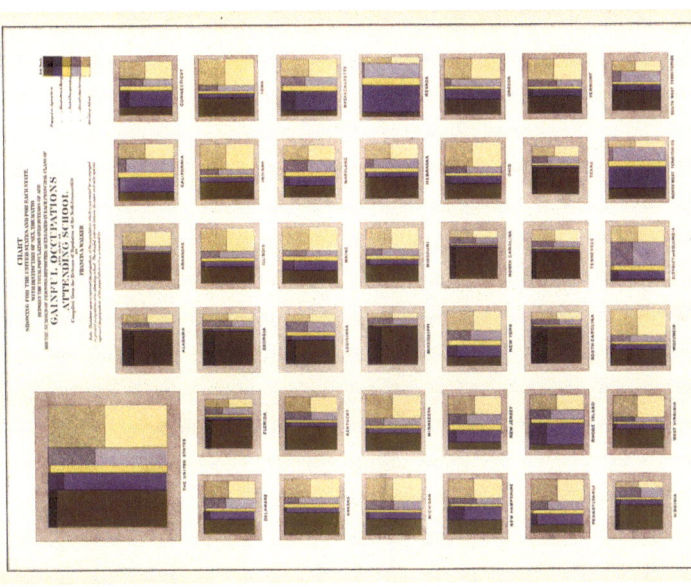

Plate 28   Mosaics from the first U.S. statistical atlas
showing occupations by states and territories

Source: Walker, Plate XXXII. Courtesy of the Library of Congress,
Geography and Map Division.

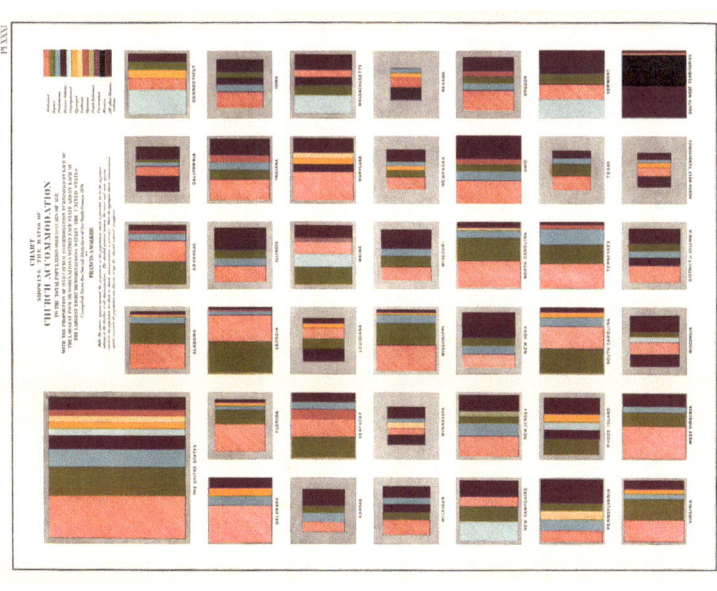

Plate 27   Mosaics from the first U.S. statistical
atlas showing church affiliations by states
and territories

Source: Walker, Plate XXXI. Courtesy of the Library of Congress,
Geography and Map Division.

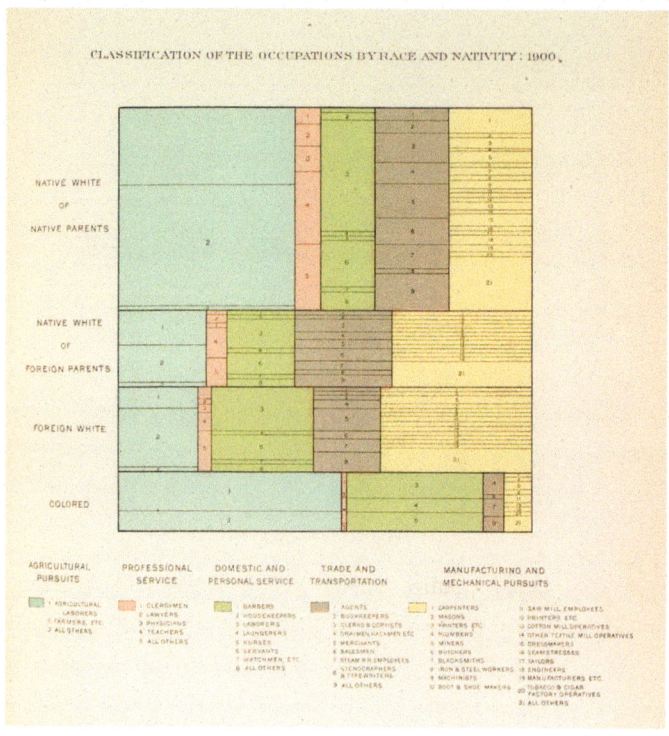

Plate 29   Mosaic from the 1903 U.S. statistical atlas showing occupations of
various segments of the U.S. population

Source: Gannett, *Statistical, Twelfth Census*, Plate 87.

Bloomberg   State Employees Cash in on Lump-Sum Retirement Checks

State Employees Cash in on Lump-Sum Retirement Checks

More than 111,000 government employees working for the 12 most-populous states collected $711 million in 2011 for unused vacation and leave accrued over their careers. A California psychiatrist took home the largest lump-sum payment.

2011 payments		Average	Top payment
Calif.	19,454	**$14,110**	$608,821
N.J.	2,277	8,539	15,000
Ohio	7,230	8,475	120,648
Ill.[1]	485	6,482	58,668
Mich.	6,684	5,228	27,257
Pa.	18,250	4,978	189,753
Texas	6,996	4,931	69,271
N.Y.	13,766	4,774	66,074
Fla.	16,074	3,812	60,946
N.C.	6,697	3,604	49,971
Ga.	7,452	3,128	65,108
Va.	5,680	3,127	44,783

California employees take home biggest lump-sum pay

Of the top 100 lump-sum payments in the 12 states in 2011, all but 10 were to California employees. The top payment went to a California psychiatrist with 2,893 hours of accrued leave.

Corrections and rehabilitation: 41
Highway patrol: 22
State police: 10
State hospitals: 7
Other: 9

Top 100 payments, by state and department
- Calif. employees
- Pa. employees

Forestry and fire protection: 5
❶ Justice: 2
❷ Developmental services: 2
❸ Public employee retirement: 2

Source: State payroll data compiled by Bloomberg
Graphic: Alex Tribou
BGOVgraphics@bloomberg.com

1 – Based on data from departments representing 90 percent of state workers.

Bloomberg GOVERNMENT

Plate 30   Bloomberg.com mosaic showing retirement payments
Source: Tribou; http://www.bloomberg.com/chart/iQ3Z7z7anAhE
© 2014 Bloomberg Finance L.P. All rights reserved. Used with permission.

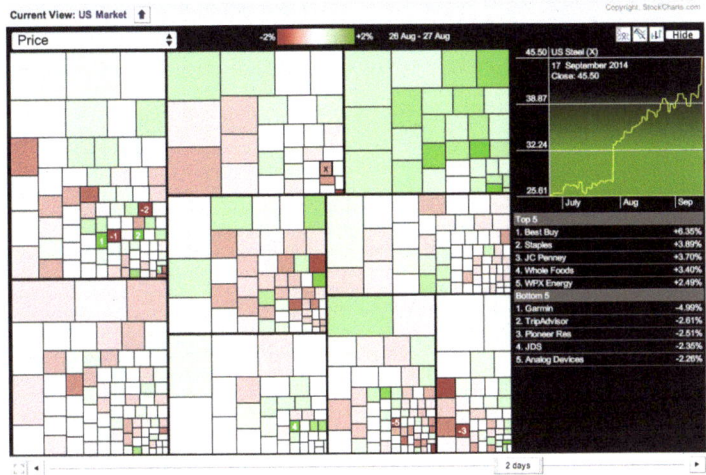

Plate 31   Online interactive treemap of U.S. stocks organized by
              investment groups

Source: StockCharts.com; http://stockcharts.com/freecharts/carpet.html
Copyright Stockcharts.com. Chart courtesy of StockCharts.com.

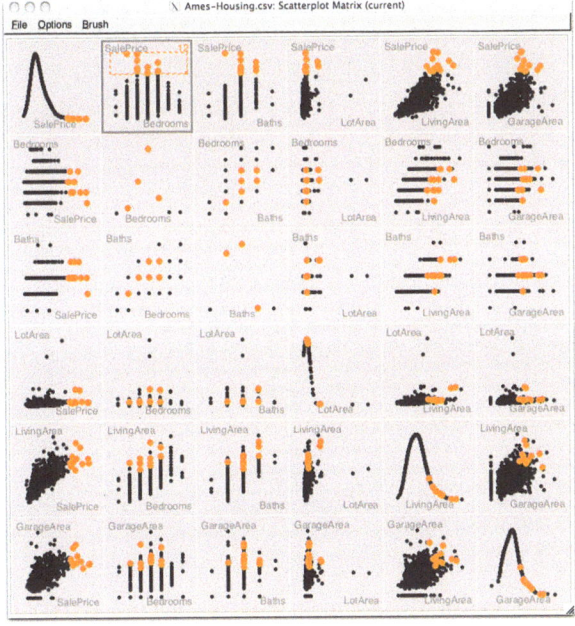

Plate 32   Brushing in a scatterplot matrix using housing data from Ames,
              Iowa, and as documented by Becker and McGill. Brushing with
              a rectangular brush is being done in the plot of sales price against
              bedrooms, with higher priced homes highlighted in orange

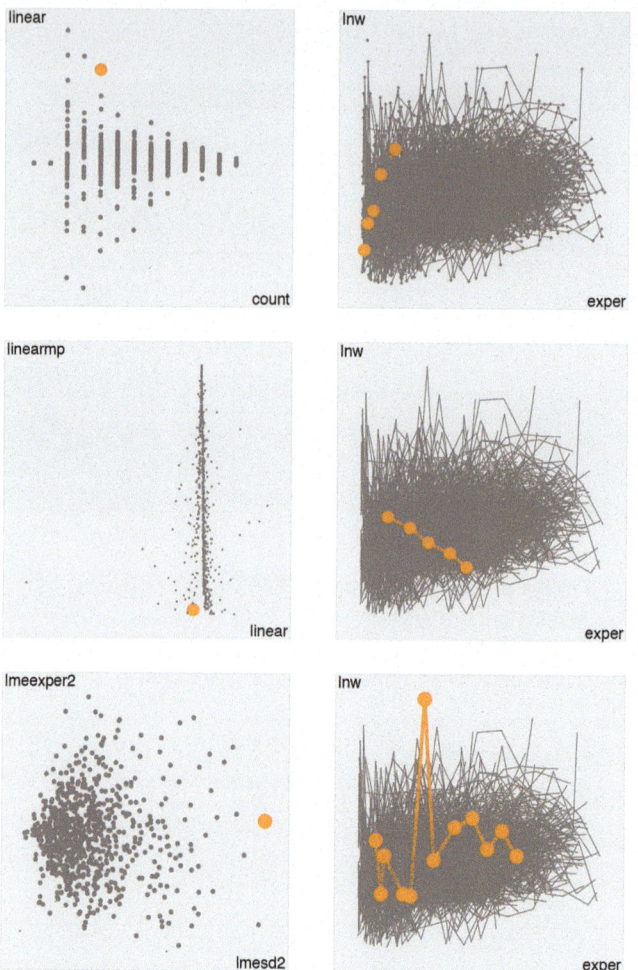

Plate 33   An illustration of brushing in multiple linked plots, used here to explore longitudinal data

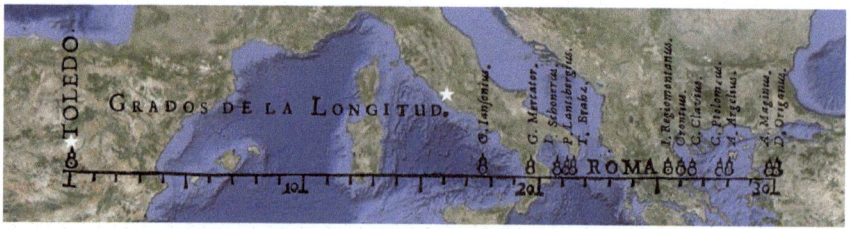

Plate 34   Van Langren's 1644 graph, re-scaled and overlaid on a modern map of Europe

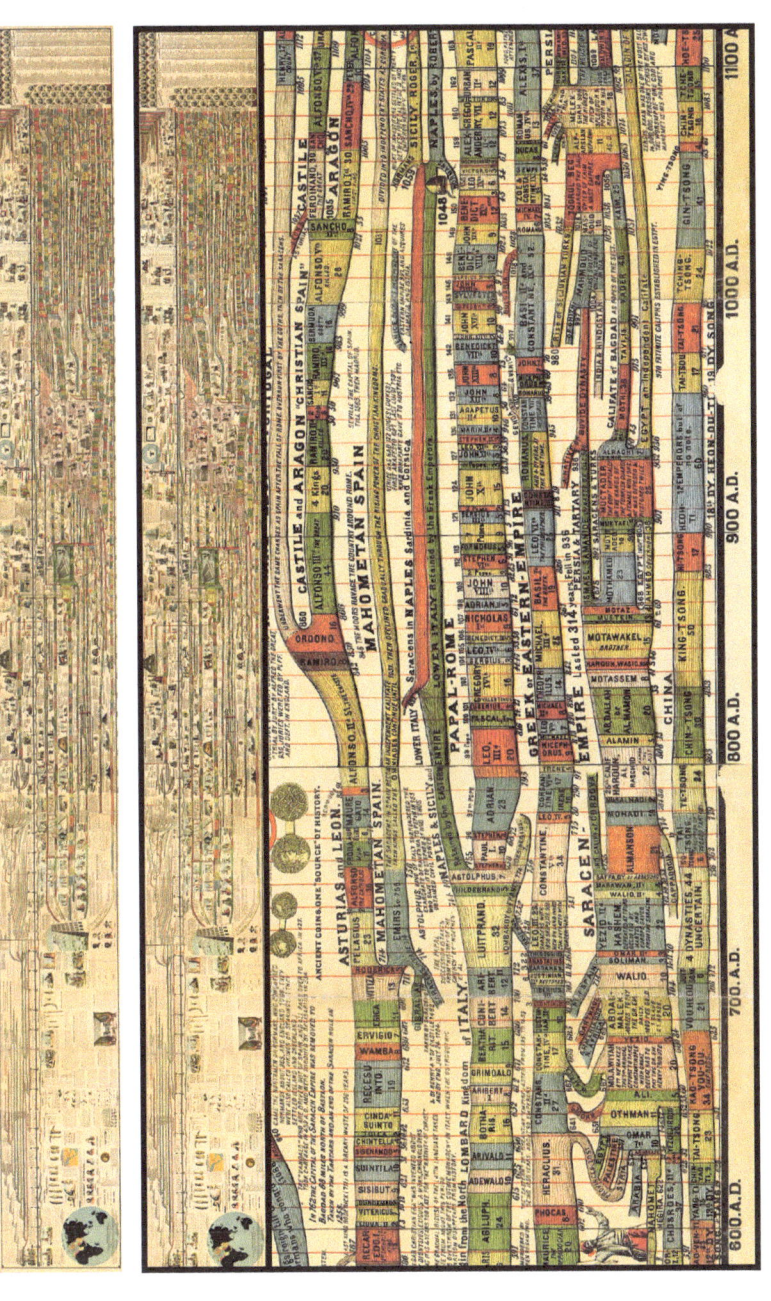

Plate 35   Above: The entirety of Sebastian Adams' Synchronological Chart of Universal History, 1881.   Below: An excerpt detailing the 600–1100 AD period. Horizontal bands trace developments in different countries, with detailed text describing significant events, and break up or merge according to political factors

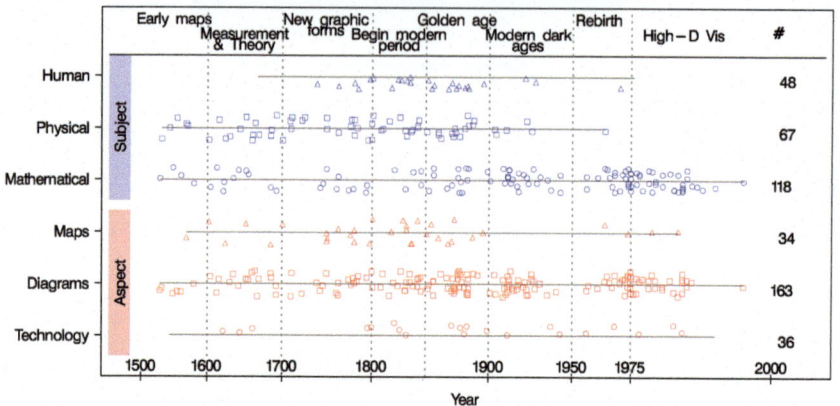

Plate 36   Sketch for a thematic timeline of milestones items, 1500–present, categorized by both the Subject (content) and Aspect (form) of the milestone item

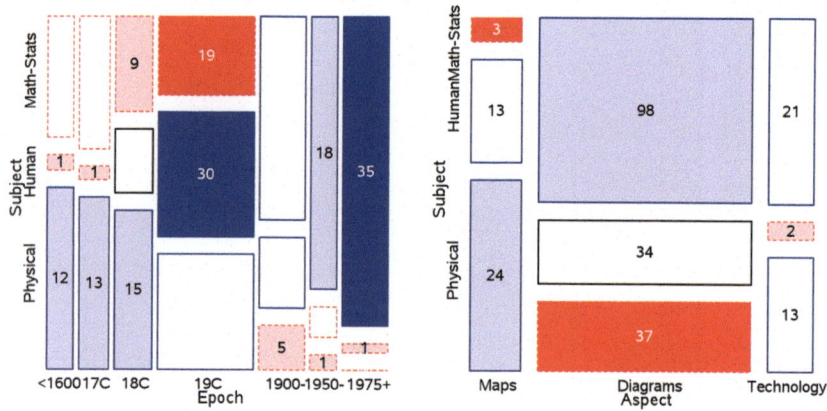

Plate 37   Mosaic displays for milestone items, classified by Epoch, Subject and Aspect. Left: mosaic for the marginal table showing differences in Subject across Epochs. Right: mosaic for the marginal table showing differences in Subject across Aspect. Numbers in the tiles give the number of milestone items

PART III
Examining Visible Numbers:
Forms, Methods, Historiographies

Studying the history of data design entails inquiry into a growing array of graphical forms, which are constantly evolving and mutating and being appropriated, deployed, and reshaped by communities of users. Establishing methods and structures within which to study this dynamic process can be challenging, both at the macro- and micro-level. This part of *Visual Numbers* features essays that explore how we understand and study the history of data visualization.

New innovations that gain popular or disciplinary status gradually evolve into stable graphical forms, the distinguishing characteristics of which Charles Kostelnick explores in his chapter on data display genres. Using the history of area charts and mosaics to illustrate this evolutionary process, he develops a model for defining and analyzing display genres based on formal and functional attributes, social and cultural factors, and technology.

The history of a field can be studied and written about only if scholars can access artifacts for analysis. The website *Milestones in the History of Thematic Cartography, Statistical Graphics, and Data Visualization* provides students and scholars with an interactive anthology of ground-breaking displays as well as sorts them by period, designer, and type. Michael Friendly, Matthew Sigal, and Derek Harnanansingh narrate the development of the Milestones website and apply graphical techniques to visualize the history of data design itself.

As a dynamic and continuous process, the history of data design continues to unfold in the recent digital revolution, which has initiated a second "golden age" of innovation and proliferation of data design. Drawing on a video library of the American Statistical Association that documents the beginnings of this revolution in second half of the twentieth century, Dianne Cook narrates the development of computer graphics in visualizing statistical data. The series of breakthroughs in graphical software that Cook analyzes created the impetus for the explosion of contemporary data visualization.

As the history of data design has received increasing scholarly attention over the past few decades, studies from several different disciplines have opened up new avenues of inquiry and methods of analysis. Kevin Van Winkle annotates a diverse cross section of scholarship in this interdisciplinary field, giving researchers new to this subject a primer on seminal studies that can help guide their research.

Chapter 8

Mosaics, Culture, and Rhetorical Resiliency: The Convoluted Genealogy of a Data Display Genre

Charles Kostelnick

Over the past two centuries, the visual language of data display has become vast and varied, encompassing an ever increasing and bewildering variety of shapes and forms. Most of these forms have been propagated in response to new visualization problems across an array of disciplines, to changing social and cultural conditions, and to innovations in technology. As they have evolved, data displays have historically been identified and categorized into what might loosely be called genres—forms that embody structural properties, achieve rhetorical purposes, and attract communities of users similar to the ways in which scholars have described the behavior of written genres (see Miller; Bazerman; Swales; Berkenkotter and Huckin; Russell; Artemeva and Freedman). Written genres supply consistency and rhetorical ballast that both engender and meet audience expectations and that, as Berkenkotter and Huckin observe, "serve to stabilize experience and give it coherence and meaning" (4). Similarly, in the realm of data displays, genres provide a reliable and relatively stable framework for deploying and interpreting the wide range of forms in which data are visualized—from pie and bar charts to line graphs, thematic maps, and scatterplots, to name some of the most prevalent (and stable) genres.

However, as data displays have proliferated, and digital and interactive displays (for example, treemaps, networks, tag clouds) have destabilized traditional designs, classifying genres has become increasingly complex and challenging. On a given day, we might encounter, in print and online, variations of existing displays, hybrids of those displays, or displays that look entirely novel and new. Nonetheless, contemporary displays cannot easily be disentangled from their predecessors in the near or even distant past: their genealogies are intertwined with those of past forms, as they are engaged in similar processes of evolution and morphology. These morphological complexities invite historical exploration and analysis that might begin with this question: how can contemporary concepts of genre explain how data displays behave over long stretches of time?

To address this question, in this chapter I will examine the factors that define, shape, and sustain visual genres as they evolve amid varying disciplinary, cultural, and political conditions. I will focus my analysis on a longstanding traditional form, mosaics, and their forerunners, rectilinear area charts. As Michael Friendly has shown, these forms emerged over two centuries ago, at the beginnings of modern data design, and have recently experienced a renaissance and reinvention ("Brief History of the Mosaic Display"). Over their history, mosaics have been deployed in a variety of forms, ranging from simple area charts representing only a handful of data to complex hierarchical diagrams with multiple levels representing dozens or hundreds of pieces of data. With the advent of contemporary software and plotting techniques, mosaics have flourished both as reincarnations of old and venerable designs and as digital innovations with interactive features. Given their long and convoluted history, mosaics provide a touchstone for studying genre by raising several key issues, ranging from their visible structure and their functionality to broader issues of cultural context, production processes, and power.

Like written genres, data display genres express and perpetuate the cultural values of the communities in which they are shaped and deployed. Historically, especially in what Friendly calls the "Golden Age of Statistical Graphics," mosaics played a key, even a dominant role, in telling complex and important (and sometimes sensitive) stories, often about people. Mosaics succeeded in doing that because of their graphical suppleness, their ability to encode complex data into multiple patterns (and both to embed and foreground that data), and the interpretive demands they made on readers to study and take them seriously. Mosaics also found a fertile aesthetic environment in the nineteenth century, when complexity and detail dominated all aspects of design before the onset of modernism. These factors enhanced the status of mosaics as a principal genre during this time, just as other confounding factors later diminished it. In these ways, examining the history of mosaics in cultural context helps us understand how genres function socially in the communities in which they live and breathe, thrive and regress, and sometimes re-emerge.

Historical Overview: The Emergence of Mosaics as a Genre

Genres are "dynamic rhetorical forms" (Berkenkotter and Huckin 4), and like written genres, visual genres are "fluid, evolving terms" (Brasseur, *Visualizing* 8) that often resist simple definitions. Genres have genealogies and histories: they undergo a process of development, evolve, morph into other forms, and sometimes decline, hibernate, or become virtually extinct. Although genres appear to be stable in the short-term, their users constantly transform them (Schryer 107–8). Like written genres that rely on the socializing force of discourse communities (Swales 9, 21–32), the morphology of data display genres depends on their standing within the communities that use them as well as the technology applied to create and proliferate them. As large communities that routinely deploy and interpret data displays, disciplines oftentimes appropriate conventional forms, repurposing them for their own data needs. Wind roses, a mid-nineteenth century invention to plot wind directions in four geographical quadrants, were appropriated by medicine and public health to visualize death and disease over time, such as Florence Nightingale's famous medical diagrams (Brasseur, "Florence"). Some genres take on new forms, like the rank chart, which in 1883 appeared prominently in *Scribner's Statistical Atlas of the United States* for comparing populations of states and cities (Hewes and Gannett, Plates 18 and 21) and which reappeared online as a digital interactive chart comparing money and performance of major league baseball teams (Fry). And some genres decline and become virtually extinct—like the train timetable charts in the second half of the nineteenth century with a web of angled lines showing how quickly (or slowly) readers might reach their destinations at varying departure times (see Marey 20, Figure 7; Tufte, *Visual Display* 31, 115–16). Today, fueled by technology and the Internet, a second "golden age" of data displays is accelerating the formation and evolution of genres.

Constantly evolving, data display genres have many progeny, some of which initiate new strains and genetic lines as their histories unfold. Some genres engender large, complex, and even flamboyant narratives, while others foster more sedate and obscure ones. Several genre narratives began in the late eighteenth and early nineteenth centuries, when data began to be visualized in an increasingly wide variety of forms, ranging from line graphs and bar charts to pie charts and data maps, each of which began with a particular exigency and evolved into different forms as designers tackled new data problems. Other displays, like area charts and mosaics, have more obscure and convoluted genealogies, often operating in the historical shadows in incipient though potent forms, awaiting their moments of glory. Any of these narratives, large or small, can tell us a great deal about genre formation and development.

The history of mosaics spans two centuries, the heir to a long-developing mélange of rectilinear area charts that flourished during the second half of the nineteenth century during the "golden age" of data display (Funkhouser, "Historical" 330; Friendly, "Golden Age"). Plate 22, for example, shows several small mosaics published in 1876 by the U.S. Department of Agriculture (Dodge) that visualize the relative land size of several states (from large to small), with the subdivisions within each mosaic showing the proportions of land surveyed (dark shading), land appropriated and presumably privately held (light shading), and public land (unshaded). Historically, this series of modest mosaics appeared amid an already long-developing genealogy. Friendly traces the origins of mosaics in the late eighteenth century and analyzes early examples from the nineteenth century ("Brief History of the Mosaic Display"). Mosaics in a variety of forms appeared in the *Statistical Atlas of the United States* (Walker; Gannett, *Statistical, Eleventh Census*), the French *Album de Statistique Graphique* (Ministère des Travaux Publics), and other government publications in the late nineteenth century, creating some of the most stunning displays of the era and legitimizing the genre by providing models and official state sanction. However, after the first few decades of the twentieth century, area charts generally, and mosaics specifically, experienced a relatively dormant period until their revival with computer technology over the past three decades. Today area charts flourish in several forms of mosaics, including treemaps and other rectilinear area displays, many of which now appear online in interactive modes.

Like other data display genres, mosaics have undergone a process of evolution, though that process has been unusually long and deliberate, spanning more than two centuries. In data display history, the beginning of that evolutionary process would be akin to a Jurassic period, when new forms suddenly emerged and subsequently morphed into a variety of additional forms. After decades of dormancy, the evolution of rectilinear area charts and mosaics has rapidly accelerated in digital form, particularly in an interactive online environment. A sophisticated taxonomy of contemporary rectilinear area charts has been developed by Hadley Wickham and Heike Hofmann, with several elements in their taxonomy having roots in nineteenth-century predecessors. Although idiosyncratic in many ways, the story of mosaics parallels those of other genres, which have been shaped and defined, often in fits and starts, over long stretches of time.

In the remainder of this chapter I will explore the history of mosaics as a genre by examining several defining characteristics—formal, functional, social, cultural, and technical—that collectively establish genre identity and boundaries, however distinct or amorphous. Although this chapter will focus primarily on the historical evolution of area charts and mosaics, particularly in the nineteenth century, the framework I use in this analysis can also be applied more broadly to other data display genres.

Formal Structures of Genres: Tiling the Mosaic

To constitute a meaningful class of objects/activities, genres must have boundaries, however sharp or murky. Among the most obvious and explicit methods for defining those edges are formal characteristics, both textual and visual, that typically describe external and easily recognizable surface features. Although those surface features lie largely outside of rhetorical exigencies and social and cultural forces, they often hint at deeper, more systemic meaning. Like many other data display genres, as mosaics emerged and evolved, the contours of these surface features underwent several transformations.

Textual Form: From Diagrams to Treemaps

Names play an important role in defining and distinguishing genres from one another and often hold significant clues about their origins and defining characteristics. In written communication monikers like *proposal, newsletter,* and *annual report* reveal aspects of their textual and visible shape as well as the purposes they serve among their communities of readers. In the realm of data design, names have become extremely prolific, with several hundred appearing in Robert Harris's *Information Graphics: A Comprehensive Illustrated Reference* even before the digital age truly blossomed. No other form of information design can claim such a rich treasury of names—full of history, nuance, design lore, and disciplinary appropriations. The sheer volume and lack of uniformity of naming can be confounding, with similar forms typically acquiring different names (bar and column charts) and with sub-genres abounding (horizontal, multiple, and divided). And of course new digital forms (treemaps, networks, tag clouds) are rapidly expanding the naming domain. So complex and tangled has naming become that charting a Linnaean taxonomy of contemporary genres would be daunting and require a high degree of contingency, cross-referencing, and interactive links.

So naming takes many different twists and turns as genres evolve—over decades and centuries—with names providing provisional lighthouses along the way marking their progress. Those lighthouses originate and function in several ways:

- Descriptive—by invoking key specific design characteristics of the genre (line graph, box plot, scatterplot, bar or divided bar chart; bubble, dot, pyramid, radar, or surface chart).
- Metaphorical—by evoking things, natural or artificial, that the genre resembles (mosaic, wind rose, spine plot; pie, donut, radar, candlestick, carpet, mountain, trellis, or stem and leaf chart).
- Purposeful—by revealing the use of the genre, often within a certain discipline (high/low chart, forest plot, isotherm map, population pyramid, rank chart, electrocardiogram).
- Eponymous—by memorializing their inventors (Mercator projection, Gantt chart, Smith chart, Pareto chart, Chernoff faces).

Descriptive names are by far the most common, and they are the easiest to invent and remember, emphasizing key design features, though sometimes they can cause ambiguity or misdirection ("bar chart" and "column chart" might not mean the same thing). And of course, often genres bear different names during their evolutionary processes: pie charts were earlier called "circles" with "sectors" (Walker 2) or as Willard Brinton put it, the "wedge or sector chart" or a chart using the "sector method" (*Graphic Methods* 6). Not surprisingly, Robert Harris's compendium of data displays contains a great deal of cross-referencing and redundancy, with alternative names for many entries.[1]

---

[1] Genre names experience varying degrees of stability and longevity. Florence Nightingale's medical appropriation of the "wind rose" (whose graphical coding resembled rose petals) was sometimes called a "coxcomb diagram" (Friendly, "Golden Age" 509), and as its uses expanded, it later acquired the name "polar diagram" and "radar chart."

Mosaics illustrate this array of naming techniques, with names ranging from the general to the specific. The term mosaic is a relatively recent moniker, generally traced to J.A. Hartigan and B. Kleiner (1981) early in the digital age (Friendly, "Brief History of the Mosaic Display" 90), and embodies the metaphor of "tiles," the pieces that collectively constitute the display. "Treemaps" also evoke a visual metaphor—branching, hierarchy, and micro-level complexity. For most of its history, however, mosaic naming was primarily descriptive. In the first U.S. statistical atlas (1874), most non-map charts, including mosaics, were called "Geometrical Illustrations" or "Charts" (Walker 1–3), with a later atlas (1898) describing them as "diagrams" (Gannett, *Statistical, Eleventh*, 3–7). German statistician Georg Von Mayr (1877) called his prototype mosaic a "Flächen-Diagramm" (79) or "area chart."

Early in the twentieth century, Karl Karsten (1923) describes two types of mosaics: "100% square" charts and "area-bar-charts," the former describing Von Mayr's purer form of mosaic, akin to the more complex genre today, and the latter describing an area chart with bars of varying width (598–612, 613–22), which Minard pioneered. Willard Brinton's 1939 *Graphic Presentation* also uses "area bar chart" to describe mosaics with variable width bars (149–51). Calvin Schmid (1954) labels this form a "one-hundred percent component-bar chart" (85, 87), given that it measures both height and width, and Harris (1996) labels it an "area column graph" (9). On the other hand, Schmid calls the more complex, contemporary form of mosaic an "Area-Bar Chart" (85–6), while Thomas Birch (1964) describes it as a rectangular form of "two-dimensional diagrams" (204–7). Finally, Harris describes it as a "mosaic graph" (242), and more recently, Leland Wilkinson (2005) refers to it by its current statistical name, a "mosaic plot" (342–3).

The term "diagram," then, evolved into combinations of "area," "square," "bar," and "100 percent," with each label—sometimes with varying combinations and meanings—capturing different aspects of the mosaic. Mosaics have recently taken root digitally in interactive forms like the "treemap," which is also sometimes called a "carpet" chart, a metaphorical name that denotes the intricate pattern of a floor covering. So a combination of descriptive and metaphoric names defines mosaics textually as a genre, revealing its design variations and documenting its evolution.

Visible Form: Composing with Rectangles

Oftentimes we define genres, both textual and graphical, in terms of the elements of their composition and the syntax and arrangement of those elements. Just as newsletters embody typical features like nameplates, multiple columns, and headings, so too do readers associate certain forms (or combination of forms) with data displays: circles with pie charts, lines/grids with line graphs, rectangles with box plots, and dots with scatterplots. Form matters as a defining characteristic of any genre because it enables us to make a quick perceptual sort based on large-scale, "supra" level features that we immediately see (Kostelnick, "Supra-

Textual"). That effect occurs, for example, when we encounter data display icons that appear in graphing programs like Microsoft Excel, which categorizes genres through shape. Even at a high level of abstraction, shape invokes genre and provides the conceptual basis for some classification systems.[2]

The wide variety of forms in which mosaics appeared suggests that they developed along several paths from the late eighteenth to the late nineteenth century. All of these forms, however, shared two traits: they were rectilinear and they represented data based on area and/or length. Beyond this simple formula, area charts during this era might be classified into the following:

- *A series of overlapping rectangles arranged progressively by size.* The earliest, incipient forms of mosaics employed this technique, whereby a series of progressively smaller rectangles are superimposed to enable readers to compare areas. In this variation of a mosaic, areas are efficiently stacked upon one another, from the largest on the bottom to the smallest on the top, rather than displayed adjacently.

- *A series of squares of different sizes, with internal subdivisions, distributed across the visual field or a map.* In the series shown in Plate 22 (Dodge, Plate 22), each area chart displays the total acres in a given state or territory, subdivided into land that has been surveyed (dark red), land that has been appropriated (light red), and unsurveyed land in the public domain (white). Western states along the top of the plate have a much higher proportion of public land compared to eastern states below, where most land is privately held. Each square and its subdivisions, then, form a separate mini-mosaic intended to be compared with others in the visual field.

- *A segmented bar, or a series of segmented bars, each of which adds up to 100 percent.* Often described as a "100 percent chart," this display looks much like a segmented bar chart, but shows proportions, rather than whole numbers, and typically has more categories per bar. The 100 percent chart still thrives today (as an option in Excel).

- *Square area charts with embedded squares proportionate to a given quantity, with the area charts often arranged in the relative order of their size.* Figure 8.1 shows examples from Clarence King's mining report that appeared in the 1881 *Second Annual Report of the United States Geological Survey to the Secretary of the Interior.* In these charts the inner shaded

---

[2]   Harris, for example, describes five different classification systems, two of which are formal systems defined by graphical elements (e.g., point, line, area, circle) or by type of axes (164–6). Leland Wilkinson uses a mathematical, "object-oriented design" system to generate a "grammar" of statistical graphics (1–3). Luc Desnoyers proposes categories of physical structure/features of what he calls "analograms," subsets of which include "punctigrams" (based on points), "curvigrams" (lines, curves), "histograms" (bars, circles), and "morphograms" (radar/radial forms) (127–30). In each of these categories, form provides the basis for grouping and parsing displays.

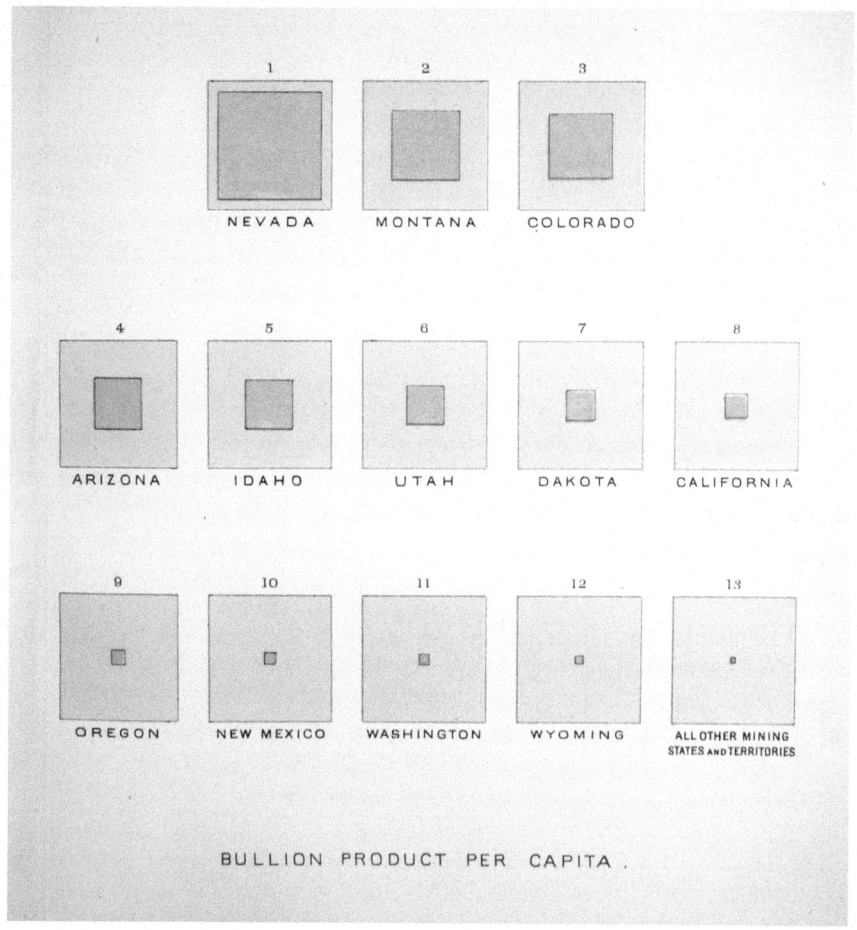

Figure 8.1   Area charts showing U.S. bullion production in states
                   and territories
Source: King, Plate XLIX.

square shows the amount of gold and silver bullion produced per capita in
several states and territories, arranged in order based on their productivity.
Compared to high-producing places like Nevada, Montana, and Colorado,
other locations shown have relatively small bullion production, including
"all other mining states and territories," visualized in the lower right corner.

- *Area bar charts showing one category of data (like a typical bar chart)
  but with different widths to represent a second category of data.* These
  charts resemble bar charts, only the bars vary in thickness to represent the
  second category. An early example appears in Plate 23, in which Charles
  Joseph Minard visualizes the freight that moves on barges along a French

canal. The height of each segment represents the tonnage of different kinds of freight (wood, coal, stone), the width shows the distance between each destination, and arrows indicate the direction the freight moved. By comparing the sizes of the rectilinear areas, readers can compare the quantities of the various goods and the distances they traveled: the larger the size of a given area, the greater the effort to transport that freight and, presumably, the greater the cost and the revenue.

- *Grids containing equal-sized squares that are apportioned according to variables.* Grids with the same scale and variables are typically arranged for comparison in the form of what Edward Tufte calls "small multiples" (*Visual Display* 170–75). In Figure 8.2 from Michael Mulhall's 1896 *Industries and Wealth of Nations*, several such grids show the distribution of occupations—agriculture, mining and manufacturing, and commerce—in several countries. With each cell (or part of a cell) graphically coded with one of the three occupations, the grids give readers a rectilinear framework within which to compare each country's occupations.

- *A collage of rectangles of varying sizes that are often clustered into subgroups, with quantities that typically add up to 100 percent.* This is perhaps the "purest" form of the contemporary mosaic. Figure 8.3 shows the "Flächen-Diagramm" (or area chart) of Von Mayr (1877), with three categories of hypothetical information, each subdivided into a, b, and c, and consisting of nine rectangles total. Von Mayr accompanies his diagram with an explanation of how the collage combines the areas to create the total amount.

These variations of mosaics are visualized conceptually in Figure 8.4 for comparison and reference. All of them embody rectilinear areas to display multiple categories of data, so in that sense they all overlap, with variation f, for example, representing a hybrid of variations c and d. And some have a stronger conventional status than others: variation a, for example, might be regarded as a short-lived prototype that never became conventional, and variation d as a longstanding conventional genre that has gradually gained in popularity. The mélange of charts in Figure 8.4. also illustrates the complications of using visual form to define genre because some forms strongly resemble those of other genres. For example, variations b and d look like segmented/divided bar charts; variation e looks like a bar chart; and variation f a simple matrix. This dilemma, of course, occurs with other forms: for example, circles provide the basic shape for several genres—pie, circle, polar, donut, and bubble charts—all of which have different genealogies and contexts of use.

Visual forms are often juxtaposed or intermingled in the same display, a factor that also affects our identification of genres. For example, Plate 24 shows a hybrid

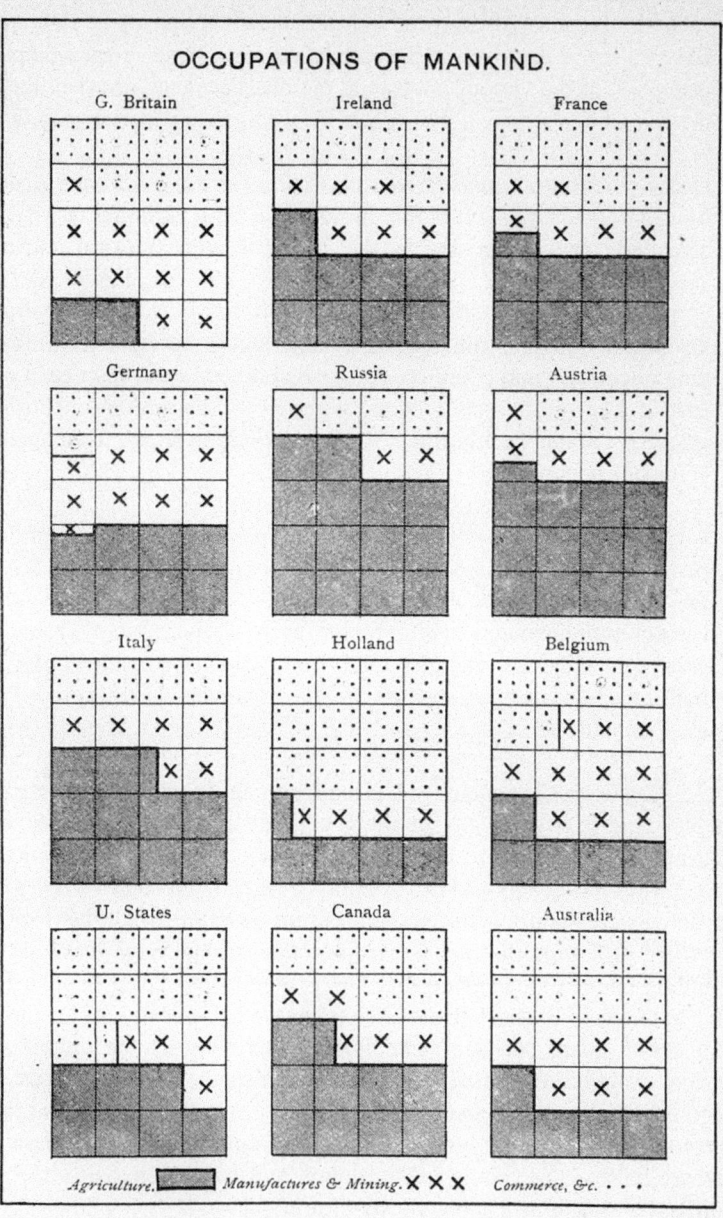

Figure 8.2 Square area diagrams showing occupations in several countries
Source: Mulhall, *Industries*, Plate III.

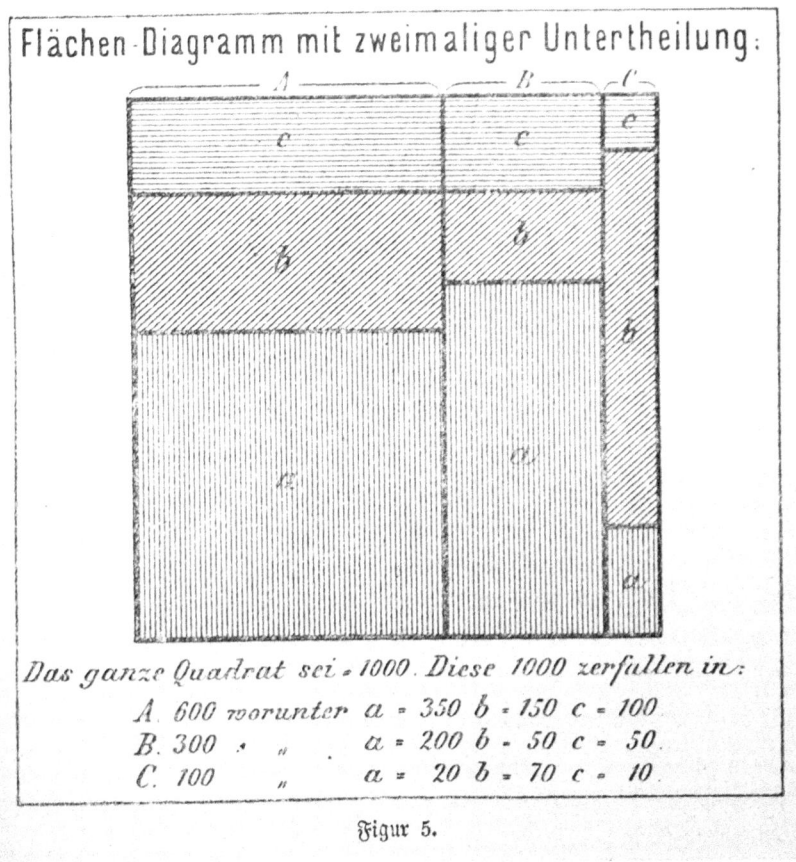

Figure 8.3   Nineteenth-century concept for a mosaic design
Source: Von Mayr 79. Courtesy University of Arkansas Libraries, Fayetteville.

of a map and mosaics from the French *Album de Statistique Graphique* of 1881.[3] Here the status of national public roads in departments across France is visualized in dozens of colorful mosaics on the map. Each mosaic displays up to three variables describing the nature and amount of road work accomplished or planned. In addition, the size of the mosaic appearing in a given region indicates the overall amount of resources the government has allocated to its roads. The map locates the

---

[3]   Friendly considers the French *Albums* among the best examples of graphical display in the later nineteenth century ("Golden Age" 517–18), and he cites a map with mosaics from the 1885 *Album* as an exemplary illustration of the genre ("Brief History of the Mosaic Display" 95–7). The 1897 Swiss atlas integrates simple area charts and maps (Statistischen Bureau, Plates XIVa, XIVb, and XVa).

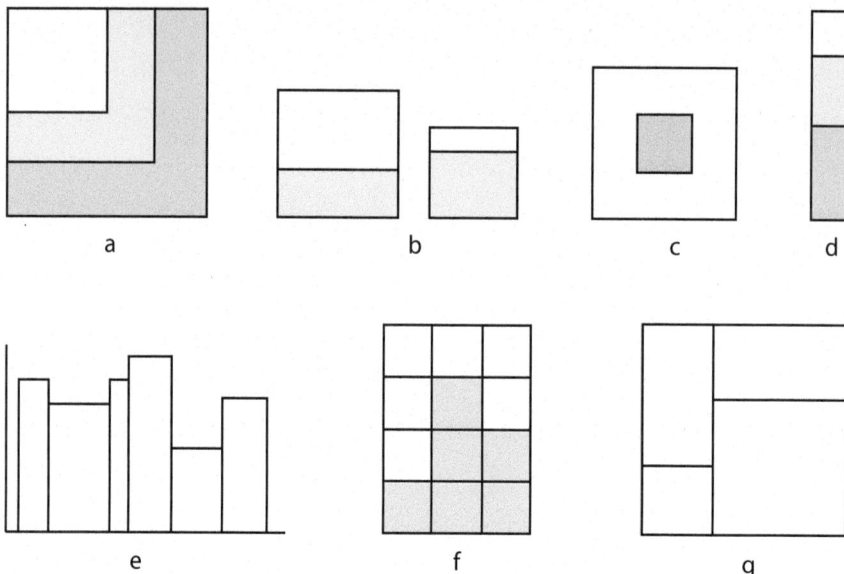

Figure 8.4   Conceptual diagram of rectilinear area charts and prototype mosaics

data geographically, and the map shading (especially in the southeast) indicates elevation, a key factor in road construction. This intermingling of map and mosaics creates a rich, hybrid display in which genres clarify and complement each other without undermining their identities or functions. Today hybrid displays abound in digital environments, where form has become highly fungible, as interactive data displays allow users to select, alter, resize, and integrate genres.

Perceptual and Rhetorical Functions: Mosaics in Action

To establish the identity of a data display genre, we also need to consider the work it does: how readers, designers, and stakeholders interact with the display and to what ends. Data displays function on a variety of levels, including their perceptual usability, how clearly and comfortably readers grasp the data envisioned, the purposes the displays serve in the situations in which they are deployed, and their rhetorical effects—persuasive, informative, emotive—in those situations. Because functionality encompasses transactions between designers and users, it opens up many new avenues for defining genre as well as addresses some of the boundary issues that arise by defining genre with names or physical shapes. However, functionality can also be slippery business because of the wide-ranging situational variables in which data displays mediate between designers and users (see Kostelnick, "Visual Rhetoric of Data Displays"). Mosaics are a good touchstone for genre functionality because their history encompasses a great deal of perceptual and rhetorical variability.

Perceptual Issues with Area Charts

Area charts and mosaics have historically raised some compelling perceptual issues: these displays can be simple and accessible, or they can be complex, perceptually challenging, and even inscrutable to the casual observer. In the abstract, perceptual criteria might be regarded as irrelevant, or even antithetical, to genre, insofar as eye, brain, and visual stimuli are deemed to operate within a sensory/cognitive system detached from social or historical context. However, perception (and beliefs and attitudes about perception) operates within an historical context. Beginning with William Playfair, data design has long been justified on perceptual grounds because it enables readers to process information quickly (*Commercial*, 3rd ed. xii), to "see" rather than to "read" information as Jacques Bertin puts it (*Graphic Information-Processing* 178–9), and to observe what Ben Barton and Marthalee Barton describe as the "synoptic" view of data ("Modes" 142–3). However, perceptual efficacy varies widely across graphical displays, as Brinton argued a century ago in *Graphic Methods for Presenting Facts* (21–9, 36–9), as early empirical research demonstrated (Macdonald-Ross), and as Tufte famously explained with his "Lie Factor" (*Visual Display* 57–8).

In theory, we might expect those genres that enable users to perform well perceptually, or are believed to do so, will evolve more successfully than those that lack perceptual efficiency—a proposition that would seem to make biological sense. However, genres that can claim superior perceptual functionality don't automatically have an evolutionary edge, as the immensely popular but perceptually deficient pie chart illustrates. Perceptually, mosaics have experienced a mixed bag. In the opening pages of *Graphic Methods for Presenting Facts*, Brinton comments on a mosaic from Gannett's 1903 U.S. statistical atlas (Gannett, *Statistical, Twelfth Census*): "This is an excellent type of chart to use if subdivisions in the component parts of any unit have to be shown" (Brinton, *Graphic Methods* 7). Brinton advocates rectilinear shapes because they can be read and measured accurately (with appropriate scales), and they are easier to label than circles, though he sharply criticizes the use of rectangles of varying areas without scales (*Graphic Methods* 4–8, 37, 69–70). Still, genres like mosaics that rely on area may have had an evolutionary disadvantage, based on early empirical research summarized by Michael Macdonald-Ross and based on William Cleveland and Robert McGill's influential taxonomy claiming that displays that arrange data points as "position along a common scale" are perceptually superior to those that rely on area, volume, or grayscales (532–6). Overall, then, in terms of perceptual functionality, mosaics have not benefited from—indeed their development in much of the twentieth century was probably stunted as a result of—the conviction that they possess perceptual deficiencies, mainly by obliging their users to make judgments about area.

That conviction, however, has waned considerably in the age of "big data" and interactive visualizations, which often employ complex rectilinear areas. Contemporary digital data design has fostered a re-examination of the perceptual weaknesses of displays, including rectangular areas. For example, Wickham and Hofmann point out that despite the inherent perceptual drawbacks of areas, rectilinear area charts are spatially flexible and can often be composed to enhance perceptual efficiency (2224–6). The perceptual constraints that earlier seemed to retard area charts and mosaics have gradually diminished, as data designers challenge (or flout) previous assumptions by creating displays with interactive features that reduce some of the guesswork entailed with static charts. So digital technology benefits a perceptually suspect genre like mosaics by unburdening them from some of the visual constraints they previously had to bear.

Continuing Functions of Mosaics

To be viable, genres must "recur" in similar situations (Miller 156). Like other kinds of genres, mosaics communicate information for purposes that recur from one situation to another, giving them rhetorical ballast. On the broadest level, from the rhetorical perspective of 10,000 feet, data displays fulfill "analytical" or "presentation" purposes (Brasseur, *Visualizing* 21–2): on the one hand, they "facilitate discovery" (Iliinsky 8) and enable "exploration" of the data (Fisher 338), and on the other hand, they convey specific information in a given situation (Iliinsky 8; Fisher 338). Mosaics have the rhetorical suppleness to serve both functions simultaneously: they are often complex and provocative enough to prompt data exploration and analysis, and their shapes are simple and accessible enough to communicate data "stories" (Tufte, *Envisioning*, 37, 108) or data arguments to readers who are motivated (and able) to interpret these displays.

Data display genres, however, typically have more specific purposes that help to clarify their identities. Sometimes those purposes are singular or even unique: for example, electrocardiograms that show data about heart function, Gantt charts that show production schedules, and isotherm maps that show wind patterns. Mosaics have the broader purposes of dividing data sets into their constituent parts and/or juxtaposing data in close proximity. However, in carrying out these functions, many types of mosaics share recurring purposes with other data display genres, which can place them in direct evolutionary competition as well as challenge genre boundaries. For example, a cascading mosaic and a circle chart with increasingly larger circles circumscribed around each other have the same purpose: to show the relative size of the data by stacking areas onto each other (see Figure 8.4, variation a). Inexplicably, history has treated circles more kindly than cascading mosaics, which have become virtually extinct. Many mosaics, particularly a 100 percent chart, function the same as a pie chart—to show parts of a whole—but use rectangular rather than curvilinear forms (see Figure 8.4, variation d). Although 100 percent charts and pie charts have appeared in the U.S. for over a century, pie charts have become the dominant genre by far.

Mosaics also serve a variety of recurring *rhetorical* purposes, which illustrate a versatility that distinguishes them from other data display genres. Some of those purposes might be described in classical terms as "judicial," "deliberative," and "epideictic" (Aristotle 47–9). For example, Mulhall's occupation charts in Figure 8.2 make the visual argument that industry largely drives Great Britain's economy, while agriculture drives those of Russia and Austria, supplying "judicial"—or "forensic"—evidence that Britain fosters a more advanced and successful economy. Graphics in a deliberative mode help readers plan and make decisions: such is probably the case for the mosaics in Plate 22, which show variances in land use across states, as well as for the mosaic map in Plate 24, which shows road conditions across France. Some graphics, like Minard's canal mosaic have the rhetorical elasticity to serve dual purposes, forensically documenting the movement of goods along a canal but also deliberatively identifying choke points and inefficiencies addressable in the future (see Plate 23). Mosaics can also invoke the epideictic function by memorializing an event, like the mosaics of the Titanic disaster that show survival rates of passengers based on age, gender, and class (Hofmann; Robert Spence 47–9; Theus and Lauer 396–8; Wilkinson 342–3).

This rhetorical flexibility springs partly from the generous affordances of mosaics, where rectangular forms can be shaped and reshaped—from short and wide to long and thin—and arranged in virtually infinite combinations. That flexibility appears in the late nineteenth-century mosaic in Plate 25 (Peale 554), which shows the distribution of major religions around the world. As a presentational graphic that appears in a reference book, it ostensibly has a purely forensic purpose—to provide facts about the subject for general American readers. However, it also functions in the epideictic mode: With the Christianity sub-chart below providing a cue, the chart tells a bleak tale, lamenting the dominance of non-Christian belief systems. And in doing so, situated in a period of intense Western colonization, it also implies a deliberative function by making a persuasive call to action to address Christianity's minority status. Today, with a more global, multicultural, and secular audience, the same chart would likely be epideictic in other ways—with many readers celebrating, while others bemoaning, its religious diversity. As this composition illustrates, the rhetorical functions of mosaics can be wide-ranging and disparate, especially when examined in specific situational contexts.

Socializing Mosaics: Conventions and Discourse Communities

Visual language is intrinsically social, as Gunther Kress and Theo van Leeuwen show, and therefore shaped by conventions that are collectively constructed by audiences of users that imitate and proliferate them (Kostelnick and Hassett). In this way, visible forms are stabilized and normalized sufficiently so that they can easily be deployed and interpreted. As social constructs that function within communities of users on a continuing basis, data display genres like mosaics both engender and meet reader expectations. In this way, a display's visible form is subsumed into its genre function and social value within a given visual discourse community and subject to the social forces that initiate and sustain the genre as well as those that destabilize, transform, or degrade it. For many genres, like mosaics, these processes unfold over long stretches of time, making genre formation inherently historical. In the context of written communication, genre evolution is typified by two related conditions: (1) as social constructs, genres require enculturation and learning; and (2) genres must gain currency by being continually deployed and interpreted by communities of users (see Swales; Bazerman). These same conditions also apply to data display genres, especially those like mosaics that have experienced long and erratic evolutionary processes.

Genre Enculturation

As social constructs, data display genres are artificial and therefore their visual languages must be acquired and learned through enculturation. Initially, then, genres that are new to readers, like mosaics, can foster both excitement and interpretive challenges. The first U.S. statistical atlas (1874) was intended for a wide public audience, as described in the Preface in a quotation from the Secretary of the Interior's 1872 *Annual Report*:

> ... for distribution to public libraries, learned societies, colleges and academies, with a view to promote that higher kind of political education which has hitherto been so greatly neglected in this country, but toward which the attention of the general public, as well as of instructors and students, is now being turned, with the most lively interest. (Walker 1)

Given this didactic purpose and large audience, the atlas needed to provide instructions on how to read several of the genres it deployed. For mosaics, these explanations appeared on the plates themselves as well as in the Preface, where specific examples were explicated (Walker 3). The French *Album de Statistique Graphique* also provided interpretive notes for mosaics, as did Minard for his canal mosaic (see plates 23 and 24). Those interpretive aids decreased in the statistical atlases of the early twentieth century (see Gannett, *Statistical*, *Twelfth Census*; Sloane, *Statistical*, 1914 and 1925), presumably because readers had by then collectively learned to read most common forms of graphical displays. Still, for some readers, mosaics must have remained relatively novel and opaque, given that Brinton himself felt compelled both to tout and explicate them (*Graphic Methods* 4–8). However, with the digital revival of mosaics in the last two decades, the learning curve has again accelerated, greatly aided by interactive displays where users can access interpretive help as well as actively customize the design.

Currency: Attracting an Audience

As a conventional construct, a viable genre has to gain a following; it must "recur" (Miller 156). Mosaics are an interesting case in terms of their currency: historically, they appeared too sparingly to achieve a high degree of recurrence, comparable to that of line or bar graphs; however, given their high-profile use in the nineteenth century in the U.S. statistical atlases and French albums and elsewhere in the U.S. and Europe, they certainly achieved status as a bona fide genre. Because state agencies like the U.S. Census Office and the Ministère des Travaux Publics sponsored their appearances, they acquired considerable political and rhetorical gravitas and ethos and on that basis achieved widespread currency. With the contemporary wave of digital mosaics and area charts, the genre is achieving currency with a much larger cohort of users, most of whom are probably oblivious to its earlier incarnations. In this way, the currency of mosaics responds to the socializing influences of both immediate conditions and invisible historical forces.

Collages of Culture: From Politics to Aesthetics

As social constructs, data display genres are situated in cultural communities and embody and propagate its values. They do so by way of their intrinsic features (Cartesian coordinate systems), metaphorical aspects (donut and pie charts), and discretionary design elements like color, emphasis, and hierarchy. In virtually any data display genre within a given cultural setting, we are likely to detect design elements that communicate that culture's values. For example, the mosaic in Plate 25 tells the story of world religions through the lens of nineteenth-century Western values: ideologically in its emphasis on Christianity (repeated and subdivided at the bottom of the chart); schematically in its rational and seemingly objective composition of angular forms, reflecting the Utilitarian and scientific values of the age; and aesthetically in its variegated color scheme and Victorian typography and page border. The same process of cultural infusion occurs throughout the history of data design and across a wide array of genres, as scholars like Ben Barton and Marthalee Barton ("Ideology"), Lee Brasseur (*Visualizing*), and Miles Kimball have demonstrated.

As Friendly demonstrates, mosaics enable readers to see relationships and patterns in complex data sets ("Brief History of the Mosaic Display" 100–105). Because mosaics can display multiple variables and categories in close proximity, they are inherently hierarchical, emphasizing some data and de-emphasizing others through grouping, spatial positioning, and graphical coding. However, mosaics are not prescriptively formulaic and therefore they afford designers considerable freedom to choose the size and shape of objects in the composition, how these objects spatially relate to one another, and the borders and graphical coding that differentiate them. This combination of complexity and flexibility of representation has particularly important consequences socially and culturally because nineteenth-century mosaics often visualize data about people—immigration patterns, nationality, religion, race, occupation, and gender.

Early Political Mosaics

Early area charts of Charles de Fourcroy (1782), August Crome (1785), and Alexander von Humboldt (1811) visualize countries, colonies, and cities—all of them politically charged tasks. Each of these charts uses the technique of overlapping rectangles, an arrangement that creates a hierarchy from smallest to largest quantities and that allows for precise comparison but that de-emphasizes entities underneath the surface, especially those beneath rectangles of only slightly smaller size, leaving minimal space between them (see Figure 8.4, variation a). De Fourcroy, for example, overlaps European cities based on population, moving from the smallest cities in the upper left to "Très Grandes Villes" in the lower right, with London, Petersberg, and Paris occupying the largest squares. Because of the variations in the size of the squares, some cities have large intervals between them, while others (like Berlin and Lyon) have extremely small intervals and are therefore scarcely visible, greatly diminishing their presence. Perhaps de Fourcroy did not intend these visual slights, and they were merely the random graphical outcomes of the display's geometry. Nonetheless, the chart invites us to read a geopolitical message, where some cities appear to have more prestige and power than others.

Crome's task was no less daunting because he visualized all European nations in a symmetrical overlapping chart titled "Groessen-Karte von Europa," with the smallest countries (by land area) in the upper center and the larger ones progressively spreading out beneath them. The result of this configuration places the smallest nations emphatically in the center—as if occupying the very heart of Europe—and the larger ones on the periphery and beneath, turning hierarchy on its head! Like de Fourcroy's chart, some rectangles overlap closely and leave only a small interval, which perceptually, and potentially politically, diminishes a given nation state. The chart's compact cascading design also creates an appearance of orderliness and connectivity, where nations are precisely stacked atop one another in harmonic equanimity. On the eve the French Revolution, and shortly before Napoleon's military conquests, such exact and symmetrical demarcations painted the illusion of political stability preceding the turbulent decades ahead.

Colonial Angst

Alexander von Humboldt's application of the overlapping area chart (see Figure 8.4, variation a) in his *Atlas Geographique et Physique de Royaume de la Nouvelle-Espange* (1811) also implicates politics and culture, particularly when considered with other charts on the same plate. The issue of colonialism on the American continents, which was particularly controversial in the early nineteenth century, permeates this series of charts. The more famous chart at the top of the plate uses de Fourcroy's overlapping technique to show the physical size of the various colonial territories of "Nouvelle-Espagne," arranged from the upper left from smallest to largest, with a dotted line showing the size of Spain—a useful benchmark for his European readers that reveals the immense disparity between colonizer and colonies (see Figure 8.5). This disparity is reinforced in the area chart below (on the left), where Spain appears in the shaded lower segment and its American colonies in the unshaded segments above. Next to Spain appear Britain and its Asian colonies, Britain and its American colonies, Turkey and its African and Asian colonies, and Russia and its Asian and American colonies, with each nation appearing in the shaded segment at the bottom and its colonies stacked in unshaded segments above. Alongside each of these area charts appear narrow bar charts showing the population of each nation (shaded) and its collective colonies above (unshaded).

These charts both quantify and question the colonial practices of the time, using incipient mosaics to tell that story in a compelling way by visualizing land areas as proportional rectangles: visual comparisons that a map—with all of its geographical complexities—could not reveal as accurately, emphatically, and (seemingly) dispassionately. Humboldt had criticized Spanish colonialism in his diaries during his expedition to South America (Rebok), and the charts subversively reveal these concerns, both implicitly and explicitly. Immediately apparent in all of the charts is the stunning magnitude of the colonies, whose land areas (and in most cases populations) are vastly larger than those of the colonizing nations. The breakaway United States segments (area and population) appear disconcerting because of their angled orientation and their placement in the center of the figure. Such emphatic dislocation may have been Humboldt's way of predicting—indeed, arguing for—similar breakaway (independence) events in the future, which later politically transformed the American continents. The exclusion of French colonies from Humboldt's charts seems odd given that France remained a major colonial power and that Humboldt published the charts in French. However, perhaps Humboldt didn't want or need to visualize France's dwindling stake in America to an audience well versed in that narrative and, more immediately, mired in Napoleonic empire-building closer to home.

These early examples of area charts and primitive mosaics tell large, complicated stories about nations, people, and politics using rectilinear forms as an efficient and seemingly objective means of comparison. As experimental prototypes, they lack conventional status and therefore have few, if any, reader expectations to meet, giving their designers the freedom to orient the data visually and culturally as they desire. These charts foreshadow the more fully developed mosaics later in the nineteenth century, establishing from the very start the cultural and political implications of visualizing statistical data.

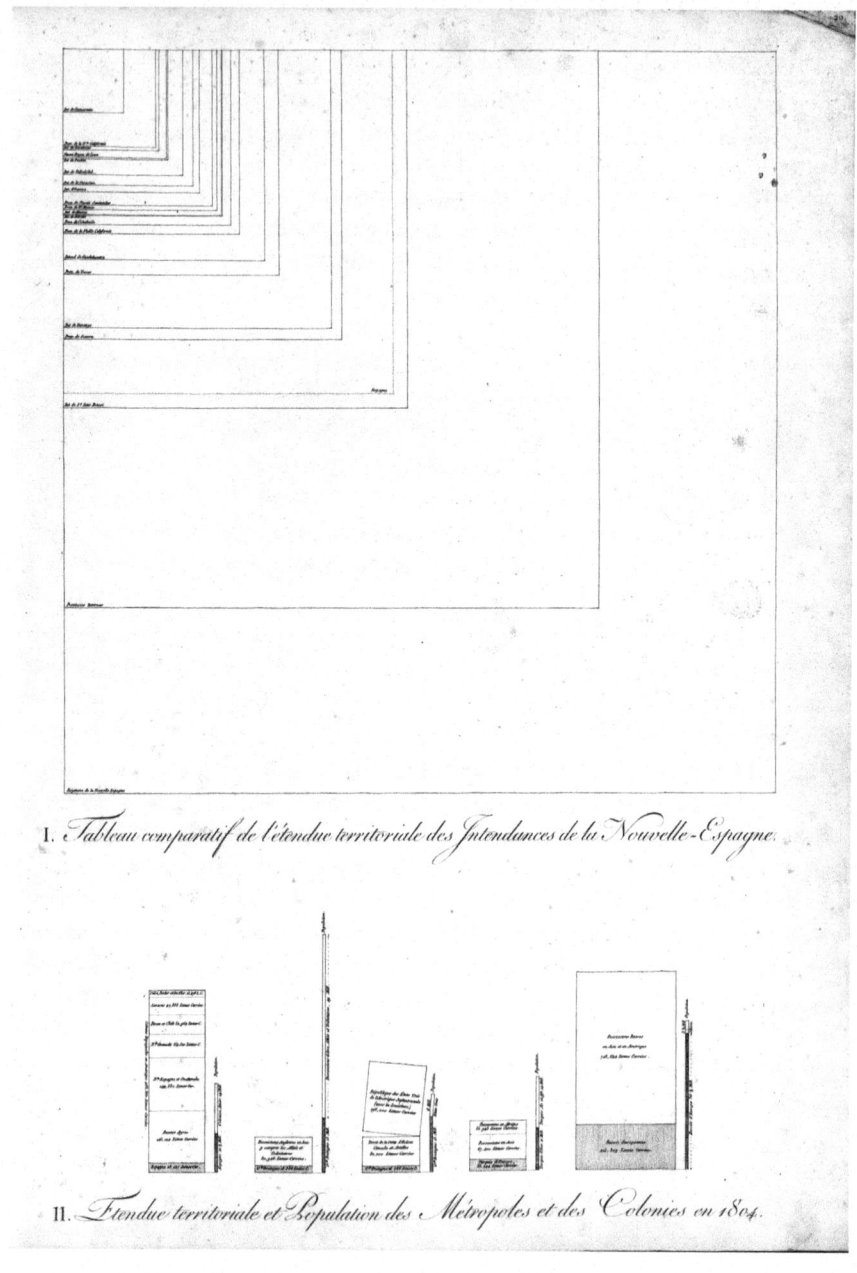

I. *Tableau comparatif de l'étendue territoriale des Intendances de la Nouvelle-Espagne.*

II. *Étendue territoriale et Population des Métropoles et des Colonies en 1804.*

Figure 8.5   Area charts showing Spanish colonies (above) and colonies of
several countries around the world (below)
Source: Humboldt, *Atlas*, Plate 20. Courtesy the David Rumsey Map Collection, http://
www.davidrumsey.com.

Graphical E Pluribus Unum

The U.S. statistical atlases needed to visualize, as efficiently as possible, complex population data showing patterns over space, time, race, gender, and nativity; mosaics met that exigency for multi-dimensional design. In the initial atlas of 1874, overseen by Francis Walker, mosaics were chosen to display data about people in three vital categories: where they reside in states across the country and their movement in and out of those states, particularly westward; which religious denominations, if any, they are affiliated with; and what kinds of occupations they engage in. *Democracy* is the pre-eminent value embedded in these charts—of showing data as equally as possible across states and population groups. Culturally and politically, mosaics in the U.S. statistical atlases became the graphical equivalent of *E pluribus unum*, representing subsumable pieces of diversity that continually shifted across population groups as the nation evolved. In the *unum* of mosaics, statistical data was literally tiled together in rectilinear fragments, visually democratizing data in complex patterns that homogenized difference (see Kostelnick, "Melting-Pot"), though hierarchical patterns also appear relative to race, gender, and nativity that reflect the growing debate over immigration.

In the states chart of the 1874 atlas (Plate 26), three key democratizing phenomena were visualized: the continuing influx of foreign immigrants, the aftermath of the Civil War and the emancipation of slaves, and the westward expansion of the population. The field of mosaics represented these narratives by visualizing the each state's population in a single square, divisible vertically and horizontally into its constituent parts. The most emphatic variable, foreigners, appears on the left of each state's square, followed by African Americans and whites, with each of these two categories divided into those born in the state (bottom) and those born in another state or territory (top). The separate rectangles to the right represent people who have left the state, African Americans appearing on the top and whites on the bottom. In terms of the their positions in the mosaics, neither foreigners nor African Americans are marginalized.[4] Nor are states themselves marginalized, as the mosaics are arranged according to their size, boustrophedon style (as the ox plows), with the *smallest* entities appearing in the top left and the largest (New York) in the less emphatic lower right. These democratizing effects are accomplished culturally and geographically despite the enormous variations across the mosaics in the composition of people and their movement into and out of each state or territory.

Religious equity is visualized in the mosaics in Plate 27, which display denominations in each state or territory, ordered largest to smallest in each mosaic from left to right. Although Methodism dominates, as seen in the national

---

[4]   On the other hand, the darkest shading represents African Americans born out of the state, though virtually the same colors are used to contrast whites born in and out of the state as are used to contrast African Americans and whites who left the state (top and bottom segments on the right).

chart in the upper left, in several states other denominations prevail, as each denomination is sequenced according to size, with "All other denominations" always appearing on the far right. With 11 church categories envisioned, this display tells a story of religious pluralism, and presumably religious tolerance, visually debunking any notion of government preference for one denomination over another. The gray area surrounding the colored swatches shows the lack of church "accommodations," or those people without access to a church, implying a gray void that needs coloring in, particularly in younger states like Nevada (middle right). Despite the pluralism and the gray voids, the plate tells a story of an overwhelmingly Christian nation, where the vast majority of citizens affiliate themselves with a church.

The mosaics from the 1874 atlas showing occupations and school attendance use a similar graphical method, with key categories on the interior and a variable gray margin surrounding them (see Plate 28). Displaying population over the age of 10, the mosaics prominently place agriculture on the left, as the country still envisions itself as a chiefly rural economy, followed on its right by "Manufactures & Mining," "Trade & Transportation," "Pers. & Profess. Services," and "Attending School." The gray margins surrounding the categories "represent the proportion of the population not so accounted for," suggesting the degree of unemployment—miniscule in Iowa (second row, right), which was mostly an agricultural state, and large in Virginia (lower left), which was undergoing reconstruction. Gender pervades the mosaic as a key variable, with females placed in the top segments (above men), an explicit nod to their growing equality, especially in education, the variable on the right. However, in occupations like Agriculture, Manufactures & Mining, and Trade & Transportation, the gender gap remains extremely large. Overall, though, the story here is muted: each mosaic occupies the same space, despite differences in population, and the color scheme alternates between cool blues and earth tones, girdled by a gray border—all of which visually homogenize variations among categories and emphasize the collective national economy over local regions.

In the occupation charts of subsequent atlases, the data categories expand from gender to include nativity and race, illustrated in Figure 8.6 from the 1898 atlas. This series of mosaics visualizes occupations in five groups, ranging from professional and agricultural to manufacturing, trades, and personal services. In the large summary chart at the top, native whites are shown at the top, followed by whites with non-native parents, non-native whites, and African Americans located at the bottom, a visual stratification that reflected the cultural orientation of the time toward race and ethnicity. The mosaics below isolate each demographic group and subdivide occupational activity by gender, with non-native whites and African Americans again located at the bottom. As they did in the 1874 mosaics, females make up a comparatively small part of these occupations, but here they appear at the bottom of each mosaic rather than the top. African American females play the largest role proportionately among females, but they appear in the least emphatic place, the far lower right.[5]

Like the occupational mosaic from the 1898 atlas, the mosaic in Plate 29, from the 1903 atlas, visualizes data about the four population groups but excludes gender as a category. This fine-grained color display of occupations contains nearly 200 tiles, revealing occupational patterns across the population groups at what Tufte calls the "macro" level and the "micro" level (Tufte, *Envisioning*, 37–51). On the macro-level, native whites are engaged in a broad spectrum of occupations, while foreign whites skew towards manufacturing, and African Americans participate largely in agriculture and domestic service, with relatively few identified as professionals.[6] On the micro-level, the mosaic subdivides occupational groups into specific vocations: within the "Professional Service" group, the proportions of clergymen (1), lawyers (2), physicians (3), and teachers (4) appear as separate tiles. These micro-level details enable readers to see specifically which occupations each group finds employment in and how the groups compare with each other. Although these small (often thinly sliced) rectangles reduce the usability of this display, their presence suggests an increasing statistical focus on who does what kind of work in a rapidly emerging industrial culture.

In France the national albums tell similarly complex stories, but with different political and cultural circumstances. The *Album Graphique de la Statistique Générale de la France* of 1907 (Ministère du Travail), for example, visualizes cultural difference explicitly, making it darker and locating it on the margins, as shown in the 100 percent charts in Figure 8.7, which display employment status based on the relative population of foreigners (top), as well as their place of birth (middle) and nationalities (bottom), divided symmetrically by gender. The bottom chart shows English foreigners in the center in near white, with the other nationalities (Belgian, German, Swiss, Spanish, Italian, and others) progressively outward—and progressively darker. What could be the rationale for these design choices? Proximity probably didn't play a role, given that France shares land borders with all of these countries (except England) and language with a few of them. One explanation might be that France and England signed the "Entente Cordiale" in 1904, engendering close relations politically and socially and presaging their strong alliance in the First World War. However, partly because of immigration issues, relations with Italy during this time were

---

[5] In the 1898 *Atlas* a more emphatic hierarchy of race and nativity appears in a series of 100 percent charts displaying the proportion of whites and African Americans living in several eastern and southern states over a century (Gannett, *Statistical, Eleventh Census* 18). African Americans appear in the hatched cells at the bottom of the charts, and whites appear in the unshaded cells above, suggesting an iconic power relationship, both spatially and graphically.

[6] See Kostelnick for additional cultural analysis of this mosaic in the context of American immigration ("Melting-Pot Ideology" 230–31).

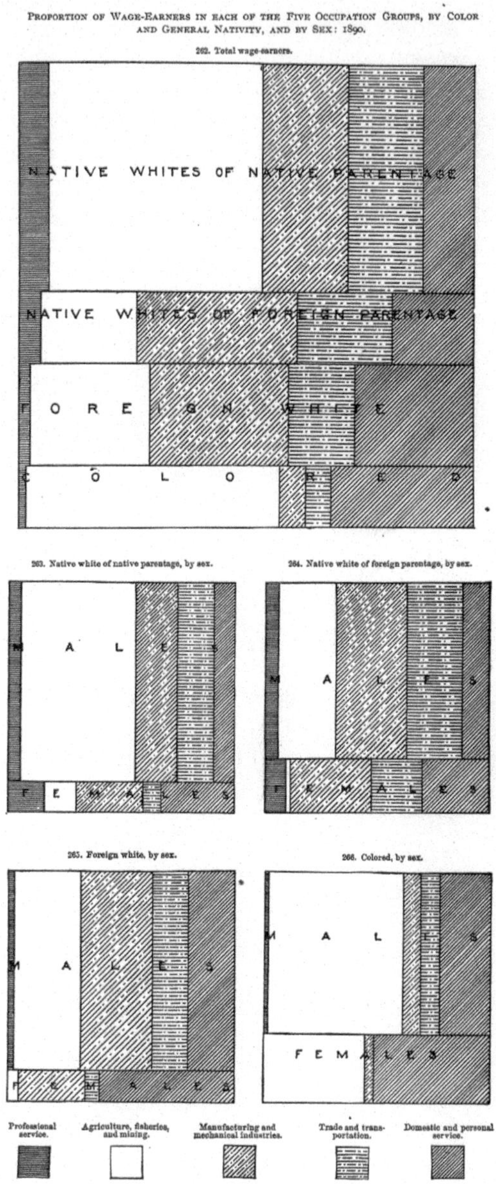

Figure 8.6   Mosaic from the 1898 U.S. statistical atlas showing occupations of various segments of the U.S. population

Source: Gannett, *Statistical, Eleventh Census*, 46. Courtesy of the Library of Congress, Geography and Map Division.

less cordial, and these political and cultural circumstances likely influenced design decisions about placement and shading.

At any historical moment, then, area charts and mosaics served as a medium for expressing and perpetuating the cultural and political values of the communities in

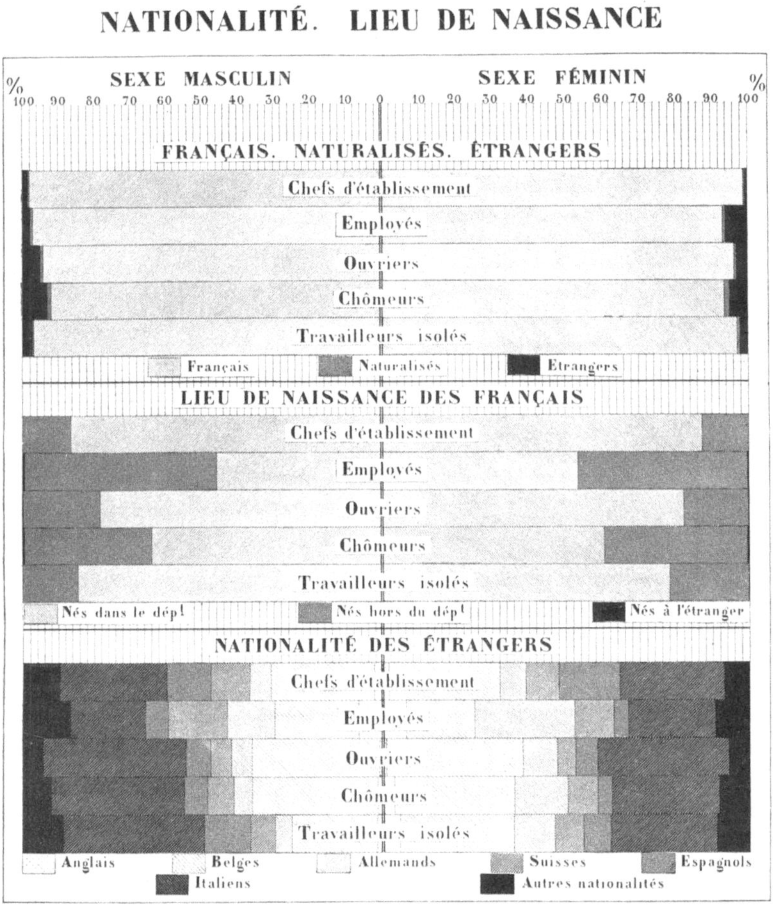

Rec. 1901, t. IV, p. 672, 673.

	PROPORTION P. 100 DES ÉTRANGERS.		PROPORTION P. 100 DES FRANÇAIS NÉS HORS DU DÉPARTEMENT.	
	MASCULIN.	FÉMININ.	MASCULIN.	FÉMININ.
Chefs d'établissement	1	1	13	12
Employés	2	6	53	46
Ouvriers	5	3	21	18
Chômeurs	7	5	35	39
Travailleurs isolés	3	2	15	22

Figure 8.7 Area charts from the 1907 *Album Graphique de la Statistique Générale de la France* showing the employment status of the native and immigrant populations

Source: Ministère du Travail 182.

which they were deployed. That behavior both typifies and distinguishes genres, as they construct and respond to the worlds in which they are situated and its exigencies, cultural conditions, and historical circumstances.

From Victorian to Postmodern Aesthetics

In visualizing statistical stories about people in the "golden age" of data graphics, mosaics also expressed cultural values aesthetically. Late nineteenth-century mosaics, particularly those that appeared in the U.S. statistical atlases, embodied a Victorian sensibility that fostered complexity, plenitude of detail, and natural color schemes. With their elaborate patterns, these mosaics conveyed the Victorian penchant for intricate designs, particularly those made by hand. Hand (as opposed to mechanical) design was promoted by the Gothic Revival movement, particularly John Ruskin, who associated hand labor with morality and imagination (151–81), principles that underpinned the Arts and Crafts movement initiated by William Morris in the 1860s. The mosaics in plates 26, 27, and 28 reflect these aesthetic impulses: lavish with color, complexity, and intricate detail; rife with subtle diversity, with each mosaic varying from the others, the plates teeming with little stories—a feast for the imagination and the senses.

In the twentieth-century atlases, however, elaborate designs like these mosaics gradually disappeared, along with the intricate tracery of Victorian porches, embellished carvings on parlor sofas, and handwriting flourishes with steel nib pens. Such extravagance simply lost its cultural capital, replaced by the high-contrast geometrical efficiency of machine-age modernism and its emphasis on perceptual immediacy and universal appeal. Not surprisingly, the waning interest in mosaics correlated with the onset of modernism—and their revival in the 1980s with the ascendancy of postmodern aesthetics, where diversity and richness of design were reaffirmed, beginning as early as Robert Venturi's 1966 *Complexity and Contradiction in Architecture*. The digital mosaic—with its array of dissimilar pieces that users can often manipulate interactively on the dynamic, ever fluctuating surface of the computer screen—epitomizes postmodern aesthetics.

Technology, Production Processes, and the Evolution of Mosaics

Genres—particularly visual genres—are defined by who can deploy and interpret them, and how a genre is designed, produced, circulated, and consumed matters a great deal to its currency and viability. The invention and proliferation of data display genres depend on technology, production processes, and resources—forces that have a particularly strong influence on mosaics. Early on, technology stabilized the growth of mosaics by restricting their currency because of the math, hand labor, and expense required to create them; recently, however, digital technology played the key role in reviving and reinventing the genre. Today mosaics are part of a larger wave of data visualizations that have collectively stretched, transformed, and destabilized genres that previously flourished in print, prompting us to question how exactly these new forms relate to traditional genres.

Manufacturing Hand-Crafted Mosaics

Like thematic maps, mosaics have not historically been the province of amateurs, largely because technology and production costs constrained their design and circulation. The first U.S. statistical atlas of 1874, which introduced mosaics to the American public, required the U.S. Congress to appropriate funds that Francis Walker, the "compiler" and chief editor of the first atlas, used to assemble a cohort of skilled labor.[7] Producing the "Geometrical Diagrams" of the 1874 atlas required numerous statisticians and draftsmen, suggesting a labor-intensive process that required considerable resources, especially for the large pages (over 20 inches in height) of first three atlases. Walker indicates that the creation and organization of the illustrations in the 1874 atlas were accomplished by a "corps of able assistants" and "several eminent gentlemen," including professors, scientists, and engineers (1). The mosaics in plates 26, 27, and 28, which were drawn by J.J. Skinner under Walker's supervision (Walker 1), surely required a great deal of industry, skill, and mastery of geometry. Although some of the technology used to create these images was simple by today's standards—rulers, compasses, pen and ink—the coloring techniques used by the printer Julius Bien to produce the plates with chromolithography were quite sophisticated.

This combination of craft, geometrical dexterity, and color reproduction coalesced in the 1903 mega-mosaic in Plate 29, which likely represented one of the most complex and interpretively demanding mosaics of the pre-digital age. The extensive scaling, coloring, and labeling must have astonished and perplexed its readers—and *demoralized* its would-be imitators. The extreme variations in the aspect ratios of the rectangles, especially the ultra-thin areas on the right representing "Manufacturing and Mechanical Pursuits," must have been challenging to execute by hand with pen and ink, utterly stretching the capabilities of the designer. Here innovation outpaces the technology available to implement it, and the genre demands new technology for its growth and survival.

In the early twentieth century, the production of mosaics (even in the larger realm of area diagrams) seems confined to a limited domain. Still, mosaics claimed some production advantages, primarily by economically visualizing several data categories in the same display. Additionally, Brinton observes that rectilinear displays like the 100 percent chart have a minor advantage over curvilinear pie charts in the ease measuring units and labeling text (*Graphic Methods* 4–7, 36).

---

[7]   An initial appropriation funded the production of several displays, which were circulated on a limited basis; a second appropriation enabled the Census Bureau, under Francis Walker's direction, to add a fuller range of displays and to publish the whole collection in the first 1874 atlas (Walker 1).

But mosaics were also constrained by production challenges posed by complex geometry and drawing, partly because they did not map as easily as other forms onto lined paper, something Brinton also notes (*Graphic Methods*, 7–8). Those production drawbacks, along with the apparent perceptual deficiencies, retarded the further growth and proliferation of the genre, which languished for decades, confined primarily to government publications, academic books, and aging atlases on library shelves.

A rare example of a twentieth-century mosaic, shown in Figure 8.8, appeared in the 1937 report of the U.S. Works Progress Administration (WPA). This chart, which was reproduced by Brinton to illustrate the "area bar chart" (*Graphic Presentation* 149) visualizes the hourly wages of workers in 10 WPA job sectors, ranging from highways to conservation to public buildings. An example of variation e in Figure 8.4, the WPA chart echoes on a smaller scale Minard's canal mosaic nearly a century earlier (Plate 23). The visual economy of the WPA mosaic also reflects the functional aesthetic of modernism, concealing the labor

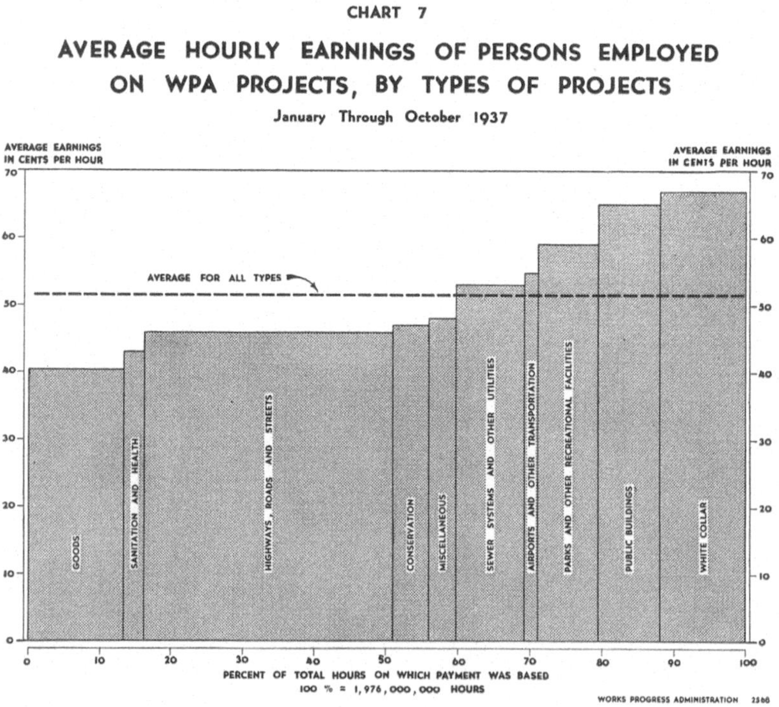

CHART 7

AVERAGE HOURLY EARNINGS OF PERSONS EMPLOYED ON WPA PROJECTS, BY TYPES OF PROJECTS

January Through October 1937

Figure 8.8   Wage chart from the December 1937 *Report on Progress of the Works Program*

Source: U.S. Works Progress Administration 52, Chart 7.

and skill required to produce it. The apparent simplicity of this mosaic also disguises its interpretive demands, which may also explain why charts like this appeared so infrequently in the WPA reports.

Digitizing a New Generation of Mosaics

With developments in computer technology in the 1970s and 1980s, a slow resurgence of mosaics occurred. Michael Friendly narrates this rebirth and development of the genre and its statistical capabilities—for example, in enabling designers to envision multiple variables in the same plot frame ("Brief History of the Mosaic Display" 99–105). Because they are generated digitally, mosaics have appeared with increasing frequency, not only in statistics but also more broadly in popular media. Plate 30, for example, shows a contemporary digital mosaic that appeared on the Bloomberg.com web site (Tribou), accompanied by an article about payments to retiring state workers (Marois and Yap). The bar chart on the right shows payments made in rank order by state, and the mosaic on the right shows the overall top 100 payments, grouping employees in proportionate areas by occupation and deploying color to differentiate Pennsylvania from California, where most payments occurred. Because the two largest groups span the top, readers can quickly see which occupations account for the most payments. Created digitally for online access, this mosaic still embodies the form and function of its ancestors—in the atlases and albums and in earlier innovations, perhaps as far back as Humboldt's designs two centuries earlier.

Contemporary technology has not only fostered a revival of the mosaic genre but also enabled a significant reinvention. For example, the mosaics that appear in Michael Friendly, Matthew Sigal, and Derek Harnanansingh's chapter in this volume visualize a wealth of data from Friendly and Denis's Milestones timeline (see Plate 37). The mosaic plot on the left of Plate 37 shows multiple categories of historical examples: three groups of disciplines (physical, human, and math/statistics) on the vertical axis and seven time zones on the horizontal axis. The resulting rectangles (tiles) show the relative number of graphical innovations in each combination of categories (discipline and era), and color indicates more (blue) or fewer (red) than expected occurrences. In these ways, the plot enables readers to identify the shifting historical patterns across disciplinary groups, as each takes its turn dominating the Milestones timeline.

Mosaics have flourished with the geometrical flexibility offered by digital technology and data visualization. Software like MANET (Hofmann and Theus), Mondrian (Theus), JMP, and R enable designers to construct mosaic plots from tables with multiple categories.[8] Today one of the most popular forms of digital mosaics is the treemap—a collage of rectangles arranged in progressively smaller

---

[8]   For example, digital mosaics visualize the Titanic passengers according to class, sex, age, and survival status (see Hofmann; Spence, 47–9; Theus and Lauer, 396–8; Wilkinson 342–3).

areas. The Stockcharts.com treemap in Plate 31 displays hundreds of rectilinear areas representing companies in the S & P stock index, with the size of each rectangle proportionate to the company's value. The treemap organizes the stocks into investment groups (for example, cyclicals, materials, and utilities across the top), with the gradient shading of each stock indicating its positive (green) or negative (red) change in value over two days. By selecting the menu in the upper left, users can choose different market variables, by moving the slider at the bottom they can review previous days, and by mousing over a company, they can access more information. Although the Stockcharts.com treemap is more data rich than most of its nineteenth-century ancestors, it performs similar rhetorical functions. Visualizing immediate and past market performance—revealed by the green and red gradients—the chart prompts users to assess what happened (forensic), plan future investments (deliberative), and celebrate gains and lament losses (epideictic). This kind of rhetorical resilience and suppleness enabled the mosaic genre to continue its long evolutionary march into the digital world.

Conclusion: Telling Genre Stories from Many Angles

Mosaics tell a long and complicated genre story that has many different angles—formal, functional, social/cultural, and technological—all of which give insights into understanding how these charts have evolved within the communities that shaped and sustained them and for which they afforded a powerful cultural lens. As we've seen with area charts generally and mosaics in particular, the process of genre formation (and transformation) can occur over long stretches of time, with history providing a window through which to observe that process as a genre evolves, stabilizes, mutates, declines, and regenerates. This story is not confined to mosaics but writ large, with any given data display genre interweaving these factors in its own way as it gains currency and conventional status.

Today this process has accelerated in digital form, particularly with interactive displays, fostering a torrent of hybridization and invention that extends and challenges historical genres and that both delights and bewilders contemporary audiences. These new developments underscore enduring principles epitomized by mosaics: that genres are vulnerable and in constant flux, but also resilient and deceptively stable. Although varieties of mosaics thrive in today's graphically dynamic environment, still uncharted are the long-term transformative effects that digital technology might have on the evolution of other genres of data display.

Acknowledgment

I am grateful to my colleague Heike Hofmann for sharing her insights about mosaics and data design during the early development of this chapter.

Chapter 9

The Twentieth-Century Computer Graphics Revolution in Statistics

Dianne Cook

There are two types of data, imperfect data and no data. If you're going to wait around for perfect data, you are going to wait around forever.

James W. Pennebaker[1]

Many statistical methods depend on the satisfaction of strict assumptions and prior probabilities in order to test hypotheses formulated in advance, but plotting data comes with few restrictions. Plots of data can be made of even the most imperfect data. That doesn't mean that insight is always guaranteed, just that data that might trip up conventional statistics can be tackled with pictures. Statistical graphics in a sense can help to generate new hypotheses and make great leaps in our fundamental understanding of the world. This is the power of the visible numbers. This is what this era of research facilitated, the tools to explore numbers.

Interactive statistical graphics, which takes plotting numbers a step further, were born in the excitement of the computer graphics revolution. The American Statistical Association Statistical Graphics Video Library documents much of the work in this new golden age, communicating to future generations the spirit of the experiments (see Table 9.1).

Interacting with Multivariate Data

In 1970 Jih-Jie Chang, a female researcher at AT&T Bell Labs, programmed the first high-dimensional rotation of a point cloud (see Figure 9.1). A 2-D grid of numbers is embedded in high dimension, and rotation of 3-D subspaces occurs until the ordered grid is revealed. This is not data. It is merely a grid of numbers, but the display helps us understand how the rotation works. Enabling the user to interact with the graphic device to rotate the data, this is the first *interactive* statistical graphic ever recorded. Prior to this, researchers had only filmed animations illustrating an algorithm; for example, Joe Kruskal's 1962 video of

---

[1]   See Hillel Italie's "Study reveals the rise of feminine pronouns." http://www.independent. co.uk/arts-entertainment/books/news/study-reveals-rise-of-feminine-pronouns-8030574.html/.

Table 9.1   Online movies available from the American Statistical
Association's Statistical Graphics Video Library

Author(s)	Title	Key descriptors
J.-J. Chang	Real-time Rotation	Dynamic graphics, rotations
J.B. Kruskal	Multidimensional Scaling	Animation, rotations
J.W. Tukey, J.H. Friedman, M.A. Fisherkeller	Prim-9	Dynamic graphics, interactive graphics, rotations, brushing, identification
R.A. Becker and R. McGill	Dynamic Data Displays	Dynamic graphics, rotations, transformations, model fitting
J.A. McDonald and S. Willis	The Grand Tour in Remote Sensing	Dynamic graphics, rotations, tours
J.A. McDonald	Antelope: Data Analysis with Object-Oriented Programming and Constraints	Dynamic graphics, interactive graphics, rotations, tours, brushing, identification
A. Buja and P.A. Tukey	Dataviewer: A Program for Looking at Data in Several Dimensions	Dynamic graphics, interactive graphics, rotations, tours, brushing, identification
R.A. Becker, W.S. Cleveland and G. Weil	Brushing and Rotation on an Iris	Dynamic graphics, interactive graphics, rotations, brushing
F.W. Young and P. Rheingans	Visualizing Multivariate Structure with VISUALS/Pxpl	Dynamic graphics, interactive graphics, rotations, brushing
D.F. Swayne, D. Cook and A. Buja	XGobi: Dynamic Graphics for Data Analysis	Dynamic graphics, interactive graphics, rotations, tours, brushing, identification
D. Cook, A. Buja, J. Cabrera and D.F. Swayne	Grand Tour and Project Pursuit	Dynamic graphics, rotations, tours
R.A. Becker and R. McGill	Brushing in a Scatterplot Matrix	Interactive graphics, brushing
C. Hurley and R.W. Oldford	Higher Hierarchical Views of Statistical Objects	Interactive graphics, linked plots, brushing
M. Koschat and D.F. Swayne	Visualizing Panel Data	Interactive graphics, brushing, identification

multidimensional scaling shows how a low-dimensional representation preserves high-dimensional inter-point distances.

The difference between an interactive graphic and an animation is huge. An animation requires that the viewer is passive and simply watches a parade of pictures. An interactive graphic, however, requires that the user actively engage in the graphic to control, change, query, or re-focus the plot. In Jih-Jie Chang's implementation, the dimension of the data is larger than the display space. A 2-D shadow of the 5-D data is shown to the user, and tools allow the user to change the viewing angle. Figure 9.2 illustrates how the rotation is achieved using a simpler object, a labeled 3-D cube. The top row shows the vertices and their labels, and the bottom row shows the full wire-frame diagram, along with a representation of the 3-D axes. The wire-frame cube shows how what seem like random ordered labels can be rotated so that they are in linear order from left to right. Jih-Jie Chang's rotation of a 5-D object is more difficult, but it is achieved the same way, by multiplying the 5-D axes by a series of 5-D rotation matrices that perform sequential 2-D rotations.[2]

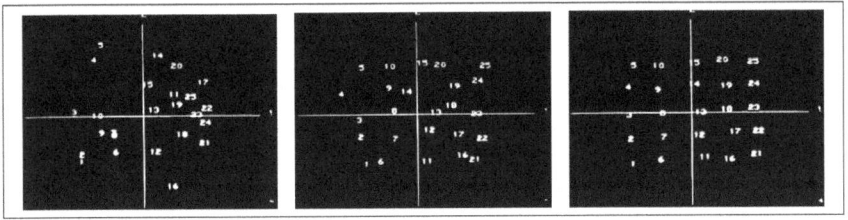

Figure 9.1   Manually rotating a 5-D point cloud to find an embedded 2-D grid pattern (http://stat-graphics.org/movies/rotation.html)

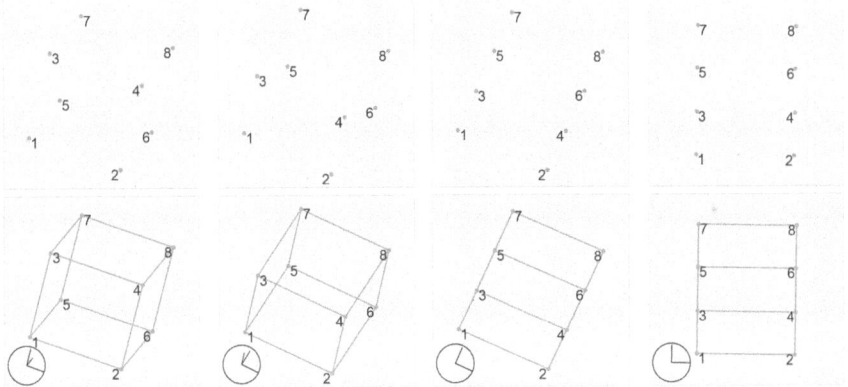

Figure 9.2   Illustration of the rotation conducted in Jih-Jie Chang's work, using a labeled 3-D cube

PRIM-9—Picturing, Rotation, Isolation, and Masking up to nine dimensions—broke barriers as the first full service system for exploring multivariate data (Fisherkeller, Friedman, and Tukey). *Picturing* means plotting the data. *Rotation* allows the exploration of high-dimensional space, to obtain alternative views of data. *Isolation* is the pre-cursor to brushing, which allows the user to focus in on subsets of the data, and *Masking* enables the removal of subsets

---

[2]   More details on this sort of methodology can be found in Buache and Asimov; Buja et al.; Dianne Cook et al.; or Cook and Swayne.

to help with the focus. PRIM-9 was driven by applications in physics from the Stanford Linear Accelerator.

Examples from the video show how the system was used to explore for structure in data. This work has been described as finding needles in the haystack of particle physics data, small signals in a cloud of noise, something that still underlies much of the analysis of physics data today. PRIM-9 was part of John Tukey's body of research in exploratory data analysis that foreshadowed today's data mining research. Tukey was a major engine of change in statistics research. Along with his experiments in interactive graphics, he produced many new pencil and paper methods for plotting data and for making quick calculations on numbers. Other statistics groups at the time were inspired to experiment with developing similar technology, named PRIM-ETH and PRIM-H.

The vocabulary introduced for interactive graphics by the PRIM-9 researchers was particularly vivid. Anyone familiar with Tukey will know that this was his norm, to coin colorful new words as he introduced new methods. His vocabulary for interactive and dynamic graphics, though, has been largely replaced by different terms now commonly used by statisticians, as described in Jürgen Symanzik. Below are some of these terms now in common use.

Brushing and Painting

What PRIM-9 referred to as masking and isolation is more commonly referred to as brushing, or painting, in statistical graphics today. Both mean to select sub-regions of the data in the display, using a geometric shape called a brush, and to change the characteristics such as color, shape, and size. This enables the user to focus on the subset of the data, and examine its distribution in relation to the rest of the data. The distinction between brushing and painting is slight. Brushing is a transient, fleeting action, and as the brush moves along, the graphical elements revert to their previous characteristics. Painting refers to a more persistent state in that the graphical elements once changed in appearance retain their appearance as the brush moves further. We typically think of color, shape, and size as the appearance, but the brush could also act like an eraser and remove data elements, or more complexly, make a complement action, where all but the selected group, have their appearances changed. These actions support what is called "drill down" in the visual data mining community.

Identification

Identification refers to how a system writes the labels of the data elements as the user points to them with the cursor. For example, if we are looking at real estate prices, the labels might be the street addresses, or if looking at a gene expression network, they might be GO annotations. If they were all drawn from the outset, labels would obscure the plotted data, in most displays. The identification enables labels to pop up and down interactively, as the user needs them. In GGobi, another early visualization application, labels could also, upon the user's request, be made *sticky* so that they remained on the plot like sticky notes (Swayne et al.).

Linking

If a user brushes in one plot, how do other plots respond? *Linking* refers to the way elements of one plot are connected with elements of another plot when multiple plots are visible. Linking was not a feature for PRIM-9, because linking is used when there are multiple plots visible, a style that emerged a decade beyond PRIM-9's days. Technically, linking is truly a property of the data being displayed, which determines the response of the other plots. For example, suppose multiple scatterplots are generated from different variables in a data table, and points in each plot correspond to rows of the table. Brushing a subset of points in one scatterplot corresponds to selecting rows of the data table, which determines which points in the other plots are highlighted accordingly. Linking can take many forms—one-to-one, one-to-many, many-to-many—and can be achieved by unique keys or categorical variables in the data. Carol Newton coined the term *linking* in 1978; the term was extended and defined in 1987 by J.A. McDonald and by Richard Becker and William Cleveland.

Tours

What PRIM-9 calls rotation is now more commonly called *tours*. Tours provide sequences of low-dimensional projections of high-dimensional data. The math behind tours is daunting, but the essence is simple. The purpose is to examine the distribution of data values in the high-dimensional space by inferring from many low-dimensional views. One might think of tours as looking at the shadows an object makes in the presence of a light source to guess what the object itself is. In PRIM-9 these were achieved by making a set of intermediate rotations in 3-D subspaces in order to navigate through high-dimensions. The development of tours allowed the reduction of visual distraction by incorporating geodesic interpolations from one low-dimensional projection plane to another (Asimov). The methods were used and extended by Andreas Buja and Daniel Asimov in 1986; McDonald and Willis in 1987; Edward Wegman in 1991; John Salch and David Scott in 1997; Buja et al. in 2005; and Dianne Cook et al. in 2007 are overview papers describing in depth the research developments on tours.

General Developments

The PRIM-9 video is piece of theatre—staged, scripted, and substantive. Subsequent videos document key developments in interactive statistical graphics research. Full systems developments, akin to PRIM-9, appeaer in the videos on DINDE, Antelope, Dataviewer, VISUALS, and XGobi (see, respectively Oldford and Peters; McDonald; Buja and Tukey; Young and Rheingans; and Swayne et al.). These mark the desire of statisticians to provide general systems that can tackle all sorts of high-dimensional data to explore the distribution of values and prepare for modeling. Missing from the collection are videos of the systems XLispStat, REGARD, QUAIL, DataDesk, MANET, Mondrian, and GGobi, which all achieved some popular support and user bases in subsequent years (see, respectively, Tierney; Unwin et al.; Oldford; Velleman; Unwin; Mondrian; and Swayne et al.). Methods and user interfaces pioneered for exploring data in these systems can often be seen in today's software systems.

Other videos in the library document the emergence of specific methods. One milestone can be seen in "Brushing a Scatterplot Matrix" (Becker and McGill). A scatterplot matrix is a layout of all pairs of plots that mirrors the arrangement in a correlation matrix. Diagonal plots may display a single variable, labels, or summary statistics. Plate 32 shows an example, made using GGobi, of housing sales data from Ames, Iowa. A sample of the data can be seen in Table 9.2. The diagonal plots display dot densities of the single variables, including sale price, number of bedrooms, bathrooms, lot area, living area, and garage area. Properties with high sales prices are selected by the brush, in the plot of sales price against bedrooms, and colored orange. These houses are then marked in the other displays.

The viewer's task is to examine the distribution of this subset in relation to the distribution of values of the remaining points. From a statistics perspective, we are comparing the 2-D marginals of the joint distribution of the subset against that of the full data. (A 2-D marginal simply means two of the variables, or two columns from the data table. The joint distribution explains the shape and relationship between all of the variables—that is, all columns of the data table.) What can we see? The highest priced properties come with a medium range of bedrooms and bathrooms. They have relatively large living areas, but lot area covers a spectrum of values that does not include the two properties with extremely large values. They also have relatively large garage areas. What is a little surprising from this inquiry is that these are not the houses that have the highest number of bedrooms and bathrooms. Actually, a few houses in the database have more bedrooms and baths but cost less. The high sales price corresponds to larger living and garage area. This illustrates the basic purpose of this type of chart and interaction—to examine the pairwise associations, and with brushing, the association in selected subsets. With this chart, the data analyst gains insight into the nature of the multivariate joint distribution.

The dashed rectangle in the SalePrice versus Bedrooms plot of Plate 32 is the brush. This is the tool by which the user interacts with the plot to select

Table 9.2   Top ten rows of the Ames housing data used in Plate 32 to illustrate linked brushing

Row No	SalePrice	Bedrooms	Baths	LotArea	LivingArea	GarageArea
1	215000	3	1	31770	1656	528
2	124500	2	1	13008	882	502
3	105000	2	1	11622	896	730
4	172000	3	1	14267	1329	312
5	176500	3	1	11029	1414	601
6	157000	4	2	10200	1434	528
7	244000	3	2	11620	2110	522
8	237500	4	2	12925	2117	550
9	206900	4	2	11075	2112	576
10	345000	4	2	13860	2704	538

graphical elements. Statistical graphics software usually provides a rectangular brush, which allows for a quick computation of the cases under the brush. The mechanism of linked brushing in the scatterplot works like this: the user employs the brush to select points in one plot, which are identified by row number, and these row numbers are used to identify the points to highlight in the other scatterplots. For example, in the plot of SalePrice versus Bedrooms one of the observations selected is row 10 in Table 9.2. The software can then easily look up the point for this row in all of the other plots and plot it accordingly in the other displays.

Brushing in a scatterplot matrix has progressed in a multi-window computing environment for multiple linked plots. Plate 33 shows an example of this technique to explore longitudinal data on wages of high school dropouts, using data from Judith Singer and John Willett. Almost 1,000 boys are tracked for as long as 12 years in the workforce, along with several demographic variables. This data is a lot to digest. The conventional statistical approach is to build a linear model with wages as the response, experience as the independent variable, and the demographic variables as covariates. It would also be usual to incorporate random effects for the individuals to quantify the variation from subject to subject. This modeling results in understanding the overall relationships, but these are actually somewhat simplistic: education improves wages, and race matters. The effect of education and race is miniscule in comparison to the variation in the data.

In Plate 33, plots of statistics calculated on longitudinal measurements are linked to a plot of the profiles of wages against workforce experience. The analyst brushes on diagnostics to reveal various characteristics in individual wages: an individual with a short but increasing wage (top), an individual with a depressingly consistent drop in wages (middle), and a person with a volatile wage experience is discovered (bottom).

This brushed linking reveals the real story: that the individual counts more than anything else: the variation from one subject to another completely dominates anything due to the demographic variables. The individual differences are staggering. Seeing this can be greatly enhanced by multiple linked plots. Plots of statistics computed for each subject—such as average wage, standard deviation of wage, and linear trend with experience—are linked with the plot of the longitudinal profiles. Plate 33 shows still frames of the linked brushing: plots on the left are the plots of the statistics, and those on the right show the longitudinal measurements. The horizontal axis of the longitudinal measurements plots ranges from 0 to 12 years of workforce experience, and the vertical axis ranges from $4 to $75 per hour on a log scale. The top row displays a plot of the slopes from linear model fits against the number of data points for each subject. One point here corresponds to one subject. Models were fit separately for each subject. The highlighted point corresponds to a subject that has a dramatic increase in wages in his or her short workforce experience history. The middle row shows slopes from a linear fit plotted horizontally and the associated p-value plotted vertically. The highlighted point corresponds to a subject that has a very steady, strong decline in wages over about six years. The bottom row shows slopes from a linear mixed effects model plotted against the standard deviation of the residuals from the fit. This point corresponds to a subject whose wages varied dramatically over about eight years.

This linked brushing technique has become very widely available in graphics software. Cook and Weisberg famously employed it in their regression diagnostics work that utilized Tierney's XLispStat. The DataDesk software (Velleman), which has been available since the early 1990s, also increased the use of interactive graphics for model diagnosis. Among psychologists, Vista (Young et al.), built with XLispStat, provided the same kind of linked model diagnostic plots. The Vista software was in common use in the mid-1990s. These systems were the first to integrate modeling and analysis with interactive graphics, something that has not yet been reproduced fully with the current technology.

Linking between multiple plots is achieved using a detect location, lookup record, and then broadcast approach. Figure 9.3 illustrates how this process works. Software systems require an event loop which provides a listener-to-user interaction with the computer—for example, mouse click, mouse drag, mouse release, key press, and key release. Functions associated with these listeners tell the computer what to do when an event occurs. For interactive graphics, typically the listener is detecting mouse click and drag events. This invokes a look up of the data to find the records that have been selected in the focus plot. Once the software finds the record, it locates associated values, and sends a "check-yourself" signal to other plots. If these plots contain the selected records, they highlight the corresponding graphical elements.

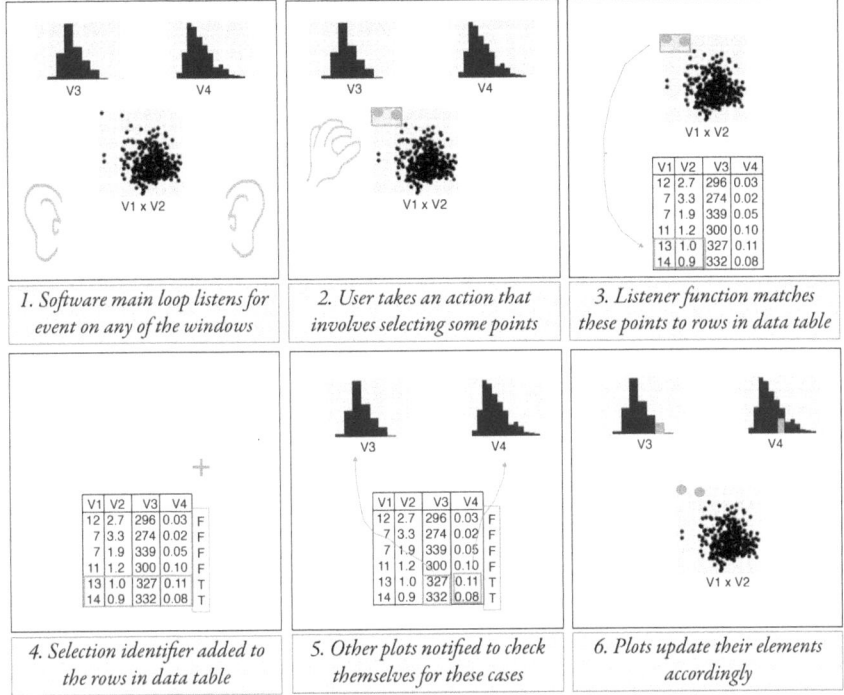

Figure 9.3   A cartoon illustrating the mechanism of linking between multiple plots

Applications

Learning how to apply new methods to analyze data is an important component of the integration of new technology into general practice. Several videos in the library document how the new interactive graphical methods can be used in practice: the incidence of the disease mumps, manufacturing process data, and display of airline traffic (see videos by Eddy and Mockus and by Eddy and Oue). These videos are examples of building animations of data, experimenting with spatial smoothing and temporal steps to obtain a smooth dynamic rendering of the data, reminding viewers of the dynamic process that gives the data context. Earlier in the video library, there are examples of interactive graphics for studying ozone, in the neighborhood of the researchers' living environment and of the exploration of environmental characteristics based on remotely sensed data (Becker et al.; McDonald and Willis). Applications drive methods development in statistics, and these videos show examples of that interplay between the practical and the abstract.

Beyond Flatland

Propelling research in interactive graphics is the need to break free from the constraints of the 2-D surface of a page to display the many dimensions of data. Data is multidimensional, in the abstract sense of Euclidean or Cartesian coordinates. In early years of school, children learn to plot pairs of numbers in a number plane. When it comes to working with data later in life, however, there are triples of numbers, or quadruples, or n-tuples, that need to be displayed. Three of the first four videos in the library operate on high-dimensional spaces to show sequences of low-dimensional views using linear combinations of the variable axes. These are mind-boggling ideas to the uninitiated, enabling the analyst to take glimpses of a high-dimensional space by looking at the shadows of the objects living in that world. It evokes Edwin Abbott's 1884 fictional story *Flatland* of 1-D and 2-D characters navigating a 3-D world. The need to show multi-dimensionality underlies tour methods: they enable the viewer to take a glimpse of high-dimensional data by examining the low-dimensional projections. See the Geometric Data web site for data, images, and movies of many standard high-dimensional shapes, as would be viewed using a tour (Schloerke).

Figure 9.4 shows a wire-frame 5-simplex being viewed using a tour, with 12 snapshots of different projections. This shape may help the reader better understand the methodology. A 5-simplex is an object that lives in 6-D space, but occupies a dimension less than the full space. It is created from six equal-length vectors, with all but one coordinate equal to zero. A 0-simplex is a point. A 1-simplex is a line segment. A 2-simplex is an equilateral triangle. A 3-simplex is a tetrahedron. In the 2-D projections the 5-simplex can appear as a triangle (top left), quadrilateral, pentagon or hexagon (bottom right). Simplexes also have a purpose in the analysis of high-dimensional compositional data. This type of data has a dependency—the proportions or percentages that are constrained by their sum. Two common examples are soil composition and opinion surveys. Soil samples are often described as a proportion of clay, sand, or silt. Opinion surveys often record the percentage of respondents answering "excellent," "very good," "average," "below average," and "poor." These types of data have values that are bounded inside a simplex. Issues of perspective and light source or shading, which are important for 3-D graphics, are irrelevant and even confusing for higher dimensions. However, because tours use projections of high-dimensional objects that are shadows, using some transparency—for example, by using wire-frames or point clouds rather than solid shadows—can help to reveal internal structure in the data.

Summary

Several names appear repeatedly in the video library, most notably that of John W. Tukey, the inspiring pioneer of interactive graphics for exploratory data

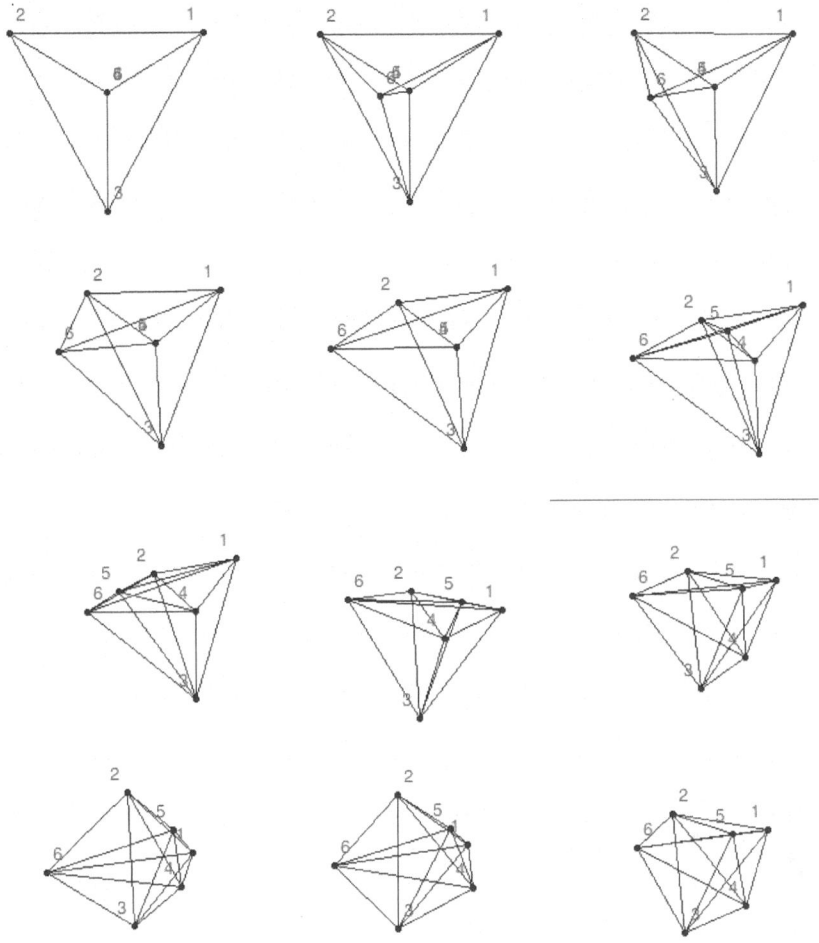

Figure 9.4   Projections of a wire-frame 5-simplex, taken from viewing the shape using a tour of 2-D projections. Views range from a triangle to a pentagon in the convex hull

analysis. Though several researchers preceded Tukey with focused experiments, his was the seminal work that provided a major impetus for others. William S. Cleveland dabbled in interactive graphics early, and he continued to work on (static) graphics developing the trellis plot ideas that form the backbone of lattice graphics, and, to a limited extent, the facetted plots of ggplot2 (see Sarkar; see Wickham, "ggplot2"). He also pioneered work in cognitive perception of statistical charts, conducting the first experiments that resulted in a hierarchy of mappings of data to graphical elements: position along a common scale, along a non-aligned scale, length/direction/angle, area, volume/curvature, and shading/color. Andreas Buja started on interactive graphics with Wolfgang Huber on PRIM-ETH, then moved on to Dataviewer, followed by XGobi and GGobi. He has continued graphics development in many ways, through applications, by developing a taxonomy of interactive tasks, and making connections between classical statistics and exploratory graphics. Many other researchers, absent from or with only fleeting appearances in this library, played important roles in the structural developments of the field.

This period was a fertile playground sparking ideas that are in common use today. Many of our common software systems employ the methods pioneered during these years, and even some aspects of the interface. Linked brushing and identification are particularly common. Some methods are less commonly available, like tours, and some have never truly emerged, such as complement brushing, what Tukey called masking. The major media outlets such as the *New York Times* (NYT), Cable News Network (CNN), and all of the major newspapers and television services commonly display real-time data analysis using techniques initiated during these years. This is especially true during events like national elections or sports broadcasts where you will see plots that have data animated over time, with linked brushing, and identification, and logical zooming. The 1970s and 1980s were exciting times for interactive graphics research.

Chapter 10

The Milestones Project: A Database for the History of Data Visualization

Michael Friendly, Matthew Sigal, and Derek Harnanansingh[1]

If you would understand anything, observe its beginning and its development

Aristotle

Introduction

Questions regarding the history of data visualization are (or at least should be) of great importance to historians of science, to current developers of graphical methods for statistical analysis and the related info-vis community, as well those just interested in the history of ideas. In the history of science, diagrams, graphs, maps, and other visualizations have often played important roles in discoveries that arguably might not have been achieved otherwise.[2] At the same time, in the fields of statistical graphics and information visualization, developers often create "new" methods without any appreciation that they have deep roots in the past.[3]

These two perspectives provided the motivation for the development of the Milestones Project. This stemmed from the fact that historical accounts of events, ideas, and techniques that relate inter alia to modern data visualization were fragmented and scattered across a wide number of fields.[4] When this work

[1] This work was supported by Grant OGP0138748 from the National Sciences and Engineering Research Council of Canada to Michael Friendly. We are grateful to Dan Denis, Stephen Stigler, and Ben Shneiderman for constructive comments on this chapter.

[2] Some salient examples are: Francis Galton's 1861 discovery of anti-cyclonic movement of wind around low-pressure areas from contour maps; Edward Maunder's "butterfly diagram" of the variation of sunspots over time leading to the discovery of the "Maunder minimum," from 1645 to 1715; and Henry Moselely's 1913 discovery of the concept of atomic number, based largely on graphical analysis (a plot of serial numbers of the elements versus square-root of frequencies from their X-ray spectra).

[3] For example, mosaic displays for frequency tables were thought to have been invented by Hartigan and Kleiner and extended to show the pattern of residuals in loglinear models by Friendly ("Mosaic Displays"). But it turns out that the essential idea behind this area-based display goes back to Georg von Mayr in 1877 (Friendly, "A Brief History of the Mosaic Display").

[4] Among these are general histories in the fields of probability (Hald), statistics (Pearson; Porter; Stigler, *History of Statistics*), astronomy (Riddell), cartography (Wallis and Robinson). More specialized accounts focus on the early history of graphic recording

began in the mid-1990s, there were no accounts or resources that spanned the entire development of visual thinking and the visual representation of data across different disciplines and perspectives. The Milestones Project began simply as an attempt to collate these diverse contributions into a single, comprehensive listing, organized chronologically, that contained representative images, references to original sources, and links to further discussion—a source for "one-stop shopping" on the history of data visualization.

In the second section of our chapter (The Milestones Project), we describe the evolution and structure of the Milestones Project. The third section of our chapter (Visualizing Time and History) presents some historical and modern approaches to one self-referential question: how can data visualization be applied to its own history? The fourth section (Using the Milestones Project for Statistical Historiography) introduces another self-referential topic we call statistical historiography, which entails the use of statistical and graphical methods for the analysis and understanding of historical innovations, developments, and trends. But first we give some brief vignettes of historical topics and questions for which the Milestones Project has proved invaluable in our own research.

The First Statistical Graph

In the history of statistical graphics, as in other artful sciences, a number of inventions and developments can be considered "firsts" in these fields (Friendly, "A Brief History of Data Visualization"). The catalog of the Milestones Project lists 70 events that can be considered to be the initial use or statement of an idea, method, or technique that is now commonplace, but there is probably no question more fundamental than that of the first visual representation of statistical data (Friendly and Denis, *Milestones*).

In Friendly, Valero-Mora, and Ulargui we argue that the one-dimensional line graph shown in Plate 34 by Michael Florent van Langen should be accorded this honour. The graph shows 12 estimates of the distance in longitude between Toledo and Rome, overlaid on a modern map. Toledo is located at lat/long (+39.86°N, −4.03°W), Rome is located at (+41.89°N, +12.5°W), both shown by markers (stars) on the map. This image makes clear what van Langren wished to communicate: the wide variability of the estimates, but also how far the estimates were biased. Van Langren used this to demonstrate that these estimates were all subject to large errors and to propose to King Phillip IV of Spain that only he had a sufficiently precise method for the determination of longitude for navigation at sea. The telling of van Langren's story not only turned out to involve astronomy, archival research, the history of patronage in the seventeenth century, and even an unsolved problem

---

(Hoff and Geddes, "Graphic Recording," "Beginnings"), statistical graphs (Funkhouser, "Note," "Historical Development"; Royston; Tilling), fitting equations to empirical data (Farebrother), cartography (Friis; Kruskal, "Visions of Maps"), thematic mapping (Friendly and Palsky; Palsky, *Des Chiffres*; Robinson, *Early Thematic Mapping*), and so forth.

of cryptography, but also serves as an example of statistical historiography. The Milestones Project provided the infrastructure for this research: through the use of a time-based, cross-referenced catalog of images, references, and links to related work, van Langren's tale was able to be studied and reported upon.

Who Invented the Scatterplot?

Although there are earlier precursors, the main graphical methods used today—pie charts, line graphs, and bar charts—are generally attributed to William Playfair in works around the beginning of the nineteenth century. All of these are essentially univariate displays of some aspect of a single variable. A logical next step would be to invent a method to reveal the relationship between two variables—what we now know as the scatterplot. By 1886, Francis Galton had utilized this truly bivariate display, which led to the discovery of correlation and regression, and ultimately to much of present multivariate statistics. However, he was not the first to use this graphical technique, and it is surprising that no one is widely credited with its invention.

In "The early origins and development of the scatterplot," we delved into this mystery. This involved tracing the early origins of ideas related to the scatterplot, which led to two compelling narratives: how, in Playfair's time, it was nearly impossible to think about and visualize bivariate relationships; and, later, how the scatterplot was essential for Galton's visual insights that would lead to the rise of modern statistics and graphics. The resources available in the Milestones Project allowed us to focus upon the events in this period and attribute the essential ideas of the scatterplot to J.F.W. Herschel and two 1832 papers. .

The Golden Age of Statistical Graphics

In our initial web presentation of the Milestones Project, it proved convenient to sub-divide the history of data visualization into epochs, each of which turned out to be describable by coherent themes. As we illustrate later, one period turned out to be particularly noteworthy, both for the sheer number of contributions, and for the beauty and elegance of their execution. We call this period, from roughly 1850 to 1900 (±10), the Golden Age of statistical graphics (Friendly, "Golden Age").

Figure 10.1 shows the time distribution of the 260 significant events that had been included in the Milestones Project database by 2007, demarcated by the labels we used for epochs. The density estimate is based on $n = 260$ significant events in the history of data visualization from 1500–present. The developments in the highlighted period, from roughly 1840 to 1910, comprise the Golden Age of statistical graphics. We traced the origin of this period in terms of the infrastructure required to produce such an explosive growth of contributions to data visualization, and found three primary sources: the systematic data collection by state agencies, the rise in popularity of statistical and visual thinking, and the enabling developments of technological innovations (see Friendly, "Golden Age").

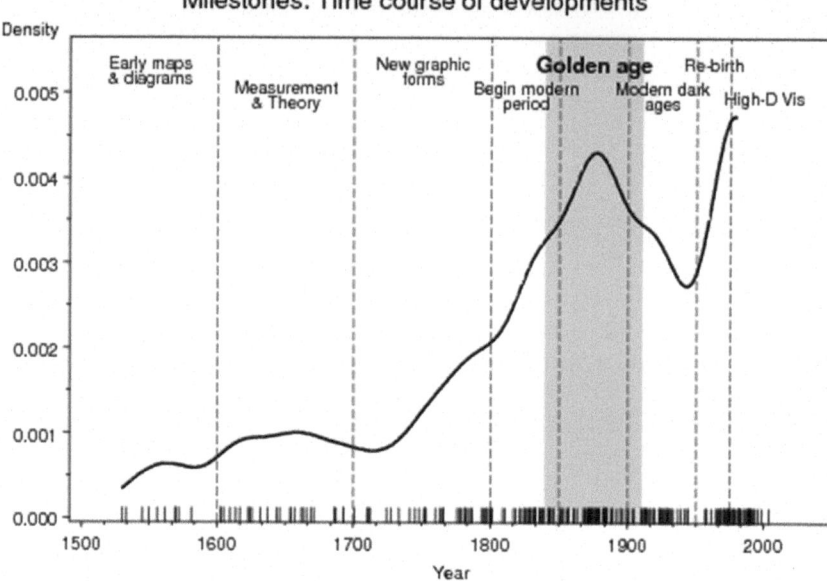

Figure 10.1   The time distribution of events considered milestones in the history
of data visualization, shown by a rug plot and density estimate

The Milestones Project

> Direction is more important than speed. We are so busy looking at our
> speedometers that we forget the milestone. (Anonymous)

An early overview of the content and aims of the Milestones Project appeared
in Friendly's "Milestones in the history of data visualization." Here we update
that description and provide a few technical details on some problems that were
encountered in attempting to make the history of data visualization convenient for
collecting, browsing, searching, and analysis.

Origin, Structure, and Evolution

The initial step in portraying the history of data visualization was to create a
simple chronological listing of milestone items with capsule descriptions;
bibliographic references; markers for date, person, and place; and links to
portraits, images, related sources, and more detailed commentaries. The initial
database contained the 105 developments listed by Beniger and Robyn in 1978,
then incorporated additional records from Hankins, Tufte (see *Visual Display of
Quantitative Information, Envisioning Information,* and *Visual Explanations*),
and Heiser, among others.

This database began as a single LaTeX file (with markup tags for all relevant bits of information), used to produce a hyper-linked PDF document. A variety of software tools (perl scripts, Unix utilities) allowed us to turn this single source directly into the web version.[5] Other custom software tools allowed us to add new milestones items from text files using a template of tags (DATE:, AUTHOR:, WHAT:, REF:, IMG:, etc.) and to extract the information about milestones items, authors, images, and the like in a variety of forms (CSV, XML, JSON) that could be used as input for analyses and graphic displays. For example, Figure 10.1 was produced in SAS software by piping the output of a LaTeX to csv translator:

```
itemdb -o milestones.csv < milestones.tex | sas -i
milestones.csv mileyears.sas
```

It soon became apparent that such a text-based representation was inadequate. Updating the milestones data required that the LaTeX file be shared among several collaborators; milestone assets, such as images, web links, and references, were not easily accessible by others, which made collaboration cumbersome. Further, each update to the website required an inefficient number of steps of verification, re-building, and synchronization with the server, meaning the website was often out of date. Around 2005 we began to make the process more dynamic, convert the flat file into a relational database, create a Milestones administrative system, and completely redesign the Milestones website. Specifically, we wanted to facilitate contributions by any number of trusted collaborators via an easy-to-use web administration area, and allow for the dissemination of milestones data via an easy-to-browse public user interface.

Migrating the data to this format posed some challenges. First, the existing milestones data needed to be restructured logically and have redundancy minimized. To do this, we partitioned the data into its relevant entities: namely the milestone itself, and its descriptors, such as its aspect, author, subject, keywords, reference, and linked media items (such as images). The aspect, author, subject, keyword, and reference descriptors exist as a many-to-many relationship between it and the milestone. The main table (milestone) contains information regarding each of the items considered a milestone in the history of data visualization, linked to other tables (for example, reference, mediaitem) by unique (primary) keys. Other supporting tables (for example, milestone2aspect) provide for convenient lookups of descriptors of these milestones items (subject, aspect, keyword). For example, an aspect can belong to one or more milestones, and the milestone can belong to one or more aspects. Media items, on the other hand, can only belong to one milestone at a time, with multiple media items possible for a single milestone. Figure 10.2 illustrates these relationships.

---

[5]   The original web version was originally shown at http://www.math.yorku.ca/SCS/ Gallery/milestone/.

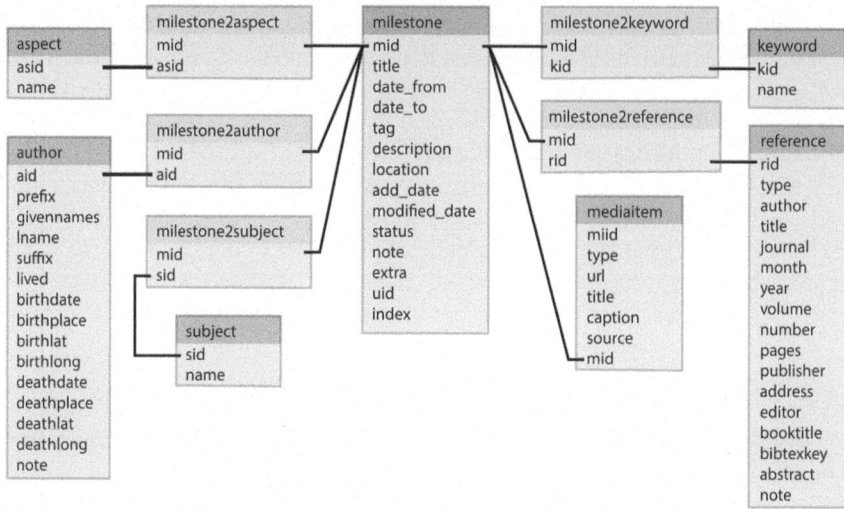

Figure 10.2   Simplified schema for the MySQL database for the
               Milestones Project

Normalizing the data in this way enabled us to free the database of modification anomalies and ensured that the database structure was scalable and could be extended with minimum modifications. Most importantly, it allows for future growth, and provides a query-neutral database model that could be used to power web presentation and customized indexed search (Codd). The last major benefit, which will be demonstrated in Section 3, is that this schema allows for any type of analysis of the Milestones data itself.

At present, the Milestones Project documents 288 contributions, with nearly 350 references, information on 336 authors, and 774 media items, made up of 371 images appearing online on the milestone site, and 403 hyperlinks to images and documents that are externally hosted. In addition, we maintain an offline image database comprising over 1,100 images collected from various sources. Over time, these too will be incorporated into the database.

User Interface

The second challenge related to how to display such a large amount of information in an easy-to-use interface that would provide overview, search, and details about these events in the history of data visualization. We decided to retain the time-based grouping of the milestones content by epochs (Pre-1600, 1600s, 1700s, etc.), each with a theme (for example, 1600–1699: Measurement and Theory) and descriptive text. The visual design of the interface adopts Ben Shneiderman's mantra: "Overview first, zoom and filter, then details on demand" (336). To do this, we added a timeline view of the milestones items displayed on the overview landing page (see Figure 10.3).

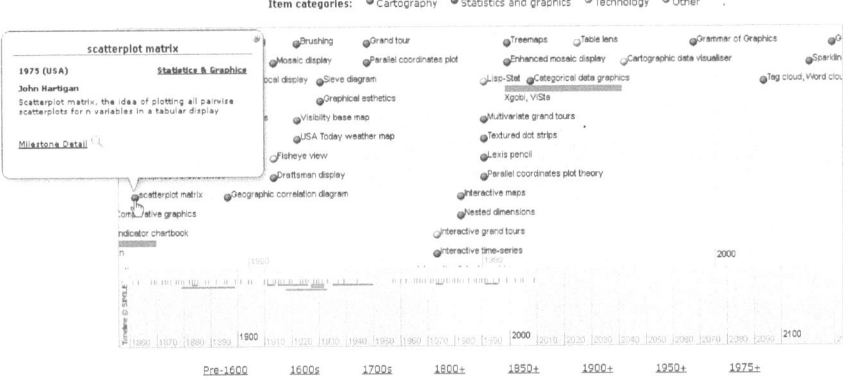

Figure 10.3   Timeline view of the Milestones Project

This timeline, based on the SIMILE Timeline Widget, allows multiple connected time bands, showing events at different resolutions. Each band can be separately panned by dragging left or right with the mouse pointer, scroll wheel, or keyboard arrow keys. In this view, the top panel shows a detailed view of the segment of history highlighted in the bottom panel, both of which can be separately scrolled. Items in the top panel show a brief text tag, color-coded by category. Clicking on an item in this panel brings up a small description, which is further linked to the details of the milestone item. The timeline view, although most obvious, is just one of several possibilities for a visual overview or interaction with the display of the milestones database. The software design of the site, using open-source tool kits, makes it relatively simple to add new ones. For example, the database can also be navigated via a list view (with drop down quick links), and in the section Milestone Authors: Geography we will illustrate how it can be explored using a map-based display.

Visualizing Time and History

What does history look like? How do you draw time? (Rosenberg and Grafton 10)

The questions in this quotation introduce an important topic in the history of data visualization: how can such a history be visualized? What methods might be called upon to detail the richness of its past?[6] Time provides an obvious dimension, but

---

[6]   Another recent book, *Visualizing Time* (Wills), discusses a range of modern graphical methods for visualizing time-based data.

what else could be included in a static display that might reveal a story previously hidden? What kinds of dynamic or interactive displays might fascinate and intrigue viewers?

An annotated visual gallery of some timeline designs and visual histories can be found in our Data Visualization Gallery. The topics covered include early visual histories, encyclopedic charts, special purpose charts, correlated histories showing events in one domain in the context of events in other areas, non-linear scales for time and space, as well as dynamic, interactive timelines. Here we present a few inventive selections from this scholarship.

The First Timelines, Reconsidered

Although there are earlier precursors, the first timelines of modern design—featuring a horizontal, linear axis for time, and vertical positions for place, theme, or category of events—were produced in the mid-1700s. Most notable of these prototypes were Jacques Barbeau-Doubourg's 1753 Carte Chronologique and Joseph Priestley's 1765 Chart of Biography. Priestley first published a small "Specimen" of this chart as a proof-of-concept, showing the lifespan of famous men in the years 600 BC to AD 1, classified as "statesmen" (from Solon to Augustus) and "men of learning" (from Thales to Ovid). Later that year, in 1765, Priestly published a detailed version that quickly became the most popular and influential timeline of the nineteenth century. The full graphic details the lifespans of more than 2,000 people from 1200 BC to AD 1750, classified by their areas of achievement (statesmen and warriors, mathematicians and physicians, artists and poets, and so on). Priestly's timeline charts can be seen on our Data Visualization Gallery, and we don't reproduce them here. Instead we show (see Figure 10.4) a re-design, in his style, of the lifespans of 79 authors from the Milestones database who were born in France or the United Kingdom between 1500 and 2000. Rosenberg and Grafton called Priestley's charts "masterpieces of visual economy" (117). Indeed, they were at the time. However, in his charts, the famous people were arranged haphazardly within category groups, so it is difficult to find specific individuals, and nearly impossible to uncover any trends, either over time or across categories.

In our version, authors are sorted by birth year within each country and the names are printed alternately at the year of birth and death. Authors are sorted within country by year of birth and labeled alternately at birth and death years, allowing better lookup and visual comparison. The result, which resembles a cumulative distribution plot, (a) allows easier visual lookup of names, (b) provides an overall "lifespan envelope," and, (c) highlights a few individuals who lived conspicuously longer or shorter lives (for example, Willam Jevons, James Maxwell, John Snow, and Phillipe Buache) than their contemporaries. Of course, to display lifespan directly requires a different kind of plot, but one that would not have been even thinkable by Priestley in 1765. We return to this question later on.

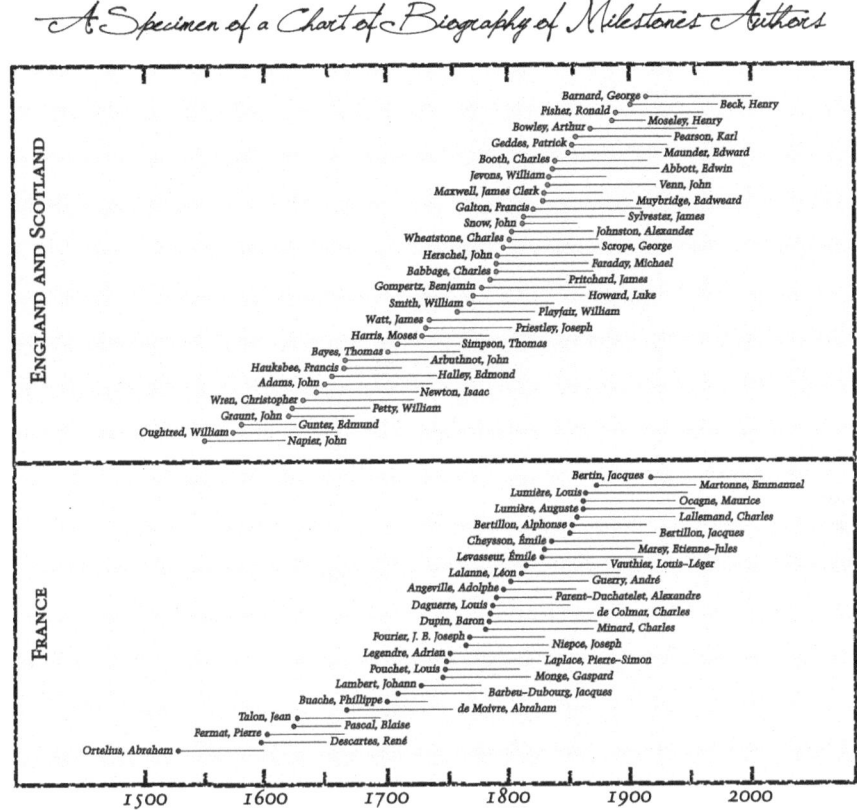

Figure 10.4   A modern re-design of Priestley's 1765 Chart of Biography, using information on authors in the Milestones database born in France or the United Kingdom

Universal Histories

In addition to unrivalled thematic maps and statistical diagrams, the Golden Age of statistical graphics also gave rise to a variety of novel attempts to visualize history in a comprehensive manner, combining parallel, intertwined time-flows, text, illustrations, maps, and other visual forms. Among the most impressive is the series of Synchronological Charts of Universal History produced by Sebastian Adams between 1871 and 1885. The 1881 version is 23 feet long and captures 5,885 years of history, from 4004 BC to AD 1881. Rosenberg and Grafton call it "nineteenth-century America's surpassing achievement in complexity and synthetic power" (172). Plate 35 shows the entire chart (note the increasing visual density towards the right), and Plate 36 shows a close-up of a small portion.[7] Adams used a linear scale for time, and so it is understandable why it took 23 linear feet to include all of recorded history.

Categorization and Non-linear Scales

Linear time scales have the advantage that they provide uniform resolution and detail across the entire time span, but events in time, or our interest in them, are rarely uniformly distributed. As exemplified by the Milestones Project, most visual histories are rather sparse at their beginning and very crowded at their end. Utilizing non-linear scales can allow resolution to vary smoothly across the range, providing greater detail in regions of interest, which are most often the recent past.[8]

Plate 36 is a proof-of-concept sketch for something that a graphic artist could use as a starting point for a chart of the history of data visualization. It uses the events from the Milestones Project, categorized by two correlated factors: subject area, in which the content has been categorized as dealing with human populations, physical properties of the world, or mathematics and statistics; and, the milestone's aspect or form, which has been categorized as dealing with cartography, graphs and diagrams, or technology. To provide greater resolution for more recent events, time (Year) is shown on a square-root scale, going backward from the year 2000. To provide greater resolution for more recent events, we have used a reverse square-root scale going backward from the year 2000. Specifically, "Year" on the horizontal time axis is actually plotted according to the formula $Year = 2*(25 - \sqrt{(2000 - Year)})$, giving the more pleasing result that the modern period 1800–2000 occupies about 60 percent of the scale, despite only comprising 40 percent of the range. This is conveyed visually by the spacing between tick marks on the X-axis.

Using the Milestones Project for Statistical Historiography

> Vision is the Art of seeing Things invisible. (Jonathan Swift)

Statistical Historiography

We use the term "statistical historiography" to refer to the use of statistical and graphical methods to explore, study, and describe historical problems and

---

[7]   The entire chart can be viewed in high-resolution at http://www.davidrumsey. com/blog/2012/3/28/timeline-maps/.

[8]   Of course, interactive graphics offer the possibility to vary resolution dynamically, by moving a "lens" across the display, as in a hyperbolic viewer.

questions.[9] This topic has a delightful self-referential quality when applied to the history of data visualization itself, since we have often found ourselves using modern methods of statistical analysis and graphics to study the development of ideas in this area. As in the Swift quotation above, one goal is to make previously hidden aspects of this history visible.

At the same time, our examination of some of the most impressive graphic works of the past sometimes left us awe-struck by their exquisite beauty and visual design.[10] On more than one occasion when looking at these elegant presentations, we wondered whether there wasn't something lost with the advent of modern software. While we can now analyze massive data sets, and generate a multitude of graphics with a simple mouse click, we still feel that designing a truly effective visual display of information requires thought and manual intervention.

For this reason, it is often quite instructive to attempt to re-create or even re-vision a graphic work from the past (Friendly, "Visions"). We can learn from this undertaking an appreciation for the insight and hard labor of our graphic heroes, and can sometimes better understand or improve on their designs by a process we call "understanding through reproduction," another facet of statistical historiography.

There is, of course, one principal requirement for statistical historiography: *data*. The Milestones Project database is the repository of all the information we have so far recorded, and modern database tools allow the possibility of simple or complex queries, limited only by the available information.[11] In related work, we have collected and disseminated data sets of historical interest on a variety of topics in statistics and data visualization, for instance via the R packages HistData and Guerry (Friendly, *HistData*; Friendly and Dray, "Guerry"). These can be considered another source for data, pictures, and stories related to statistical

---

[9]   As far as we know, the initial expression of this idea appeared in a paper by Rubin discussing various ways in which statistical methods could be applied to historical topics. These included the use of sampling methods to test historical theories, statistical distributions applied to historical data, and the use of time series graphs with smoothed curves to study historical trends. More recently, many examples of the application of these ideas to statistical topics can be found in Stigler (*History; Statistics*), as well as our own papers on the history of data visualization, cited inter alia.

[10]   Some examples are Charles Joseph Minard's famous depiction of Napoleon's March on Moscow (Friendly, "Visions and Re-Visions"), Francis Galton's detailed study of weather patterns in Europe (see: Friendly, "Golden Age"), and André-Michel Guerry's (Guerry, Plate 17) semi-graphic table depicting the relations of occurrence of crimes to a wide variety of social and demographic factors (see Friendly, "A.-M Guerry's").

[11]   It should be noted that, beyond the basics of recording milestones items, images and references, inputting the other meta-data (content and form categories, keywords, etc.) is highly labor-intensive. Thanks are due to many research assistants and graduate students who have and continue to work on the Milestones Project, including Dan Denis, Matt Dubins, Yvonne Lai, Avi Lipton, and Carolina Patryluk.

historiography, and understanding through reproduction. This is the essence of the motto on the datavis.ca website: looking back, going forward.

In the subsections below, we describe a few applications of these ideas using the Milestones Project database and case studies that arose from this work. There is an interesting interplay between such historical analyses and these data collections. Some studies called for us to find and incorporate new data sources, such as our paper on Guerry's moral statistics of France and the Guerry package, to which we added Angeville's extensive 1836 data on social and economic characteristics of France (Friendly, "Visualizing"). In other cases, our analyses suggested new or different ways to visualize historical data.

Milestone Authors: Lifespan

As noted earlier, we record information relevant to the contributors of milestones events in an author table in the database. For most of these individuals, Internet and biographical searches allowed us to determine the dates and places of their birth and death. One simple question that can be posed using this information is how long did these contributors live? As illustrated earlier in Figure 10.4, Joseph Priestley was the first to develop the idea of using a graphic representation to show the lifespan of famous men. His "charts of biography" did this in a particularly evocative form, representing each person by a line segment whose length was defined by the individual's lifespan and then grouped by occupational category.

These "timespan" charts tell an interesting story, but they do not provide an answer to the question of how long, in general, these individuals lived. However, with the author table from the Milestones Project, it is a simple matter to calculate lifespan and to obtain a direct answer to this query.

Figure 10.5 shows one display of this information, using a combined density plot and rug plot, similar to the one used in Figure 10.1. Individual observations are shown by a (jittered) rug plot, and the three extremes on each end are identified by name. The red vertical lines show the quartiles of the distribution. Several features of this plot deserve comment, and also invite further inquiry: most notable is that, by and large, milestones authors generally lived to a ripe old age—the median lifespan is 73.0, but the density plot peaks at around 79. This contrasts with a detailed study on famous people between 2400 BC and AD 1880 by David de la Croix and Omar Licandro. In their research, the typical lifespan fluctuated around a mean of 61 years for four millennia; it only gradually reached 69 toward the end of their sample. Such a discrepancy between the two studies might warrant further investigation—for instance, by classifying the individuals into occupational, locational, or otherwise more delineated groups, and looking for trends. Another interesting feature that becomes apparent in this graphic is the noticeable bump in the distribution around 45 years. This occurrence calls for some attempt at further explanation. We don't pursue this here, but again note that such graphs often suggest further analyses (breakdowns by region or time period), or cry out for the collection of more data.

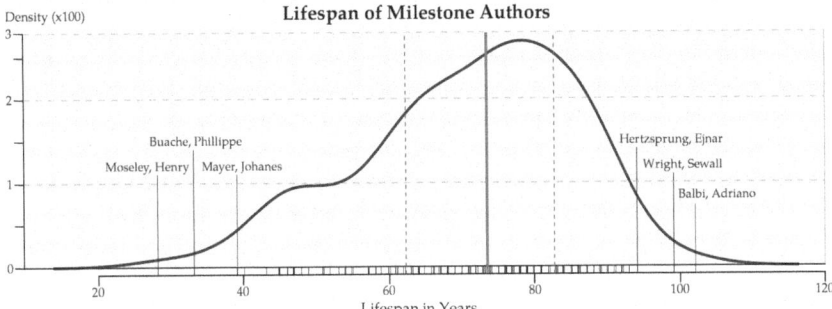

Figure 10.5   Density plot of the lifespan of the 172 authors in the Milestones
                Project database who were born after 1500 and for whom lifespan
                can be determined

Finally, although Figure 10.5 is just a summary graph, we have labeled a few
extreme observations on each end, which may relate to telling parts of the story
of the history of data visualization. Among these, Henry Moseley, who is known
for the discovery of atomic number from a graphical display, died the youngest,
as a consequence of serving in the British Army during the First World War. But,
we were surprised to see the noted and prolific French cartographer Phillippe
Buache, and the German physicist and astronomer Johann Tobias Mayer, show up
in positions two and three. On the other end, we were delighted to see that Adriano
Balbi, a Venetian geographer and early collaborator of André-Michel Guerry had
the longest lifespan, just exceeding the population geneticist, Sewall Wright, who
invented path analysis and the path diagram around 1920 (Balbi and Guerry).
By incorporating these details, the visualization is able to reveal narratives that
otherwise would have been concealed.

Milestone Authors: Geography

The Milestones Project website provides an initial page showing an interactive
timeline of the events in this history as a visual overview (Figure 10.1). A long-
term goal has been to provide other views of this history and other tools for
searching and exploring the database. With recent technological developments,
it became evident to us that one such method would be through the use of
geographical data.

So far, the primary geographic information we have encoded in the database
refers to the birth and death place of the milestone authors. This is an imperfect
representation, as these locations may not accurately represent the author's
primary residence. For instance, Charles Joseph Minard was born in Dijon, and
died in Bordeaux, but all of his work was done in Paris while he worked at the
École Nationale des Ponts et Chaussées. Nevertheless, a geographic view of the

available information is potentially useful. In this regard, we used the Google geocoding tools to provide latitude and longitude for the locations listed in the author table. Using this and the R package googleVis created by Gesmann and de Castillo, we developed the interactive map shown in Figure 10.6.

Like other instances of Google Maps, this graphic can be panned and zoomed using mouse controls. The place markers display tool tips when hovered over and, when clicked, link to a search page that details all of the Milestone items that are related to that author. This interesting visualization will soon be revealed on the Milestones Project website, with future work planned to incorporate other types of data in addition to the birth and death locations.

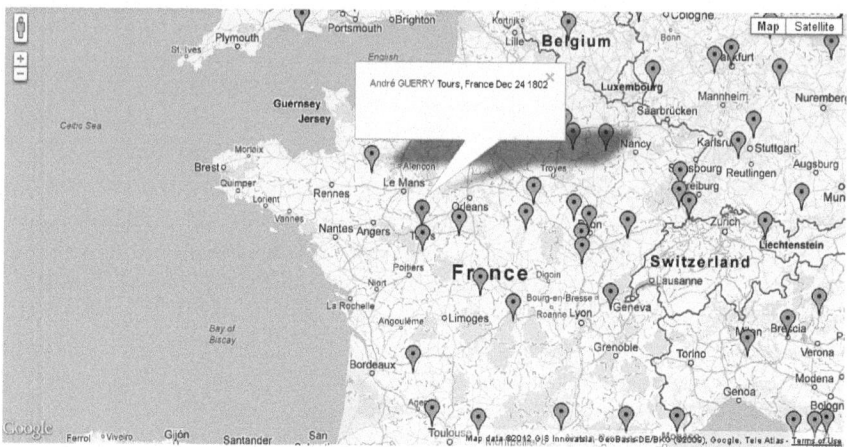

Figure 10.6   Birthplaces of 188 milestone authors, shown on an interactive Google map, centered on France. Each geographic marker is linked to an author query on the datavis.ca website that lists the contributions by that individual

Milestones: Themes and Trends

The records in the Milestones Project database also feature various text fields for each logged event. These include a brief item tag, a full description of the event, and relevant keywords, as well as categorical codes for the content (Subject), and form (Aspect) of the item. Treating this information as "data" allows us and others to study themes and trends in these developments. Modern methods of text mining and data visualization can provide insights into this history not available through other means.

As one simple illustration of this approach, Plate 37 shows two mosaic displays[12] that explore the relationships among Epoch, Subject, and Aspect. The left panel shows changes in the distributions of milestone events by Subject over time. It can readily be seen that while most of the milestone innovations up to the end of the eighteenth century were about the physical world (astronomy, geodetic measurement, weather, etc.), this trend changed in the nineteenth century, where there was a large shift toward problems that related to human populations (for example, pertaining to mortality, births, disease, crime). Beginning in the early 1900s, the pattern changes again, with advances in mathematics and statistics becoming the dominating force.

The right panel shows the association between Subject and Aspect, pooled over Epoch. As is not surprising, maps and other cartographical representations were most often used to show data of the physical world, while graphs and diagrams were most often associated with mathematical and statistical subjects. Other statistical graphs and analyses could be used to explore these and other relationships in more detail. The key to this is of course the existence and availability of data—in this case reflected by the coding of graphical milestones in our database.

Conclusion and Future Directions

The Milestones Project began as a simple attempt to collect a comprehensive history of innovations and developments in data visualization in a single location. Like Topsy, it grew over time, with images, historical papers and references, suggestions, and other contributions graciously provided by friends and collaborators, most notably from the members of Les Chevaliers des Albums de Statistique Graphique.

In this chapter, our primary goal was to introduce the second and latest iteration of this project. The redesign was undertaken to make this history more accessible for browsing and searching, and to attempt to make the database more amenable to additions, edits, and extensions among collaborators. However, we find that the most exciting aspect of the new structure is its flexibility in terms of data retrieval, and our newfound ability to use and manipulate the data for graphic-based statistical historiography.

One goal for the future, as we suggested earlier, is to extend the user interface to provide multiple views and advanced text search and filtering capabilities. One convenient path for this development is provided by the SIMILE Exhibit framework. This provides web software libraries (Ajax, JavaScript, CSS) for

---

[12]   Mosaic displays show the frequencies in cells of a cross-classified table by the area of each tile. The tiles are shaded according to departure from a null model of no-association, using blue for cells with frequencies substantially greater than chance, and red for cells with frequencies that are lower than expected.

timelines, interactive maps, tabular displays, image "tiles" and other visualizations. Various views can be composed for browsing as tabbed, alternatives or as faceted displays, showing for example an interactive timeline and a map.

Equally important, the Exhibit framework allows us to present some of the milestones tables to be used as filters for the items displayed in these views. Tables for subject, aspect, keywords, location, epoch, and so on would allow the user to select milestone events based on some or all of these criteria, providing a way to ask such questions as "what milestones events between 1700 and 1900 involving social science occurred in Europe?"

Finally, we would like to make the milestones database more publicly accessible for use by others on the history of data visualization. For the examples we have shown here, we connect to the milestones database directly via MySQL or ODBC interfaces to SAS and R, but this presents security risks. Happily, the Exhibit framework also provides methods for data export from various views, using JSON or CSV formats. In addition, we contemplate adding facilities for users in the data visualization community to add comments, notes, references and links to milestones items. These extensions will comprise the Milestones Project 3.0.

Chapter 11

Annotated Bibliography of Scholarship on the History of Data Graphics

Kevin Van Winkle

One of the goals for *Visible Numbers* is to inspire other researchers to investigate and contribute to the developing field of data visualization scholarship. This annotated bibliography was created to assist scholars in such an endeavor. Like the field of data visualization itself, the items included here are wide-ranging. They span historical eras, cross national borders, and intersect many academic disciplines. Readers will find here articles and books that conduct historiographical scholarship on data visualization similar to that found in the chapters of *Visible Numbers*. They will also find theoretical and methodological works helpful in fashioning future studies on data visualization. Readers will likewise recognize some well-known works whose influence on the field was just too profound not to include. As with any good visualization, data reveals itself to the viewer in a way that might have been imperceptible in another form. I hope that the 42 items collected here will have a similar effect, creating a clearer picture of what has been done, how it was done, and most importantly, what comes next.

Barton, Ben F., and Marthalee S. Barton. "Ideology and the Map: Toward a Postmodern Visual Design Practice." *Professional Communication: The Social Perspective*. Edited by Nancy Roundy Blyler and Charlotte Thralls. Newbury Park, CA: SAGE Publications, 1993. 49–78.

Barton and Barton begin this important article on the ideological underpinnings of seemingly objective data graphics by addressing the duplicitous nature of ideology. They assert that ideology has a Janus face, privileging some meanings and practices at the same time that it suppresses evidence of its own partiality. The authors examine the creation and use of maps to show how this dual mode of inclusion and exclusion works to create ideological dominance. For instance, some Iraqi maps have included Kuwait as the nineteenth province of Iraq rather than as an independent nation. Other forms of inclusion relate to the size or placement of information, a practice the authors label "privileged positioning." This is seen most notably in Mercator's frequently used fifteenth-century map of the world that centers on the cartographer's homeland of Germany. Working in tandem with these methods of inclusion are methods of exclusion. Some things, such as slums and other "unsightly" geographical features, are not often found on a map. Another form of exclusion is the use of exonyms by the cartographer. The result of these and other dubious methods is the creation of a hegemony that seems "natural." In order to denaturalize these mapping practices, the authors propose a postmodern design practice. This "new politics of design" valorizes differences and works to create a heterogeneous cartography (70). Specifically, the authors advocate the use of maps as collages or palimpsests to help ensure that future map making is more inclusive and more easily recognized as an interpretive practice subject to human bias.

Barton, Ben F., and Marthalee S. Barton. "Modes of Power in Technical and Professional Visuals." *Journal of Business and Technical Communication* **7.1 (1993): 138–62.**

Using the two modes of panoptic power, synoptic and analytic, established by Michel Foucault in his renowned book *Discipline and Punish,* Barton and Barton examine the power that visuals have to communicate both macro and micro information. The synoptic power of visuals, the authors explain, is the ability to show multivariable data in an instant. The analytic mode of power is the capacity of visuals to showcase the particulars, such as individual data points. In Foucault's conception of the panopticon, both the synoptic and the analytic are at work, allowing the tower guard to see the global picture of the prison at the same time it makes possible individual scrutiny of the prisoners. By applying this taxonomy of power modes to technical communication visuals, the authors show that visuals offering only one mode or the other stifle viewers' empowerment. For instance, during the disaster that occurred at the Three Mile Island nuclear facility, control room operators could see only individual data points. This analytic mode of surveillance made it difficult to see the larger event: a facility meltdown. Conversely, the much-admired London Underground Diagram, with its synoptic view of London's subway system, appears to be empowering to users, but is actually the opposite. With so many travel options available at a glance, the viewer's individual needs are left unsatisfied. Barton and Barton come to the conclusion that by using both synoptic and analytic modes of surveillance, technical communicators, among others, can create more useful and empowering visuals.

Beniger, James R., and Dorothy L. Robyn. "Quantitative Graphics in Statistics: A Brief History." *The American Statistician* **32.1 (1978): 1–11.**

Beniger and Robyn outline the history of quantitative graphics as they were used and developed in the fields of science and statistics. They structure their article around the concept that the field of quantitative graphics advanced as solutions to one of four problems displaying data. Scientists and analysts in the seventeenth and eighteenth centuries, tasked with plotting data spatially, turned to coordinate systems and tables. In the eighteenth and nineteenth centuries, the problem was displaying the links between data sets, a quandary that led to the development of graphics of discrete comparison, such as Playfair's bar chart. Later in the nineteenth century, the need for quantitative graphics showing a continuous distribution of data gave rise to the ogive and the histogram. Later still, and into the twentieth century, researchers needed to show the relationship between multivariate data, hence the creation of contour maps and stereograms. The authors conclude their article with the prediction that the need to address problems and show data is ongoing, and therefore the innovation of quantitative graphics is likely to evolve in the future the same way that it has in the past. While the history Beniger and Robyn provide is valuable, perhaps the most useful features of their article are its appendix and reference lists. In the former, the authors provide a timeline of statistical graphics that begins as far back as 3800 BC. In the latter, there is an exhaustive list of books and articles concentrated on statistical graphics.

Bertin, J. *Semiology of Graphics: Diagrams, Networks, Maps*. Trans. W.J. Berg. Madison, WI: University of Wisconsin Press, 1983.

In his foreword to Bertin's book, Howard Wainer writes, "*Sémiologie graphique* is the most important work on graphics since the publication of Playfair's Atlas" (ix). One feature of the book that supports Wainer's claim is the exhaustive level of definition, explanation, and example that Bertin uses to discuss the creation and utilization of graphic sign-systems. This extensive amount of information is broken down into two parts: semiology of the graphic sign-system and utilization of the graphic sign-system. In the first section, Bertin argues that visualization is a fundamental language of the human mind. Furthermore, graphic representation allows us to both store and research information with eight different content representative variables, which Bertin calls "retinal variables." These variables are the x and y dimensions of the plane, along with the size, value, texture, color, orientation, and shape of the "visual marks" used on the page. For Bertin, this multiplicity of signs combined with our intuitive sense of their use justifies his semiological approach to their study at the same time that it underscores the superiority of graphics for communication. From this, Bertin establishes a system for creating efficient images, one that allows for quick and accurate perception of graphically represented data. In the second part of his text, Bertin refers to his previously established system to discuss and correct the problems that can arise when one creates or uses different types of diagrams, networks, and maps. The scope and detail of Bertin's book is difficult to summarize succinctly, but, suffice it to say, those interested in the design, history, or use of data graphics will likely find it as valuable as Wainer does.

Biderman, Albert D. "The Playfair Enigma: The Development of the Schematic Representation of Statistics." *Information Design Journal* **6.1 (1990): 3–25.**

Given the frequent use of statistics and Cartesian coordinate systems for the plotting of empirical data by scientists, why is it that William Playfair, a political pamphleteer and sometime scoundrel, is accredited with the invention of statistical graphics? In this article Biderman attempts to answer this question. According to the author, and contrary to the viewpoint of many, Playfair just happened to be in the right place at the right time. Prior to and for many decades after Playfair's invention there existed a resistance to the spatial representation of non-spatial data. Additionally, Biderman claims that the zeitgeist of the scientific community during the seventeenth century was one that sought to identify and order the work of God in scientific terms, most identifiable in mathematics and geometry. From these conditions, Biderman concludes that there were many other scientists and scholars who could have, or got close to, making inventions similar to Playfair's; however, these would-be inventors were either hindered by the rationalistic orientation of their times or simply uninterested in the kind of socio-political and economic data Playfair used in his graphics. Not being a member of the scientific community, combined with his pragmatism and technological expertise in drafting, created a perfect storm for Playfair, one that secured both the creator's and his inventions' exalted status in the history of data graphics.

Brasseur, Lee E. "Florence Nightingale's Visual Rhetoric in the Rose Diagrams." *Technical Communication Quarterly* **14.2 (2005): 161–82.**

In this article Brasseur tells the story of Florence Nightingale's rose diagrams. She does so in order to show "an important example of how visual abstraction of data can help further an argument" and incite people to action (180). Concerned with the high mortality rates of its soldiers fighting in the Crimean War, the British government sent a team of nurses led by Nightingale to Scutari to help field doctors treat the wounded. What they encountered there was a highly dysfunctional and unsanitary healthcare system, one that contributed to soldier deaths more than actual combat. After Nightingale and her team instituted sanitary reforms, there was a precipitous drop in the number of British casualties. Correctly assuming other British field hospitals had similar unsanitary conditions, Nightingale produced a report for the British government and Army leaders outlining the changes she made at Scutari and advocating for their implementation army-wide. This report was largely ignored, however, leading Nightingale to publish and disseminate an addendum. In it she included three graphs, now known as the rose diagrams. These diagrams, named for the way in which blocks of data line up next to each other like the pedals on a rose, were Nightingale's unique invention. According to Brasseur, these diagrams represent a significant achievement in information graphics and a stroke of rhetorical genius. Knowing that the government was unwilling to accept her statistics and suggestions for reform, and that the general public would be unable to understand complicated statistical data, Nightingale invented the rose diagrams to display the data in a way that was both irrefutable and easy to comprehend. As a result of her information graphics, Nightingale's reforms were eventually implemented throughout the British Army.

Brasseur, Lee E. *Visualizing Technical Information: A Cultural Critique.* Amityville, NY: Baywood, 2003.

In an effort to correct for the favoritism shown towards cognitive-based theory in the scholarship of technical visuals, Brasseur examines five types of data graphics—graphs, technical illustrations, charts and diagrams, tables, and information visualizations—through genre theory. She acknowledges that genre theory is a fluid and evolving theory, but unlike the popular cognitive-based theoretical approach to examining data visualizations, it does not separate users from their contexts. Instead, genre theory treats data visualizations as cultural artifacts, and in doing so, provides a more thorough understanding of how data visualizations work. For each type of data visualization listed above, the author asks seven questions: 1) What is the genre?; 2) How has the genre developed?; 3) What categories exist within it?; 4) How is the genre written about and taught?; 5) What is the relationship of text to the genre?; 6) What theory underlies the genre?; and 7) How is the genre designed? (12). After asking these questions of each type of data visualization, Brasseur comes to several related conclusions. First, she reasons that pragmatism must be central to the future study of the different genres of technical visuals. Second, the author advocates for a departure, or at least an addition to, the cognitive-based research typically conducted on visuals. Lastly, and most importantly, Brasseur concludes that technical communicators must begin to consider the influence of society on the use and production of data visualizations.

Camerini, Jane R. "Heinrich Berghaus's Map of Human Diseases." *Medical History* **44 Supplement 20: Medical Geography in Historical Perspective (January 2000): 186–208.**

Camerini states that the "intent of this essay is to contextualize Berghaus's 1848 map of diseases, the earliest world map in an atlas showing the geographical distribution of epidemic and endemic human diseases" (186). She does this to argue that Berghaus's map, along with other maps made around the same time, were not necessarily influenced by his teacher and friend Alexander von Humbodlt quite as much as some believe. Camerini points to the scholarly work of Susan

Cannon for this mistaken attribution of Humboldt's influence. Cannon, Camerini argues, created the phrase "Humboldtian science" as a way to describe Humboldt's prominence in the zeitgeist of nineteenth-century medical cartography. In doing so, she overreached, thereby reducing the significance of several contextual phenomena, such as society's view of science and their expectations for the rhetorical and graphical practices used to communicate scientific information during the last half of the nineteenth century. To make her case, Camerini provides a detailed analysis of Berhaus's thematic world map of human diseases and an account of how he came to create it. Published in 1845 and then again in 1848, the atlases that included Berghaus's map were not as detailed or as accurate as Humboldt's. Furthermore, the theme of the map had more to do with Britain's colonialism, the rise of statistics in Europe at the time, and the public's demand for accessible and affordable atlases than it did with Humboldt's more purely academic endeavor to visualize the interrelationships between the world and the people who populate it. While the exact balance of influence "Humboldtian science" and political and societal factors had on Berghaus's map is incalculable, it is clear from Camerini's work that, much like the entire history of data visualization, what gets invented and why is often the result of a confluence of several factors.

Crawley, Charles R. "From Charts to Glyphs: Rudolf Modley's Contribution to Visual Communication." *Technical Communication* **41.1 (1994): 20–26.**

From 1933 to 1945, Rudolf Modley was the creator and president of the Pictograph Corporation, located in New York City. This agency created and sold pictorial statistics in a style similar to Otto Neurath's Isotype to a variety of businesses, government agencies, and periodicals. It is not surprising that Modley's pictographs were so similar to Neurath's, for Modley had once been a student of the famed designer while still in his home country of Vienna during the 1920s. Crawley shows that, through example and prescribed principles of design, Modley managed to exert an incredible and long-lasting influence on the field of data visualization. Like his predecessor Neurath, Modley also aimed for a language of pictures universally recognized and understood. He worked on this sign-system, which came to be called "glyphs," with famed anthropologist Margaret Mead. Even though no such system of universally understood symbols exists, it is hard to say that Modley failed, for as Crawley concludes, the effect of Modley's work continues to be found in contemporary examples of visual communication.

Dragga, Sam, and Dan Voss. "Cruel Pies: The Inhumanity of Technical Illustrations." *Technical Communication* **48.3 (2001): 265–74.**

According to Dragga and Voss, technical communicators too frequently sacrifice humanity for objectivity when creating data visualizations. The authors include several examples of data visualizations that fail to represent the human element of the data displayed, such as a pie chart that reports the horrifying ways lumberjacks have died, or a bar graph that displays the instances of baby-walker-related injuries. These graphs, and ones like them, are more than unethical; they are inhumane. In what proved to be a provocative remedy for this condition, Dragga and Voss suggest "semantic fusion" or "hybrid literacy." These approaches to visual literacy describe a juxtaposition of visuals with words or other visuals in ways that highlight, rather than obscure, the humanity connected to the data. One prominent example created by the authors is a reiteration of Charles Joseph Minard's famous Napoleon's March on Moscow data graphic that features pictographs representative of the soldiers who lost their lives during the march. By doing this, Dragga and Voss highlight the humanity obscured in Minard's original version. This practice, they believe, should be more frequently employed by technical communicators working with data graphics. Chart-junk, distractions, creator subjectivity, and the necessary level of humanization are all concerns that come with humanizing data graphics, but as the authors point out, "Technical communicators, schooled in rhetoric and trained in the humanities, are in a unique position to help raise the level of ethics in the visual display of technical information" (272).

Friendly, Michael. "A.-M Guerry's 'Moral Statistics of France': Challenges for Multivariable Spatial Analysis." *Statistical Science* **22.3 (2007): 368–99.**

As he has done for many others, Friendly works here to contextualize and describe the contributions of a data visualization pioneer: André-Michel Guerry. Published in 1833, during the "Golden Age of Statistics," Guerry's *Essai sur la Statistique Morale de la France* was a response to the ever increasing amount of data on social and moral statistics being gathered by the state. The corpus for the *Essai* was 25 years of crime and suicide data, along with the places such events occurred. Guerry arranged this "avalanche of numbers" into several striking maps and tables for the *Essai*, a task for which he won many awards and high praise. Friendly uses the same data set as Guerry to produce a number of modern graphics, ones not yet invented in Guerry's time. Friendly's scatterplots, biplots, HE plots, geovisualizations, and other updated graphics provide him and readers with some insight into the benefits and challenges of visualizing such a massive amount of multivariate data. He calls this process of updating and revisioning statistics and statistical graphics *statistical historiography*. This process proves to be very valuable at creating a better understanding of information graphics for both past and present data visualizations.

Friendly, Michael. "The Golden Age of Statistical Graphics." *Statistical Science* **23.4 (2008): 502–35.**

Friendly's article traces the rapid expansion of statistical graphics that occurred during the last half of the nineteenth century, a period of time he labels the "Golden Age of Statistical Graphics." These 50 years saw a growth in the use of statistics in both public and private spheres, the advent of new ways to graph and display statistical data, and technological advances that enabled the publication and distribution of these new data visualizations to a greater number of people. The individuals working during this time in the nascent field of data visualizations should be familiar to readers of *Visible Numbers*; Friendly discusses the contributions of Baron Charles Dupin, Francis Galton, Charles Joseph Minard, André-Michel Guerry, Florence Nightingale, Adolphe Quetelet, and many others to show how the practices and use of data visualizations progressed during this time. Friendly states that his goal for the article was to "tell the story of a particular slice of the history of statistics and data visualization"; however, he also wants to remind readers of the interocularity—that is, the visual impact—of such graphics so that we can keep it in mind while viewing or creating contemporary data visualizations (532).

Friendly, Michael. "Visions and re-visions of Charles Joseph Minard." *Journal of Educational and Behavioral Statistics* **27.1 (2002): 31–51.**

Re-vision, Friendly explains, can mean both to look at again and to revise and make new. With this dual definition as a preface to his article, Friendly reviews the work of one of the central figures in the history and development of thematic cartography and the visual display of data: Charles Joseph Minard. By using it as an important figure in his influential book *The Visual Display of Quantitative Data* (1983), Edward R. Tufte played a significant role in making Minard's *Carte pagne de Russe 1812–1813* (Napoleon's March on Moscow graphic) a well-known example of excellence in the visual display of data; however, Friendly contends that the popularity of this single graph distracts from a lifetime of elegant and effective data visualizations. Consequently, Friendly "re-visions" Minard's full body of work by discussing and showcasing other examples of Minard's catalog, such as his center-of-gravity map of Paris and his divided-circle map of butcher's meats supplied to Paris markets. Friendly contends that these figures, along with several others, show that Minard gave much more to the field of data visualization than a single graphic. In the sense of the second meaning of "re-vision," Friendly discusses recent attempts by contemporary graphics designers to reproduce and make new the Napoleon's March on Moscow graphic. The results are interesting, but more important for Friendly is that such attempts show how Minard still has valuable lessons to teach contemporary statisticians and cartographers.

Friendly, Michael, and Daniel J. Denis. "The Early Origins and Development of the Scatterplot." *Journal of the History of the Behavioral Sciences* **41.2 (Spring 2005): 103–30.**

The scatterplot "may be considered the most versatile, polymorphic, and generally useful invention in the entire history of statistical graphics" (103). Oddly, though, there is little known about its origins. Friendly and Denis address this gap in the history of information graphing by tracing the scatterplot's development. While graphs similar to the scatterplot appeared in William Playfair's work on economics and Francis Galton's graphs of the relationships between parents and their children's physiological development, neither completely fulfilled the criteria to be called a scatterplot, that is "a plot of two variables, x and y, measured independently to produce bivariate pairs (x_i, y_i), and displayed as individual points on a coordinate grid typically defined by horizontal and vertical axes, where there is no necessary functional relation between x and y" (103). It was not until 1833, when the astronomer John Herschel used a graph to show the orbits of double stars, that the first scatterplot appeared. Once others began to realize the scatterplot's value, its use grew quickly. By the first half of the twentieth century the graph became mainstream. Presently, the scatterplot is still used often to show the complex relationships between bivariate data.

Friendly, Michael, Pedro Valero-Mora, and Joaquin Ibáñez Ulargui. "The First (Known) Statistical Graph: Michael Florent van Langren and the 'Secret' of Longitude. *The American Statistician* **64.2 (2010): 174–84.**

Correctly calculating longitude, especially at sea, proved to be a seriously difficult problem for the cartographers of the sixteenth and seventeenth centuries. It was so difficult, in fact, that Michael Florent van Langren made correctly calculating longitude his life's work. Unfortunately for sailors and other explorers of the time, he failed in his endeavor. Fortunately for data visualizations scholars, though, is that in doing so van Langren created the first known statistical graph. This graph was a visual depiction of previous estimations of the longitude between Toledo and Rome. Although it was an attempt to calculate a more precise measurement of the distance between these two cities, it failed to show the correct measurement, 16.5°, which wouldn't be known for almost another century. Nonetheless, the authors contend that it was still "remarkable" for several reasons. First, showing the data in tabular form would have been easy and expected, yet van Langren chose to display it in graphic form. He did this in order to make the data more dramatic and more perceptible to the viewers. Second, the graph showed the geographical bias of previous longitude estimators starkly, always an important feature of a good data graphic. And, lastly, the graph displays van Langren's rhetorical concern for data ordered for maximum effect. All of these factors, say the authors, make the graph "right," if only in terms of data visualization.

Funkhouser, Howard Gray. "Historical Development of the Graphical Representation of Statistical Data." *Osiris* **3 (1937): 269–404.**

This prominent history of data graphics comprises several distinct chapters. In the early chapters, Funkhouser details the circumstances and events that led to the common use of graphical representations to display statistical data. Funkhouser's account is comprehensive up to the point of its publication in 1937. He begins his history of data graphics in antiquity, pointing to rudimentary visualizations of data found in ancient Egypt and Rome, and traces it through to the early part of the twentieth century. While Funkhouser attributes the most credit to William Playfair and Charles Joseph Minard, he still provides an exhaustive list of others whose inventions and actions encouraged the advancement of data graphics. The most important of these is Adolphe Quetelet, the initial advocate for the International Statistical Congress. Responding to a new enthusiasm for statistics in the nineteenth century, the Congress was an attempt to standardize and otherwise respond to issues concerning statistics. Despite lasting only a couple of decades, this group did much to further the use of graphical representations in statistics. In the latter chapters, Funkhouser goes on to discuss contemporary—for the time—data graphics like the nomogram and the possibility of data graphics as pedagogical tools. Funkhouser concludes with a glossary of terms and an extensive annotated bibliography. Taken together, these chapters make for a text that is closer to a book than a journal article, but in either case a highly important text in the field of data graphics.

Gilbert, E.W. "Pioneer Maps of Health and Disease in England." *The Geographical Journal* **124.2 (1958): 172–83.**

In all of medical geography, there might not be a map more well known than that of Dr. John Snow's *Deaths from Cholera in the Broad Street Area of London in September 1854*. Using dots, Snow's map made visible the deaths caused by cholera infection occurring in the Broad Street area at the time. Snow's visualizing led him to identify correctly that cholera is a water borne pathogen and was being spread in the Broad Street area via a community water pump. Aside from saving numerous lives, Snow's map was a milestone of both epidemiology and data visualization. Yet, as Gilbert shows in his article, Snow's was not the first or the last map to be made concerning outbreaks of cholera in the United Kingdom. In

1833 Dr. Robert Baker made a choropleth map of cholera infections for Leeds. In 1848 W.P. Ormerod mapped cholera infection rates for Oxford. In 1849 Dr. Thomas Shapter created a dot map of cholera deaths, very similar to what Snow would later develop, for Exeter. Others, such as Augustus Petermann and Dr. Henry Wentworth Acland, made similar maps around the same time, but theirs were not published until after Snow's. While not directly stated, one can infer from Gilbert's article that these other maps failed to achieve the same stature as Snow's because they did not make the same connection between drinking water and cholera. In other words, his was "right" while theirs were not.

Godlewska, A.M.C. "From Enlightenment Vision to Modern Science? Humboldt's Visual Thinking." *Geography and Enlightenment.* **Edited by D.N. Livingstone and C.W.J. Withers. Chicago: University of Chicago Press, 1999. 236–80.**

In this chapter, Godlewska discusses the role of Alexander von Humboldt in the history of geography; however, she characterizes him more as an innovator of data graphics, someone on the margins of the map-making field, than an actual cartographer. Nevertheless, his goal of creating data graphics that combined analytical depth, rigorous research, and holistic vision led to some basic but useful ways of representing landscapes and data. The unity between these things—depth, rigor, and holistic vision—was at the crux of Humboldt's work. Godlewska contends that this makes him representative of the evolution of scientific thought and visual depiction that occurred post Enlightenment. This evolution was the move from an explanatory science that most often practiced mimesis of its subjects, to theoretically driven science that sought to detail the connections between the interiority of natural and social phenomena. The author characterizes the difference between these two approaches as the difference between the study of static forms and phenomena and the study of dynamic ones. In searching for these connections, Humboldt engendered a form of mutlidisplinarity in cartography and data graphics that persists to this day, particularly in the field of thematic mapping.

Hankins, Thomas L. "A 'large and graceful sinuosity': John Herschel's Graphical Method." *Isis* **97.4 (2006): 605–33.**

In this article, Hankins details the use of graphs in scientific research during the nineteenth century by way of their strongest advocate, John Herschel, a mid-nineteenth-century English philosopher and scientist. Hankins points out that graphs were not always believed to be a good way to show data. There were many scientists and researchers during Herschel's time who thought that numbers alone were the correct way to display data; however, in his article "On the Investigation of the Orbits of Revolving Double Stars: Being a Supplement to a Paper Entitled

'Micrometrical Measures of 364 Double Stars &c, &c.," Herschel showed how a good data graphic can reveal complexities and relationships between data sets that would otherwise be obscured or unnoticed in numbers alone. Herschel drew by hand a line through his plotted data on the orbits of double stars. In doing so, he showed that the data, like the stars' orbit, was in the shape of an ellipsis, one with a "large and graceful sinuosity." Not surprisingly, other scientists balked at Herschel's use of such a subjective method. Hankins goes on to explain thoroughly the ensuing controversy, which ultimately did not diminish the value of data graphics in scientific research.

Hankins, Thomas L. "Blood, Dirt, and Nomograms: A Particular History of Graphs." *Isis* 90.1 (1999): 50–80.

Originally a lecture delivered to the History of Science Society in 1997, Hankins' article details the invention and early uses of a "peculiar diagram" known as the nomogram, a computing graph that allows users to arrive quickly at the solution to complex mathematical problems or to see easily the relationship between three-variable data. While the work of William Playfair, James Watt, and Charles Joseph Minard paved the way for the nomogram, Hankins shows that it was Maurice d'Ocagne who deserves credit for inventing this particular graph in the last decade of the nineteenth century. During d'Ocagne's lifetime, military strategists and civic engineers each had a need to reshape the terrain, or, as Hankins puts it, to "move dirt." This task proved to be more difficult than it sounds; the refiguring of three-dimensional terrains into two-dimensional data figures was a complicated process. The nomogram, however, allowed users to arrive at the answers to these sorts of complicated problems by simply drawing a line between two points of data in order to identify a third variable. So simple and accurate was the nomogram that it became a popular computation tool for years afterwards, sometimes even used outside of its original field. In the early part of the twentieth century, for instance, scientist L.J. Henderson used the graph to diagram the components and correlations of human blood. While Henderson's blood diagrams, with their multiple variables and busy display, were difficult to interpret, the system they revealed would have been impossible without the nomogram. Hankins contends that it is this combination of revelatory ability and flexibility that make the nomogram such a valuable tool and important information graphic.

Harley, J.B. "Deconstructing the Map." *Cartographica* 26.2 (1989): 1–20.

Harley uses the theories of postmodernists Michel Foucault and Jacques Derrida to create a broad approach for deconstructing maps. Doing so, he contends, helps us identify the often ethnocentric and culturally subjective underpinnings influencing their creation. His analytical approach is tripartite. Using Foucault's theories of discourse analysis, Harley shows how maps are not simply transparent images of scientific measurement, but instead are artifacts imbibed with the values and rules of the society in which they were created. Per Derrida, Harley's second analysis considers the map as "text," and shows that the "steps in making a map—selection, omission, simplification, classification, the creation of hierarchies, and 'symbolization'—are all inherently rhetorical" (11). For the third and final part of his analysis, Harley returns to Foucault to demonstrate how the map becomes a nexus of political and technological power, features he labels external and internal respectively. In his conclusion, Harley rationalizes a deconstructive approach to map-reading by reminding readers that maps are never neutral, and that "[b]y dismantling we rebuild" (5). Furthermore, deconstructing maps encourages us to deconstruct other "texts," thereby achieving a richer understanding of our social reality and its artifacts.

Kimball, Miles A. "London through Rose-Colored Graphics: Visual Rhetoric and Information Graphic Design in Charles Booth's Maps of London Poverty." *Journal of Technical Writing and Communication* **36 (2006): 353–81.**

Between the years of 1886 and 1902, Charles Booth, a shipping magnate and social reformer, created a series of thematic maps depicting the state of economic conditions and social classes in Victorian London. With these graphics Booth created an eight-part class structure, ranging from poorest to richest, and then used a color code to signify the habitats where these groups lived on a street-level map of London. Kimball contends that Booth's maps were far from objective, and instead were very much a product of the rhetorical, ideological, and cultural milieu they were created in. For instance, the poorer echelons of society were represented by black and dark blue, colors that implied their "hidden" status in society. The wealthier classes were represented with cheerier colors, such as pinks and yellows. The consequences of Booth's design choices had significant and longlasting effects. By creating a hierarchy of economic classes, Booth created the "poverty line," a concept that persists to this day. Moreover, Booth's maps literally changed the way society viewed the poor. Many once believed that poverty was a "hidden" problem, but with Booth's maps the viewer could plainly see the small dark spots of poverty peppering the larger map of London. The issue of poverty became manifestly visible and therefore seemingly manageable. As a result, social programs were created to assist the indigent. While the intent and reactions to Booth's maps may have been somewhat positive, what is worrisome is how viewers took them as fact, never considering their limitations or rhetorical effects. Kimball concludes that current information graphics are little different and should always be viewed with a critical eye.

Koch, Tom. "The Map as Intent: Variations on the Theme of John Snow." *Cartographica* **39.4 (2004): 1–14.**

Using John Snow's map of the 1854 cholera outbreak in the Soho area of London and Roland Barthes's semiological concept of myth, Koch makes the case for authorial intent as the prevailing reason for why maps appear as they do. Koch uses the reprinted cholera maps of Snow's made by Gilbert, Tufte, Monmonier, and the Center for Disease Control to show that none is true to the creator's original. Some have changed the markings from lines to dots; some have altered the symbols used, while others have purposefully left out data. Their motivations for doing so may be reasonable, such as to demonstrate a more profound point, but Koch has difficulty with the fact that none specifically stated they were doing so. Regardless of whether one finds this problematic or not is secondary to the larger theme of Koch's article: maps, even those maps used as examples to discuss maps, are always value laden and always reflect the intent of the cartographer as much as they do the content of the map.

Kostelnick, Charles. "Melting-Pot Ideology, Modernist Aesthetics, and the Emergence of Graphical Conventions: The Statistical Atlases of the United States, 1874–1925." *Defining Visual Rhetorics.* **Edited by Charles A. Hill and Marguerite Helmers. Mahwah, NJ: Lawrence Erlbaum Associates, 2004. 215–42.**

In an attempt to make census data more accessible to the public, the U.S. government published a series of six atlases between the years of 1874 and 1925. These atlases were a mix of maps, graphs, and charts that showed how the country was taking shape both geographically and demographically. Kostelnick uses these artifacts to demonstrate how social and cultural context can influence the aesthetic features and rhetorical effects of data graphics. For instance, following the aesthetic trends of the times, the first atlases displayed an ornate, Victorian style. Later iterations conveyed a more pared down and distinctly modern look. More than just appealing to the viewers of the times, though, these atlases were exceptionally persuasive. Depicting the west as a vast expanse of land to be settled became a powerful visual argument to move west. And graphing the multiple ethnicities of the people working in and populating the country caused many to believe America truly was the "melting-pot" of the world. While these atlases had a profound effect on the audience, influencing their understanding of public policy issues, such as immigration, they also help contemporary audiences understand the rhetoric of our present design ideals. The confluence of the atlases, their design, and their rhetorical effect reinforces an important feature of visual rhetoric and information design for Kostelnick: it is "an intensely social process" (239).

Kostelnick, Charles, and Michael Hassett. *Shaping Information: The Rhetoric of Visual Conventions.* **Carbondale, IL: Southern Illinois University Press, 2003.**

In their now well-known and oft-used book, Kostelnick and Hassett set out to "build a framework for structuring visual language around a wide range of conventional practices" (5). For them, visual language comprises socially constructed conventions ranging from pie charts to architecture. Visual conventions can last unchanged for centuries, or they can develop into something very different from their origins in only a few years. To show how this evolutionary process occurs, or does not for that matter, the authors consider a number of historical visualizations. Using exploded view diagrams, cut away drawings, charts, graphs, and a variety of other visualizations, Kostelnick and Hassett consider the types of discourse communities in which visual conventions are born and used, the rhetorical functions of such visual conventions, and the external factors, like technology, that also influence their inherently mutable nature. For those interested in information graphics and other forms of visual communication, there is much to be gained from *Shaping Information*—too much to list here. Underpinning it all, though, is what might be best a called the moral of the book: despite how ostensibly transparent some visual conventions seem, they are never *arhetorical*.

Lupton, Ellen. "Reading Isotype." *Design Issues* **3.2 (1986): 47–58.**

In this article, Lupton blends culture, philosophy, and aesthetics to create a detailed history and analysis of Isotype. Created by Otto Neurath, Isotype is a system of communication that "uses simplified pictures to convey social and economic information to a general public" (47). Neurath, a philosopher and social scientist, developed Isotype in an attempt to create a visual language that could transcend language barriers and unify disparate international societies. Primary to Neurath's motivations for doing so was his status as a logical positivist. Logical positivism, Lupton explains, is the merging of two seemingly incongruous philosophical stances, a mix of rationalism with empiricism. The combination of the two provided a foundation to analyze language and search for connections between abstract terms and direct experience. Lupton describes how Isotype characters are both positive and logical: "An Isotype character is *positive* because, as a picture, it claims a base in observation; it is *logical* because it concentrates experienced detail into a schematic, repeatable sign" (49). Isotype, in its effort to supplant interpretation with perception, advanced similar themes in art and design. Furthermore, Isotype influenced the design of pictorial statistics, allowing for a substitution of pictures for numbers while adding layers of meaning to a graphic. Although noble in his pursuit for a consistent language that could transcend national boundaries, Neurath's Isotype characters, like all images, are not immune from cultural bias. Nevertheless, as

Lupton concludes, Neurath's work has had lasting and significant influence over theorists like Barthes and Derrida, as well as the overall field of design.

Monmonier, Mark. *How to Lie with Maps*. 2nd ed. Chicago: University of Chicago Press, 1996.

Whether it is in the process of converting expansive three-dimensional land features into flat two-dimensional surfaces or the way populations of people and the environments they exist in are reduced into simple symbols, maps must distort reality. Or, as Monmonier puts it, they must lie. In his book, Monmonier uses both theoretical and historical examples to show how maps do so, why they do so, and to what degree. The ways maps lie can range from simple mistakes and "white lies," such as the distortions that come from adjusting the scale, projection, or symbols on a map, to full on propaganda, as in the case of empires who want their domain to seem larger or their enemies more threatening. Regardless whether the reason for the mendacity is ignorance or manipulation, Monmonier wants his audience to understand that maps, often viewed as authoritative and infallible, are man-made and therefore can be biased and flawed. "The wise map user is thus a skeptic, ever wary of confusing or misleading distortions conceived by ignorant or diabolical map authors" (184).

Monmonier, Mark. "The Rise of the National Atlas." *Cartographica* 31.1 (1994): 1–15.

Which country had the first national atlas? For Monmonier, it depends on how you define "atlas." In 1960 the International Geographic Union's (IGU) Commission on National Atlases outlined three criteria for defining an atlas: it must cover a single country or nation; it must provide multiple thematic maps depicting demographic and other pertinent data for the area; and it must be endorsed by the nation's government and scientific communities. By these standards, Finland's 1899 national atlas was the first. However, as Monmonier points out, this is only one organization's view, and a rather problematic one at that. By failing to consider things like geographic knowledge, technological affordances, cartographic intent, and even the processes of publication, the IGU Commission creates a false historical demarcation that fails to recognize many earlier documents as atlases. By ignoring this limited definition, Monmonier shows how Christopher Saxton's 1579 atlas of England and Wales, Maurice Bouguereau's 1594 atlas of France, and Francis Walker's 1874 *Statistical Atlas of the United States* are all "atlases" as well. Furthermore, compared against the IGU Commission's criteria, many documents created in the late nineteenth and throughout the twentieth century, documents that are clearly atlases, would also fail to be identified as such. This issue of what is or what is not an atlas is not simply one of semantics. With the progression of technology and the ever-changing nature of a nation and its people, it seems likely to Monmonier that digital atlases are certain to be the future of such documents. It is important, then, that we recognize them as such and not limit ourselves to narrowly defined criteria.

Palsky, Gilles. "The Debate on the Standardization of Statistical Maps and Diagrams (1857–1901). Elements for the History of Graphical Language." *Cybergeo: European Journal of Geography* **(1999). http://cybergeo.revues. org/148**

The increasing use of data graphics to show statistics in the early and middle nineteenth century resulted in heterogeneous conventions for data display. This inconsistency had many in the scientific and statistics communities concerned. Some turned to the creation of governing bodies that could debate and codify rules governing the use of data graphics. Palsky's article details the creation of these international scientific congresses, their debates over and attempts to regulate this ever growing visual idiom, and their ultimate failure to do so. The first issue the international congresses set out to solve was classification. They attempted to prescribe a specific type of data graphic to certain data content, but before this problem was adequately resolved, future congresses moved on to the question of standardization. This second issue was also left unresolved, due to an inability to reach agreement on the rules governing choropleth maps. There were simply too many factors—shape, color, class intervals, etc.—to regulate in the diverse use of such maps. For Palsky, what these aborted debates represent is the inability to mitigate the differences between the artistic aspect of creating data graphics and the statisticians' needs to communicate clearly in unequivocal terms. This debate between style and content, Palsky concludes, echoes through current debates over information graphics, their design, and use.

Palsky, Gilles. "Emmanuel de Martonne and the Ethnographical Cartography of Central Europe (1917–1920)." *Imago Mundi* **54.1 (2002): 111–19.**

In support of the Paris Peace Conference following the end of the First World War, France established a committee of expert cartographers to help institute and display new national boundaries. Among others, this group included Emmanuel de Martonne, an expert in physical geography and a progenitor of geomorphology. Able to speak the language and especially familiar with the area, de Martonne was responsible for mapping Romania, including its geographical boundaries and its inhabitants. In order to show the landscape of the area, its population, and the ethnic breakdown of this population, de Martonne created a new sort of map that included several different types of information graphics. Despite his contention that the outcome was an objective picture of the land and its inhabitants, Palsky finds several elements of these graphics to be prejudiced, evidence of de Martonne's cultural biases and favoritism towards Romania. For instance, the graphic's title, which translates into English as "Distribution of Nationalities in Regions Dominated by Romanians," is far from impartial. Furthermore, de Martonne's choice of red, a color that stands out among the others, to represent the Romanians made the group appear larger and more significant than others in the region. These design choices influenced viewers' perception and made the Romanian population in the area seem predominate and of more consequence than other ethnicities. Perhaps the most egregious of de Martonne's offenses was the way in which he shaped the new country in response to a geographical ideal. Romania, he felt, should have a harmonious shape, one that fit within the geomorphic features, so he made it that way. Palsky concludes his article by noting that there was much more going on in the shaping of national boundaries during the Peace Conference than this single map, and so it is therefore hard to show exactly how much influence de Martonne's ethnographic maps had on the outcome of national boundaries. Nevertheless, the final shape given to Romania was generous, leading Palsky to speculate that de Martonne's efforts must have been influential.

Robinson, Arthur H. *Early Thematic Mapping in the History of Cartography*. Chicago: University of Chicago Press, 1982.

Robinson defines a thematic map as a map that "focuses on the differences from place to place of one class of feature, that class being the subject or 'theme' of the map" (16). He positions the development of thematic mapping inside the larger history of cartography, but his book is mostly focused on the years between 1800 and 1860, during which time the use and creation of thematic maps grew rapidly. Robinson attributes this rapid growth to developments in intellectual, social, and economic interests and attitudes, along with industrial and commercial developments. As society grew curious about the physical world and the relationships between that world and its inhabitants, thematic maps offered up new ways to display and understand this information. Using specific examples of early thematic maps from Edmond Halley, Alexander von Humboldt, Charles Joseph Minard, William Playfair, and many others, Robinson details the first appearances and subsequent developments of thematic mapping. He shows how these men's often pragmatic concerns for displaying correlations between different types of data led to a revolution in cartography, one that still influences thematic mapping today. This connection to the present is the author's stated impetus for writing the book. He concludes that it is "worthwhile to look back" because "[o]ften the basic truths and concepts of a field are made clearer by observing how the innovators coped with problems no one else ever faced" (219).

Spence, Ian. "No Humble Pie: The Origins and Usage of a Statistical Chart." *Journal of Educational and Behavioral Statistics* **30.4 (2005): 353–68.**

The field of information graphics owes much to the Scottish engineer and economist William Playfair. As Spence notes, Playfair was responsible for the creation and/or the popularization of several now common forms of data visualization, such as the bar chart, the line graph, and the pie chart. It is this last one that Spence explores in his article. In his 1801 book, *The Statistical Breviary*, Playfair included several graphs using circles to depict statistical data for European countries. In doing so he "introduced three new forms of the statistical graph: the circle diagram, the pie chart, and the Venn-like diagram" (356). According to Spence, it was typical for Playfair to name and explain how he developed his graphs, yet with the pie chart he was curiously silent. Spence speculates that this is because Playfair likely did not consider these types of charts his own and were instead influenced by the studies of his older brother and guardian, John Playfair. As a prominent mathematician and an ordained minister in the Church of Scotland, John had most likely read the works of previous centuries' mathematicians and Christian philosophers who used circle diagrams that could be considered rudimentary pie charts or Venn diagrams. Spence suggests it is likely Playfair adapted these earlier information graphics to suit his own needs. Regardless where they came from, though, these types of circle graphs, especially the pie chart, did not catch on during Playfair's time and in fact still remain a dubious way to display data. Critics contend that the way a person perceives and compares shapes can influence how accurately they assess the data in a pie chart. Spence closes his article with a rejoinder to such critics, pointing out that, when used properly, the pie chart can be a useful and appealing graphic, a fact attested to by its longevity and frequency of use.

Spence, Ian and Robert F. Garrison. "A Remarkable Scatterplot." *The American Statistician* **47.1 (1993): 12–19.**

Noting the pedagogical importance of graphs and their ability to show the unexpected, Spence and Garrison provide the history of an important but often overlooked one: the Hertzsprung-Russel diagram. Prior to the introduction of this diagram circa 1912, it was difficult for astronomers to fully comprehend the evolution of stars; there were just too many variables in size, position, and trajectory to form a concept of a star's evolution with any accuracy. The Hertzsprung-Russel diagram is a scatterplot of this multivariable data, one that makes clear just how a star lives and dies. This scatterplot, the authors explain, had a profound effect on the field of astronomy, one that echoes into the present-day. Of course, there are many more known stars today than there were at the time of the scatterplot's introduction. So, in order to show just how useful the scatterplot can be for visualizing such data, the authors modernize

Hertzspring-Russel's diagram with this expanded and updated data to show how the scatterplot, and other data visualizations like it, can provide tremendous insight into the movements of celestial objects and even the make-up of the universe itself.

Stigler, S.M. "Galton Visualizing Bayesian inference." *Chance* **24.1 (2011): 8–10.**

Bayes' Theorem, well known to statisticians, is a way of calculating conditional probability. In this article, Stigler wonders what this theorem looks like. He does not mean in the traditional, algebraic equation sense, but rather he wonders what a visualization of this theorem in calculating action would *physically* look like. For his answer, Stigler turns to Francis Galton, the nineteenth-century polymath who actually built a machine that did just this. Stigler acknowledges that Galton did not purposefully set out to visualize the Bayes Theorem, but the result of his attempts to visualize "the action of natural selection in a model for inheritance of quantitative characteristics" did just that (10). Since this sort of conditional probability calculating is what Bayes Theorem does, Stigler retroactively repurposed Galton's device to answer his questions about visual and physical representation. More important than the actual process of physically visualizing Baryes Theorem is the lesson Stigler learns from doing so: complex data of all types, including complicated mathematical calculations, is made more understandable through visualization.

Tilling, Laura. "Early Experimental Graphs." *The British Journal for the History of Science* **8.3 (1975): 193–213.**

Noting the usefulness of graphs to scientific experimenters and their audiences, Tilling sets out to trace the evolution of experimental graphs. Her search began with the gathering of numerous journals and textbooks published in the eighteenth century. Within this corpus, Tilling found only a few examples of experimental graphs. Oddly, though, Tilling also discovered that a number of scientific experiments conducted in the eighteenth century utilized measuring devices that produced graphical representations of data innately. Leopold's Universal Wind Machine and Watts's indicator of variations in pressure relative to the volume in a steam engine were devices that naturally delivered results in graphical form. Yet, Tilling claims that these men, like other scientists with similar devices, were not convinced of the value of data graphics and therefore published their results numerically and in tabulated form. There is one exception, however: J.H. Lambert. A philosopher and a scientist, Lambert published a number of graphs to explicate and substantiate his findings from experimental testing. Lambert's results from testing the periodic variation in temperature and rates of evaporation relative to temperature were all recorded and displayed in scientific journals of the time using graphs. Tilling acknowledges that Lambert's use of experimental graphs went largely unnoticed by his contemporaries and that it was not until some 50 years later that men like William Herschel and John Dalton revived their use. Tilling found no explanation for the gap in the use of data graphics, but she does remind us that despite a choppy start, the use of data graphics as part of scientific research is now so ubiquitous that they are often taken for granted.

Tufte, Edward R. *Envisioning Information*. Cheshire, CT: Graphics Press, 1990.

In this follow-up to his influential 1983 book *The Visual Display of Quantitative Information*, Tufte continues to lay out the strategies for excellence, accuracy, and beauty in information graphics. Using both historical and contemporary examples, he shows how graph makers "escape the flatland" of two dimensionality to create information graphics with depth and density, graphs that truly can be called excellent. A large part of Tufte's criteria for graphical excellence relies on the relationship between the micro and macro data presented in the graphic. In other words, the small details and the larger visual composition of an information graphic should strike an effective balance. A viewer should get a sense of the data being shown, regardless how complex, just by looking at the information graphic. A closer inspection, however, should provide the viewer a deeper understanding, perhaps even a uniquely personal one, of the data presented. When this balance between the micro and the macro elements of the display is off, clutter and confusion occur, distorting the graphic's purpose. Layering, separation, and color, when done well, are ways designers can ensure that data, no matter how complex, are not obscured. Another way the designer can accomplish a clear and useful presentation of data is by using small multiples, the numerous displays of data variations designed and exhibited in ways that make comparison easier. The issue of comparison is crucial to the effective information graphic for Tufte. In fact, he states that "[a]t the heart of quantitative reasoning is a single question: *Compared to what?*" (67). Tufte concludes his book with a discussion of information graphics that must display multivariate data, such as time and space, in the flatland. Timetables are his main focus because the issues they present—to display large amounts of multivariate data—represent a culmination of all the needs and goals of information graphics: density, complexity, and beauty.

Tufte, Edward R. *The Visual Display of Quantitative Information.* **Cheshire, CT: Graphics Press, 1983.**

A highly influential work in the field of data graphics, Tufte's book details the principles of graphical excellence and the theories for creating beautiful and effective graphical designs. Tufte uses the work of data graphics inventors and innovators—such as J.H. Lambert, William Playfair, and Charles Joseph Minard—to provide a history lesson on the creation and development of data graphics, while also creating the historical foundations for his own principles of graphical excellence. Excellent data graphics, according to Tufte, provide complex data in easily interpretable formats with "clarity, precision, and efficiency" (p. 51). They also have graphical integrity, a phrase Tufte uses to define data graphics that tell the truth. The level of graphical integrity has much to do with context, meaning whether a graph could be considered deceptive or not relies heavily on the numbers, words, and visual elements of the graph that accompany the data. But a graph's integrity is also connected to the designers who created it. Tufte argues that bad graphics are the result of graphic designers who know too little about statistics, or statisticians who know too little about graphic design. Those who hope to make effective graphs, free from error or deception, should be competent at striking a balance between form and content. Perhaps the greatest value of Tufte's book is his creation of an entire lexicon for discussing data graphics. Chart-junk (the superfluous visual elements of a graph that can cloud the data), data-ink (the essential content of a graphic), and multifunctioning (the use of a graphical element to show more than one aspect of the data) are some of the many terms that Tufte uses to explain his theories of good graphical design. These terms, and indeed Tufte's whole body of work, continue to shape and influence the discussions and research on data visualizations.

Tukey, John W. "Data-based Graphics: Visual Display in the Decades to Come." *Statistical Science* **6.3 (1990): 327–59.**

In this article, prominent statistician and expert on the visual display of data John Wilder Tukey outlines 19 ideas and questions that will shape the future of data visualization. Despite being over two decades old, Tukey's article, indeed his whole body of work, has proven to be both prescient and influential to the field. For Tukey, much of how good data graphics work relies on their ability to go beyond just the display of numbers and to reveal the phenomena hidden in the data. When revealed, such phenomena can be impactful, enabling viewers to recognize patterns and trends previously invisible. This, Tukey maintains, is the greatest possibility of data visualization. Color, flow, comparison, and the influence of both cartographers and graphic designers are discussed in order for Tukey to explain what works and what does not when it comes to creating clear data graphics. Furthermore, he maintains that visual data display experienced a deep transition caused by the work of Jacques Bertin, whose descriptions of effective data graphics modernized the field. The future, Tukey predicts, will include even more changes. The progression of computers, their use to compute previously incomputable numbers, and the possibility of dynamic graphics will all change the design and use of visual data display greatly. Hindsight allows us to see just how right Tukey was, while also reminding us that more changes are likely to come. As such, it would be wise to follow the moral of Tukey's article: let us go forward with an understanding of our past, so that we can be active in the shaping of our future.

Wainer, Howard. *Graphic Discovery: A Trout in the Milk and other Visual Adventures.* **Princeton, NJ: Princeton University Press, 2005.**

Wainer, a scientist and statistician, uses history, humor, and personal anecdotes to show how data graphing began, when it works best, and where it is likely to go in the future. The book is divided into three sections, the first being a detailed biography of William Playfair. Unlike many other writers on Playfair, Wainer does not limit his discussion to the man's data graphics inventions. Playfair was, among many things, an engineer, an accountant, a silversmith, a pamphleteer, and a publicist. He was also a blackmailer, a one-time convict, and a participant in the storming of the Bastille. For Wainer, above all other jobs or titles, and for the way he revolutionized the design and display of data in his visual inventions, Playfair is the "father of modern graphical display" (5). The second part of *Graphic Discovery* presents numerous examples of data visualizations to show how discovery is facilitated, or sometimes suppressed, by their use. Using personal examples, such as the graphs he frequently finds in *The New York Times* or in the junk mail for life insurance he receives, Wainer shows how order, scale, sampling, and size can all influence the way data is perceived. In the final section, Wainer details conversations he held with the late John Wilder Tukey, a genius in the field of statistical graphing and data visualizations. From their conversations, Wainer predicts that, with the aid of computers, data graphics will become even more revelatory. Dynamic displays of information will offer new ways to display data, which, in turn, will provide users even more ways to understand and extrapolate data. Before concluding his book, Wainer provides a list of the dramatis personae for the history of data graphics. This list, spanning several centuries, delivers brief bios and contributions of important people in the field, rounding out an interesting and informative discussion on data visualizations.

Wainer, Howard. "Graphical Visions from William Playfair to John Tukey."
***Statistical Science* 5.6 (1990): 340–46.**

Wainer compares three points of agreement and one difference of opinion
between statisticians and information graphics designers William Playfair
and John Tukey. The first point on which these two historically separated yet
philosophically close designers agree upon, according to Wainer, is that an
information graphic must reveal the data vividly, in ways that have an impact on
the viewer. Even though many, namely Edward R. Tufte, have long advocated
for clear, simple designs that lend themselves to transparency, that Playfair
and Tukey agree on the occasional value of a dramatic display tells us that we
should perhaps not always be so quick to seek maximization of the data-ink
ratio. Another point on which Playfair and Tukey seem to agree is that users
must understand the graph. This may mean the inclusion of a useful legend or
a lengthy preface. Regardless the method, though, without user understanding,
a graph can have little value, or worse could be misleading. Both Playfair and
Tukey also appreciated and valued the use of graphs to show things that might
have otherwise gone unnoticed. Indeed, it could be said that showing what was
previously unseen is the whole purpose of information graphics. Wainer admits
that he had to search hard to find a difference between Playfair's and Tukey's
principles of information graphics, but he does point to a very specific issue that
distinguishes the two men's ideas: the use of curved lines. Tukey found them to
be "always poor graphics," while Playfair did not, as evidenced in his famous
"Chart of Exports and Imports to and from the East Indies." Wainer's article is
not intended to identify only the similarities and a difference between these two
men's ideas of graphical display. Instead, Wainer's juxtaposition of Playfair and
Tukey shows us how prescient Playfair was and, consequently, how prescient
Tukey may prove to be.

**Yates, JoAnne. "Graphs as a Managerial Tool: A Case Study of Du Pont's
Use of Graphs in the Early Twentieth Century."** *Journal of Business
Communication* **22.1 (January 1985): 5–33.**

Given the current ubiquity of data visualizations, especially in business, it
can be difficult for contemporary audiences to appreciate that at one time they
were new and therefore not widely understood devices for communicating
information. Perhaps even more surprising, some business executives of the
past argued *against* the use of data visualizations, labeling them as overly
simplistic and functionless. Such was the case at Du Pont during the very early
part of the twentieth century. Yet, as Yates shows through her extensive research
of the Du Pont Corporation, once employees and executives began to use data
visualizations, they caught on quickly and spread rapidly. Citing the recorded
minutes of meetings and other relevant historical documents, Yates shows that

in 1904 Du Pont and its executives used very few data visualizations; however, by 1915 they were an intrinsic part of the corporation's operations and ensuing success. While Yates shows that there was a confluence of factors contributing to the rapid rise of data visualizations at Du Pont, she gives the most credit to Irenee Du Pont. In 1909, Irenee worked as the Director of the Development Department. Years later he would become CEO. During his time at Du Pont, Irenee strongly advocated for the use of data visualizations. His advocacy eventually culminated in the advent of the chart room. The chart room, a mechanized room filled with hundreds of different charts that could be brought before the viewer, enabled Du Pont executives to keep track of the diverse departments of the corporation without being directly involved in the daily operation of them or suffering from information overload. The chart room was so successful that it remained unchanged and secret from competitors for almost 50 years. The lesson to be learned from Yates's article is one of symbiosis. Data visualizations, she writes, are "most valuable when they work closely with organizational needs" (29). According to her, both business managers and educators should remember this lesson.

Works Cited

A Square [Abbott, Edwin]. *Flatland: A Romance of Many Dimensions*. London: Seeley & Co., 1884.

Adams, Dean C., F. James Rohlf, and Dennis E. Slice. "Geometric Morphometrics: Ten Years of Progress Following the 'Revolution.'" *Italian Journal of Zoology* 71.1 (2004): 5–16.

Advertisement. "Mulhall's Dictionary of Statistics." *The Athenaeum* 3298 (1891): 66.

Alborn, Timothy L. "Fletcher, Joseph (1813–1852)." *Oxford Dictionary of National Biography*. Oxford: Oxford University Press, 2004.

Amaral, Ana Rita, Maria M. Coelho, Jesus Marugán-Lobón, and F. James Rohlf. "Cranial Shape Differentiation in Three Closely Related Delphinid Cetacean Species: Insights into Evolutionary History." *Zoology* 112 (2009): 38–47.

American Statistical Association, Statistical Computing. *Statistical Graphics Video Library*. Web. http://stat-graphics.org/movies/

Angoff, Charles, and H.L. Mencken. "The Worst American State, Part I." *The American Mercury* 24.93 (September 1931): 1–16.

____. "The Worst American State, Part II." *The American Mercury* 24.94 (October 1931): 175–88.

____. "The Worst American State, Part III." *The American Mercury* 24.95 (November 1931): 355–71.

Arbuthnot, John. "An Argument for Divine Providence Taken from the Constant Regularity Observ'd in the Births of both Sexes." *Philosophical Transactions of the Royal Society* 27 (1710): 186–90.

Aristotle. *On Rhetoric: A Theory of Civic Discourse*. Trans. George A. Kennedy. 2nd ed. New York: Oxford University Press, 2007.

Arnheim, Rudolf. *Art and Visual Perception: A Psychology of the Creative Eye: The New Version*. Berkeley, CA: University of California Press, 1974.

____. *Entropy and Art: An Essay on Disorder and Order*. Berkeley, CA: University of California Press, 1974.

____. Personal Communication to Lee Brasseur. 1988.

____. *The Power of the Center: A Study of Composition in the Visual Arts, the New Version*. Berkeley, CA: University of California Press, 1988.

Artemeva, Natasha, and Aviva Freedman, eds. *Rhetorical Genre Studies and Beyond*. Winnipeg, Manitoba, CA: Inkshed, 2008. Web. http://http-server.carleton. ca/~nartemev/Artemeva%20&%20Freedman%20Rhetorical%20Genre%20 Studies%20and%20beyond.pdf

Asimov, Daniel. "The Grand Tour: A Tool for viewing Multidimensional Data." *SIAM Journal of Scientific and Statistical Computing* 6.1 (1985): 128–43.

"Australasian Statistics." *Journal of the Statistical Society of London* 46.1 (1883): 128–45.

Balbi, Adriano, and André-Michel Guerry. *Statistique Comparée de l'État de l'Instruction et du Nombre des Crimes dans les Divers Arrondissements des Académies et des Cours Royales de France*. Paris: Jules Renouard, 1829.

Barton, Ben F., and Marthalee S. Barton. "Ideology and the Map: Toward a Postmodern Visual Design Practice." *Professional Communication: The Social Perspective*. Edited by Nancy Roundy Blyler and Charlotte Thralls. Newbury Park, CA: SAGE Publications, 1993. 49–78.

____. "Modes of Power in Technical and Professional Visuals." *Journal of Business and Technical Communication* 7.1 (1993): 138–62.

Bate, Jonathan. *Romantic Ecology: Wordsworth and the Environmental Tradition*. London: Routledge, 1991.

Bazerman, Charles. *Shaping Written Knowledge: The Genre and Activity of the Experimental Article in Science*. Madison, WI: University of Wisconsin Press, 1988.

Beck, Henry. *The London Underground Map*. London, 1931.

Becker, Richard A., and William S. Cleveland. "Brushing Scatterplots." Cleveland and McGill 201–24.

Becker, Richard A., William S. Cleveland, B. Kleiner, and J.L. Warner. *Ozone in the Northeast*. ASA Statistical Graphics Video Lending Library, 1978. Web. http://stat-graphics.org/movies/ozone.html

Becker, Richard A., and Robert McGill. "Brushing a Scatter Plot Matrix." ASA Statistical Graphics Video Lending Library. Web. http://stat-graphics.org/movies/brushing.html

____. *Dynamic Displays of Data*. ASA Statistical Graphics Video Lending Library, 1985. Web. http://stat-graphics.org/movies/dynamic-displays.html

Beniger, James R., and Dorothy L. Robyn. "Quantitative Graphics in Statistics: A Brief History." *The American Statistician* 32.1 (1978): 1–11.

Berkenkotter, Carol, and Thomas N. Huckin. *Genre Knowledge in Disciplinary Communication: Cognition/Culture/Power*. Hillsdale, NJ: Lawrence Erlbaum Associates, 1995.

Bertin, Jacques. *Graphics and Graphic Information-Processing*. Trans. William J. Berg and Paul Scott. New York: Walter De Gruyter, 1981.

____. *Semiology of Graphics: Diagrams, Networks, Maps*. Trans. William J. Berg. Madison, WI: University of Wisconsin Press, 1983.

Bessant, Kenneth C., and Eric D. MacPherson. "Thoughts on the Origins, Concepts, and Pedagogy of Statistics as a Separate Discipline." *The American Statistician* 56.1 (2002): 22–8.

Biderman, Albert D. "The Playfair Enigma: The Development of the Schematic Representation of Statistics." *Information Design Journal* 6.1 (1990): 3–25.

Birch, T.W. *Maps, Topographical and Statistical*. 2nd ed. Oxford: Oxford University Press, 1964.

Blunden, Edmund. *Undertones of War*. Garden City, NJ: Doubleday, 1929.

Bookstein, Fred L. Email to Alan Gross. 16 June 2011.

____. "A Hundred Years of Morphometrics." *Acta Zoologica Academiae Scientiarum Hungaricae* 44.1–2 (1998): 7–59.

____. *The Measurement of Biological Shape and Shape Change: Lecture Notes in Biomathematics*. Vol. 24. Edited by S. Levin. Berlin: Springer-Verlag, 1978.

Bose, K.C. "The Prevalence of Smallpox in Calcutta and How to Arrest its Spread." *Indian Medical Gazette* 25 (1890): 77–85.

Bradley, Leslie, ed. *Smallpox Inoculation: An Eighteenth-Century Mathematical Controversy*. Nottingham: University of Nottingham Press, 1971.

Brandes, H.W. "Einige Resultate aus der Witterungs-Geschichte des Jahres 1783, und Bitte um Nachrichten aus jener Zeit; aus einem Schreiben des Professor Brandes an Gilbert." *Annalen der Physik* 61 (1819): 421–6.

Brasseur, Lee E. "Florence Nightingale's Visual Rhetoric in the Rose Diagrams." *Technical Communication Quarterly* 14.2 (2005): 161–82.

____. *Visualizing Technical Information: A Cultural Critique*. Amityville, NY: Baywood, 2003.

Brewer, Cynthia A., and Trudy A. Suchan. *Mapping Census 2000: The Geography of U.S. Diversity*. Washington, DC: U.S. Government Printing Office, 2001.

Brinton, Willard C. *Graphic Methods for Presenting Facts*. New York: The Engineering Magazine Co., 1914.

____. *Graphic Presentation*. New York: Brinton Associates, 1939.

Brown, Kevin. "The Evolution of the Comparative Mountains and Rivers Chart in the 19th Century." May 12, 2010. *Geographicus Antique Map Blog*. Web. http://www.geographicus.com/blog/rare-and-antique-maps/the-evolution-of-the-comparative-mountains-and-rivers-chart-in-the-19th-century/

Buache, Philippe. *Carte Minéralogique où l'on Quoit la Nature et la Situation des Terreins qui Traversent la France et l'Angleterre*. Paris, France, 1746.

____. "Essai de Géographie Physique." *Mémoires de L'Académie Royale des Sciences*, 1752: 399–416.

Buja, Andreas, and Daniel Asimov. "Grand Tour Methods: An Outline." *Computing Science and Statistics* 17 (1986): 63–7.

Buja, Andreas, Dianne Cook, Daniel Asimov, and Catherine Hurley. "Computational Methods for High-dimensional Rotations in Data Visualization." Rao, Wegman, and Solka 391–413.

Buja, Andreas, and Paul A. Tukey. *Dataviewer: A Program for Looking at Data in Several Dimensions*. ASA Statistical Graphics Video Lending Library, 1987. Web. http://stat-graphics.org/movies/dataviewer.html

Büsching, Anton Friedrich. *Neue Erdbeschreibung*. Vol. 1. Hamburg, Germany, 1754.

Byatt, A.S. "Trench Names." *The New Yorker* 6 April 2009. Web. http://www.newyorker.com/fiction/poetry/2009/04/06/090406po_poem_byatt

Caan, Woody, and Dawn Hillier. "How Do We Perceive Risks?" *Communicating Health Risks to the Public: A Global Perspective.* Edited by Dawn Hillier. Aldershot: Gower, 2006. 33–46.

Cairo, Heriberto. "The Field of Mars: Heterotopias of Territory and War." *Political Geography* 23 (2004): 1009–36.

Camerini, Jane R. "Heinrich Berghaus's Map of Human Diseases." *Medical History* 44 Supplement 20: Medical Geography in Historical Perspective (January 2000): 186–208.

Card, Stuart K., Jock D. Mackinlay, and Ben Shneiderman. *Readings in Information Visualization: Using Vision to Think.* San Francisco, CA: Morgan Kaufmann Publishers, 1999.

Carter, James R. "The Map Viewing Environment: A Significant Factor in Cartographic Design." *American Cartographer* 15 (1988): 379–85.

_____. "Weather Maps on Television in the USA." *Proceedings of the Sixteenth International Cartographic Conference, Cologne, Germany,* 1993. 244–54.

Chang, Jih-Jie. *Real-time Rotation.* 1970. ASA Statistical Graphics Video Lending Library. Web. http://stat-graphics.org/movies/rotation.html

Chasseaud, *Rats Alley: British Trench Names of the Western Front 1914–18.* Staplehurst, UK: Spellmount Ltd. Publishers, 2006.

_____. *Topography of Armageddon: British Trench Map Atlas of the Western Front, 1914–18.* Lewes, UK: Mapbooks, 1991.

Cifelli, Robert, Nolan Doesken, Patrick Kennedy, Lawrence D. Carey, Steven A. Rutledge, Chad Gimmestad, and Tracy Depue. "The Community Collaborative Rain, Hail, and Snow Network: Informal Education for Sand Citizens." *Bulletin of the American Meteorological Society* 86.8 (August 2005): 1069–77.

Cleveland, William S., and Marylyn E. McGill, eds. *Dynamic Graphics for Statistics.* Monterey, CA: Wadsworth, 1988.

Cleveland, William S., and Robert McGill. "Graphic Perception: Theory, Experimentation, and Application to the Development of Graphical Methods." *Journal of the American Statistical Association* 79.387 (1984): 531–54.

Codd, E.F. "Further Normalization of the Data Base Relational Model." IBM Research Report, San Jose, CA, RJ909. 1971.

Cohen, I. Bernard. "Florence Nightingale." *Scientific American* 250.3 (1984): 128–37.

Comte, Auguste. *Cours de Philosophie Positive.* Vol. 4. Paris: Bachelier, 1839.

Cook, Dianne, Andreas Buja, Eun-Kyung Lee, and Hadley Wickham. "Grand Tours, Projection Pursuit Guided Tours and Manual Controls." *Handbook of Data Visualization.* Edited by Chun-houh Chen, Wolfgang Härdle, and Antony Unwin. Vol. 3 of *Handbook of Computational Statistics.* Heidelberg: Springer, 2008. 295–314.

Cook, Dianne, and Deborah F. Swayne. *Interactive and Dynamic Graphics for Data Analysis with Examples Using R and GGobi.* New York: Springer, 2007.

Cook, Edward Tyas. *The Life of Florence Nightingale.* Vol. 1: *1820–1861.* London: Macmillan and Co., 1914.

Cook, R. Dennis, and Sanford Weisberg. *An Introduction to Regression Graphics.* New York: Wiley, 1994.

Cook, Robert, and Howard Wainer. "A Century and a Half of Moral Statistics in the United Kingdom: Variations on Joseph Fletcher's Thematic Maps." *Significance* 9.3 (2012): 31–6.

Coquerelle, Michelle, Fred L. Bookstein, Jose Braga, and Demetrios J. Halazonetis. "Fetal and Infant Growth Patterns of the Mandibular Symphysis in Modern Humans and Chimpanzees *(Pan troglodytes)*." *Journal of Anatomy* 217 (2010): 507–20.

Cottereau, Alain. "The Fate of Collective Manufactures in the Industrial World: The Silk Industries of Lyons and London, 1800–1850." *World of Possibilities: Flexibility and Mass Production in Western Industrialization.* Trans. C. Blamires. Edited by Charles F. Sabel and Jonathan Zeitlin. Cambridge: Cambridge University Press 2002. 75–152.

Covello, Vincent T., and Jeryl Mumpower. "Risk Analysis and Risk Management: An Historical Perspective." *Risk Analysis* 5.2 (1985): 103–20.

Cozzens, Suzan. "Comparing the Sciences: Citation Context Analysis of Papers from Neuropharmacology and the Sociology of Science." *Social Studies of Science* 15.1 (1985): 27–53.

Crampton, Jeremy W., and John Krygier. "An Introduction to Critical Cartography." *ACME: An International E–journal for Critical Geographies* 4 (2006): 11–33.

Crawley, Charles R. "From Charts to Glyphs: Rudolf Modley's Contribution to Visual Communication." *Technical Communication* 41.1 (1994): 20–26.

Croix, David de la, and Omar Licandro. *The Mean Life of Famous People from Hammurabi to Einstein.* Web. http://www.fcs.edu.uy/archivos/BCU_clebrities.pdf

Crome, August F.W. *Über die Grösse und Bevölkerung der Sämtlichen Europäischen Staaten.* Leipzig: Weygand, 1785.

Cumpston, J.H.L. *The History of Smallpox in Australia, 1788–1908.* Melbourne: Commonwealth of Australia, Quarantine Service. Service Publication #3. 1914.

Damasio, Antonio R. *Descartes' Error: Emotion, Reason and the Human Brain.* New York: Avon Books, 1994.

Darton, William, and W.R. Gardner. *New and Improved View of the Comparative Heights of the Principal Mountains and Lengths of the Principal Rivers in the World, the Whole Judiciously arranged from the Various Authorities extant.* London: William Darton, 1823. *David Rumsey Map Collection.* Web. http://www.davidrumsey.com/luna/servlet/detail/RUMSEY~8~1~240~20049:New-and-Improved-View-of-the-Compar?trs=57&mi=34&qvq=mgid%3A2151

Das, Santanu. "Slimescapes." *World War I Centenary: Continuations and Beginnings.* Web. http://ww1centenary.oucs.ox.ac.uk/body-and-mind/slimescapes/

Desnoyers, Luc. "Toward a Taxonomy of Visuals in Science Communication." *Technical Communication* 58.2 (2011): 119–34.

Diamond, Marion, and Mervyn Stone. "Nightingale on Quetelet." *Journal of the Royal Statistical Society.* Series A (General). 144.1 (1981): 66–79.

Dilke, O.A.W. "Cartography in the Ancient World: An Introduction." *The History of Cartography*. Vol. 1: *Cartography in Prehistoric, Ancient, and Medieval Europe and the Mediterranean*. Edited by J.B. Harley and David Woodward. Chicago: University of Chicago Press, 1987. 105–6.

Dodge, J.R. *Centennial Album of Agricultural Statistics, Including Maps, Charts, Diagrams, Illustrations of Industrial Colleges, and Type Specimens of Breeds of Farm Animals*. Washington, DC: U.S. Department of Agriculture, 1876.

Doesken, Nolan, and Henry Reges. "The Value of the Citizen Weather Observer." *Weatherwise* 63.6 (2010): 30–37.

Dorling, Daniel. *Human Geography of the UK*. London: SAGE Publications, 2005.

Dorling, Daniel, and Bethan Thomas. *Bankrupt Britain: An Atlas of Social Change*. Bristol: Policy Press, 2011.

Dossey, Barbara Montgomery. *Florence Nightingale: Mystic, Visionary, Healer*. Springhouse, PA: Springhouse Corporation, 1999.

Doubleday, Thomas. *True Law of Population*. New York: A.M. Kelley, 1842.

Dragga, Sam, and Dan Voss. "Cruel Pies: The Inhumanity of Technical Illustrations." *Technical Communication* 48.3 (2001): 265–74.

Duncan, W., and D. Duncan. *Comparative Heights of Mountains*. Glasgow: Blackie and Son, 1853.

Dupin, Charles. *Carte Figurative de l'Instruction Populaire de la France*. Paris, 1826.

Eagleton, Terry. *Literary Theory*. Minneapolis, MN: University of Minnesota Press, 1983.

Eddy, William F., and Audris Mockus. *Incidence of Disease Mumps*. ASA Statistical Graphics Video Lending Library, 1994. Web. http://stat-graphics. org/movies/mumps.html

____. *Manufacturing Process Data*. ASA Statistical Graphics Video Lending Library, 1993. Web. http://stat-graphics.org/movies/manufacturing-process. html

Eddy, William F., and Shingo Oue. *Display of U.S. Air Traffic*. ASA Statistical Graphics Video Lending Library, 1994. Web. http://stat-graphics.org/movies/ air-traffic.html

Edison, William G. "Confusion, Controversy, and Quarantine: The Muncie Smallpox Epidemic of 1893." *Indiana Magazine of History* 86.4 (1990): 374–98.

Edney, Matthew H. "Cartography without 'Progress': Reinterpreting the Nature and Historical Development of Map Making." *Cartographica* 30.2/3 (1993): 54–68.

Elkins, James. *The Domain of Images*. Ithaca, NY: Cornell University Press, 2001.

Epidemic of Smallpox in Muncie, 1893. Indianapolis, IN: W.B. Burford, 1894.

Farebrother, Richard William. *Fitting Linear Relationships: A History of the Calculus of Observations 1750–1900*. New York: Springer, 1999.

Farr, William F., ed. "Diagram Representing the Force of Mortality in Small Pox. Days 18–35." *British Medical Almanack; With a Supplement*. London: John

Churchill, 1838. Web. http://wellcomeimages.org/indexplus/image/L0069605. html

Faulkner, Ronnie W. "Hinton Rowan Helper (1829–1909)." North Carolina History Project, 2012. Web. http://www.northcarolinahistory.org/commentary/345/entry

Faure, Fernand. "The Development and Progress of Statistics in France." *The History of Statistics, Their Development and Progress in Many Countries.* Edited by John Koren. New York: Macmillan, 1918. 217–329.

Ferraris, Victor A., and Suellen P. Ferraris. "Risk Stratification and Comorbidity." *Cardiac Surgery in the Adult.* Edited by Lawrence H. Cohn and L. Henry Edmunds, Jr. New York: McGraw-Hill, 2003. 187–224.

Feynman, Richard. *The Character of Physical Law.* Messenger Lectures, 1964. Cambridge: MIT Press, 2001.

Fiebrich, Christopher A. "History of Surface Weather Observations in the United States." *Earth Science Reviews* 93 (2008): 77–84.

Findlen, Paula, Dan Edelstein, and Nicole Coleman. *Mapping the Republic of Letters.* 2008. Web. <https://republicofletters.stanford.edu/

Fisher, Danyel. "Animation for Visualization: Opportunities and Drawbacks." Steele and Iliinsky 329–52.

Fisherkeller, M., Jerome H. Friedman, and John W. Tukey. "PRIM-9: An Interactive Multidimensional Data Display and Analysis System." ASA Statistical Graphics Video Lending Library, 1974. Web. http://stat-graphics. org/movies/prim9.htmlhttp://stat-graphics.org/movies/prim9.html

Flandreau, Marc. "Caveat Emptor: Coping with Sovereign Risk under the International Gold Standard, 1871–1913." *International Financial History in the Twentieth Century: System and Anarchy.* Edited by Marc Flandreau, Carl-Ludwig Holtferich, and Harold James. Cambridge: Cambridge University Press, 2003. 17–50

Fletcher, Joseph. *Education: National, Voluntary, and Free.* London: James Ridgway, 1851.

_____. "Moral and Educational Statistics of England and Wales." *Quarterly Journal of the Statistical Society of London* 10.3 (September 1847): 193–233.

_____. "Moral and Educational Statistics of England and Wales." *Quarterly Journal of the Statistical Society of London* 12 (August 1849): 151–76, 189–335.

_____. "Progress of Crime in the United Kingdom: Abstracted from the Criminal Returns for 1842, and the Prison Returns for the Year Ended at Michaelmas, 1841." *Quarterly Journal of the Statistical Society of London* 6 (August 1843): 218–40.

_____. "Reports from Assistant Hand-Loom Weavers Commissioners, Part IV." *The Sessional Papers of the House of Lords* 38 (April 1840): 1–372.

_____. *Summary of the Moral Statistics of England and Wales.* Privately printed and distributed. N.d.

Foege, William H. *House on Fire: The Fight to Eradicate Smallpox.* Berkeley, CA: University of California Press, 2011.

_____. "Plagues: Perceptions of Risk and Social Responses." *In Time of Plague: The History and Social Consequences of Lethal Epidemic Disease.* Edited by Arien Mack. New York: New York University Press, 1991. 9–20.

Ford, E.B., and J.S. Huxley. "Genetic Rate Factors in Gammarus." *Archiv für Entwicklungsmechanik der Orgsanismen* 117 (1929): 67–79.

Foucault, Michel. "Of Other Spaces." *Diacritics* 16 (1986): 22–7.

Fourcroy, Charles de. *Essai d'Une Table Poléométrique, ou Amusement d'un Amateur de Plans sur les Grandeurs de Quelques Villes.* Paris: Dupain-Triel, 1782.

Foville, Achille-Louis de. "The Abuse of Statistics." Trans. M. de Foville. *Journal of the Royal Statistical Society* 50.4 (1887): 703–8.

Frere de Montizon, Armant J. *Carte Philosophique Figurant la Population de la France.* 1830.

Friendly, Michael. "A Brief History of Data Visualization." *Handbook of Data Visualization.* Edited by Chun-houh Chen, Wolfgang Härdle, and Antony Unwin. Vol. 3 of *Handbook of Computational Statistics.* Heidelberg: Springer, 2008. 15–56.

_____. "A Brief History of the Mosaic Display." *Journal of Computational and Graphical Statistics* 11.1 (2002): 89–107.

_____. *A Brief History of the Mosaic Display.* 2001. Web. http://www.datavis.ca/papers/moshist.pdf

_____. "A.-M. Guerry's 'Moral Statistics of France': Challenges for Multivariable Spatial Analysis." *Statistical Science* 22.3 (2007): 368–99.

_____. *Data Visualization Gallery.* Web. http://datavis.ca/gallery/timelines.php

_____. *Gallery of Data Visualization.* Web. http://datavis.ca/gallery/

_____. "The Golden Age of Statistical Graphics." *Statistical Science* 23.4 (2008): 502–35.

_____. *Histdata: Data Sets from the History of Statistics and Data Visualization.* R package version. 2011.

_____. "Milestones in the History of Data Visualization: A Case Study in Statistical Historiography." *Classification: The Ubiquitous Challenge.* Edited by C. Weihs and W. Gaul. New York: Springer, 2005. 34–52.

_____. "Mosaic Displays for Multi-way Contingency Tables." *Journal of the American Statistical Association* 89 (1994): 190–200.

_____. "Visions and Re-visions of Charles Joseph Minard." *Journal of Educational and Behavioral Statistics* 27.1 (2002): 31–51.

Friendly, Michael, and Daniel J. Denis. "The Early Origins and Development of the Scatterplot." *Journal of the History of the Behavioral Sciences* 41.2 (Spring 2005): 103–30.

_____. *Milestones in the History of Thematic Cartography, Statistical Graphics, and Data Visualization.* Web. http://www.datavis.ca/milestones/

Friendly, Michael, and S. Dray. "Guerry: Maps, Data and Methods Related to Guerry (1833) 'Moral Statistics of France.'" R package version 1.4. 2010.

Friendly, Michael, and Gilles Palsky. "Visualizing Nature and Society." *Maps: Finding our Place in the World.* Edited by James R. Akerman and Robert W. Karrow, Jr. Chicago: University of Chicago Press, 2007. 205–51.

Friendly, Michael, Pedro Valero-Mora, and Joaquin Ibáñez Ulargui. "The First (Known) Statistical Graph: Michael Florent van Langren and the 'Secret' of Longitude." *The American Statistician* 64.2 (2010): 174–84.

Friis, Herman R. "Statistical Cartography in the United States Prior to 1870 and the Role of Joseph C.G. Kennedy and the U.S. Census Office." *The American Cartographer* 1 (1974): 131–57.

Fry, Ben. "Salary vs. Performance: What Baseball Teams are Spending Their Money Well, and How Does It Change over the Course of the Season?" 2014. Web. http://benfry.com/salaryper/

Funkhouser, H[oward] Gray. "Historical Development of the Graphical Representation of Statistical Data." *Osiris* 3 (1937): 269–404.

____. "A Note on a Tenth-Century Graph." *Osiris* 1 (1936): 260–62.

Galton, Francis. "Regression towards Mediocrity in Hereditary Stature." *Journal of the Anthropological Institute* 15 (1886): 246–63.

Gannett, Henry, and the U.S. Census Office. *Statistical Atlas of the United States, Based upon Results of the Eleventh Census.* Washington, DC: Government Printing Office, 1898. Library of Congress, Geography and Map Division: Cultural Landscapes. Web. http://www.loc.gov/item/07019233/

____. *Statistical Atlas: Prepared under the Supervision of Henry Gannett, Geographer of the Twelfth Census.* Washington, DC: U.S. Census Office, 1903.

Gehan, Edmund A., and Noreen A. Lemak. *Statistics in Medical Research: Developments in Clinical Trials.* New York: Plenum Medical Book Company, 1994.

Gelman, Andrew. *Red State, Blue State, Rich State, Poor State: Why Americans Vote the Way they Do.* Princeton, NJ: Princeton University Press, 2008.

Gentle, James E., Wolfgang Härdle, and Yuichi Mori, eds. *Handbook of Computational Statistics: Concepts and Methods.* Berlin/Heidelberg, Germany: Springer, 2004. Web. http://sfb649.wiwi.hu-berlin.de/fedc_homepage/xplore/ebooks/html/csa/node81.html

German Empire. *Vaccination Law of April 8th 1874.* B. Paul Berlin SW 48, 1904.

Gesmann, Markus, and Diego de Castillo. *googleVis: Interface between R and the Google Visualisation API.* R package version 0.2.12. 2011.

Gilbert, E.W. "Pioneer Maps of Health and Disease in England." *The Geographical Journal* 124.2 (1958): 172–83.

Gliddon, Gerald. *Somme 1916: A Battlefield Guide.* Dover, NH: Sutton Publishing, 1995.

Godlewska, Anne Marie Claire. "From Enlightenment Vision to Modern Science? Humboldt's Visual Thinking." *Geography and Enlightenment.* Edited by David N. Livingstone and Charles W.J. Withers. Chicago: University of Chicago Press, 1999. 236–80.

Goldie, Sue M. *Florence Nightingale: Letters from the Crimea 1854–1856.* Manchester: Manchester University Press, 1997.

Grabill, Jeffrey T., and W. Michele Simmons. "Toward a Critical Rhetoric of Risk Communication: Producing Citizens and the Role of Technical Communicators." *Technical Communication Quarterly* 7.4 (1998): 415–41.

Graunt, John. *Natural and Political Observations on the London Bills of Mortality.* London: John Martyn and James Allestry, 1662.

Great Britain War Office. Geographical Section. *Report on Survey on the Western Front, 1914–1918.* London: His Majesty's Stationery Office, 1920. Web. http://discovery.nationalarchives.gov.uk/SearchUI/details?uri=C2357603

Gregg, W.R., and I.R. Tannehill. "International Standard Projections for Meteorological Charts." *Monthly Weather Review* 65 (1937): 411–15.

Griggs, A.L. "The Background and Development of Weather Charts." *Bulletin, Geography and Map Division, Special Libraries Association* no. 21 (1955): 10–13.

Guerry, A[ndre]-M[ichel]. *Statistique morale de l'Angleterre Comparée avec la Statistique Morale de la France, d'après les Comptes de l'administration de la Justice Cen Angleterre et en France, etc.* Paris: J.-B. Baillière et fils, 1864.

Guettard, Jean-Etienne. "Mémoire et Carte Minéralogique sur la Nature et la Situation des Terreins qui Traversent la France and l'Angleterre." *Histoire de l'Académie Royale des Sciences.* Paris, 1751. 363–92.

Hald, Anders. *A History of Probability and Statistics and their Application before 1750.* New York: John Wiley and Sons, 1990.

Halley, E[dmond]. "An Historical Account of the Trade Winds, and Monsoons, Observable in the Seas between and near the Tropicks; with an Attempt to Assign the Physical Cause of Said Winds." *Philosophical Transactions of the Royal Society* 16.183 (1688): 153–68.

Handa, Carolyn. *Visual Rhetoric in a Digital World: A Critical Sourcebook.* Boston, MA: Bedford/St. Martin's, 2004.

Hankins, Thomas L. "A 'large and graceful sinuosity': John Herschel's Graphical Method." *Isis* 97.4 (2006): 605–33.

———. "Blood, Dirt, and Nomograms: A Particular History of Graphs." *Isis* 90.1 (1999): 50–80.

Harley, J.B. "Deconstructing the Map." *Cartographica: The International Journal for Geographic Information and Geovisualization* 26.2 (1989): 1–20.

Harness, Henry Drury. *Atlas to Accompany the Second Report of the Railway Commissioners, Ireland.* Dublin: Irish Railway Commission, 1837.

Harris, Robert. L. *Information Graphics: A Comprehensive Illustrated Reference: Visual Tools for Analyzing, Managing, and Communicating.* Atlanta, GA: Management Graphics, 1996.

Hartigan, J.A., and B. Kleiner. "Mosaics for Contingency Tables." *Computer Science and Statistics: Proceedings of the 13th Symposium on the Interface.* Edited by William F. Eddy. New York: Springer-Verlag, 1981. 268–73.

Haygarth, John. *An Inquiry How to Prevent the Smallpox and Proceedings of a Society for Promoting General Inoculation at Stated Periods, and Preventing the Natural Smallpox in Chester*. London: Cadell and Davies, 1801.

Heiser, W.J. "Early Roots of Statistical Modelling." *Social Science Methodology in the New Millennium: Proceedings of the Fifth International Conference on Logic and Methodology*. Edited by Jörg Blasius, J. Hox, E. de Leeuw, and P. Schmidt. Amsterdam: TT-Publikaties, 2000. (CD-ROM publication).

Heiskell, Henry L. "The Commercial Weather Map of the United States Weather Bureau." *Yearbook of the Department of Agriculture* (1912): 537–9.

Helper, Hinton Rowan. *The Impending Crisis of the South: How to Meet it*. New York: Burdick Brothers, 1857.

Hewes, Fletcher W., and Henry Gannett. *Scribner's Statistical Atlas of the United States*. New York: Charles Scribner's Sons, 1883. Web. Library of Congress, Geography and Map Division: Cultural Landscapes. Web. http://www.loc.gov/item/a40001834/

Hilgard, J[ulius] E[rasmus]. "The Advance of Population in the United States." *Scribner's Monthly* 4.2 (1872): 214–18.

Hoff, Hebbel E., and Leslie A. Geddes. "The Beginnings of Graphic Recording." *Isis* 53 (1962): 287–324. Pt. 3.

____. "Graphic Recording before Carl Ludwig: An Historical Summary." *Archives Internationales d'Histoire des Sciences* 12 (1959): 3–25.

Hofmann, Heike. "Mosaic Plots." *Encyclopedia of Measurement and Statistics*. Edited by Neil J. Salkind and Kristin Rasmussen. Vol. 2. Thousand Oaks, CA: SAGE Publications, 2007. 631–3.

Hofmann, Heike, and Martin Theus. *MANET: Missings Are Now Equally Treated*. 2000. Web. http://rosuda.org/MANET/

Hopkins, Donald R. *The Greatest Killer: Smallpox in History*. Chicago: University of Chicago Press, 2002.

Hopkins, Eric. *Birmingham: The First Manufacturing Town in the World, 1760–1840*. London: Weidenfeld and Nicolson, 1989.

Horn, Robert. *Visual Language: Global Communication in the 21st Century*. Bainbridge Island, WA: Macrovu, 1998.

Howat, Jeremy. "Edward T. Mulhall and Michael G. Mulhall." 2002. Web. http://www.argbrit.org/Mul1863/MulBros.htm

Humboldt, Alexander von. *Atlas Géographique et Physique du Royaume de la Nouvelle-Espange*. Paris: F. Schoell, 1811. *David Rumsey Map Collection*. Web. http://www.davidrumsey.com/luna/servlet/view/search?QuickSearchA= QuickSearchA&q=Atlas+Géographique+et+Physique+du+Royaume+Humb oldt&sort=pub_list_no_initialsort%2Cpub_date%2Cpub_list_no%2Cseries_ no&search=Search

____. "Sur Les Lignes Isothermes." *Annales de Chimie et de Physique* 5 (1817): 102–12.

Hunt, Stephen E. *Green Romanticism*. Saarbrücken: VDM Verlag, 2011. Web. http://www.barnesandnoble.com/w/green-romanticism-stephen-e-hunt/1105318874

Hurley, Catherine B., and R.W. Oldford. "Statistical Graphics in QUAIL: An Overview." 1997. Web. https://www.stat.fi/isi99/proceedings/arkisto/varasto/hurl0937.pdf

Huxley, Julian S. "Appendix." *Problems of Relative Growth*. New York: Dover, 1932.

____. *Problems of Relative Growth*. New York: Dover, 1932.

____. *Problems of Relative Growth*. 2nd ed. New York: Dial Press, 1972.

Huygens, Christiaan. *Oeuvres completes*, Tome Sixieme Correspondance. The Hague: Martinus Nijhoff, 1895. 515–18, 526–39.

IBM. *Many Eyes*. Web. http://www.manyeyes.com/software/analytics/manyeyes/

Iliinsky, Noah. "On Beauty." Steele and Iliinsky 1–13.

Italie, Hillel. "Study Reveals the Rise of Feminine Pronouns." *The Independent* 10 August, 2012. Web. http://www.independent.co.uk/arts-entertainment/books/news/study-reveals-rise-of-feminine-pronouns-8030574.html

Jenner, Edward. *An Inquiry into the Causes and Effects of the Variolæ Vaccinæ, A Disease Discovered in Some of the Western Counties of England, Particularly Gloucestershire, and Known by the Name of the Cow Pox*. 3rd ed. London: Printed for the Author by D.N. Shury, 1801.

Johnson, A.J. *Johnson's Chart of Comparative Heights of Mountains, and Lengths of Rivers of Africa ... Asia ... Europe ... South America ... North America.* Edited by A.J. Johnson. *Johnson's New Illustrated Family Atlas of the World.* New York: A.J. Johnson, 1874.

Johnson, Steven B. *The Ghost Map: The Story of London's Most Terrifying Epidemic and How It Changed Science, Cities, and the Modern World*. New York: Riverhead Books, 2006.

Kannel, William B., Thomas R. Dawber, Abraham Kagan, Nicholas Revotskie, and Joseph Stokes III. "Factors of Risk in the Development of Coronary Heart Disease: Six-Year Follow-up Experience—the Framingham Study." *Annals of Internal Medicine* 55 (1961): 33–50.

Karsten, Karl G. *Charts and Graphs: An Introduction to Graphic Methods in the Control and Analysis of Statistics*. New York: Prentice-Hall, 1923.

Keegan, John. *The First World War*. New York: Knopf, 1999.

Keith, Jocelyn M. "Florence Nightingale: Statistician and Consultant Epidemiologist." *International Nursing Review* 35.5 (1988): 147–50.

Kendall, David. Comment on "Size and Shape Spaces for Landmark Data in Two Dimensions." *Statistical Science* 2 (1986): 222–6.

____. "The Diffusion of Shape." *Advances in Applied Probability* 9 (1977): 428–30.

Kimball, Miles A. "London through Rose-Colored Graphics: Visual Rhetoric and Information Graphic Design in Charles Booth's Maps of London Poverty." *Journal of Technical Writing and Communication* 36.4 (2006): 353–81.

King, Clarence. "Production of the Precious Metals in the United States." *Second Annual Report of the United States Geological Survey to the Secretary of the Interior 1880-'81*. Comp. J.W. Powell. Washington, DC: Government Printing Office, 1882. 331–401.

Kircher, Athanasius. *Magnes Sive de Arte Magnetica*. Rome, 1641.

____. *Mundus Subterraneus, quo Universae Denique Naturae Divitiae.* Rome, 1665.

Klein, Judy I. *Statistical Visions in Time: A History of Time Series Analysis, 1662–1938.* Cambridge: Cambridge University Press, 1997.

Koch, Tom. "The Map as Intent: Variations on the Theme of John Snow." *Cartographica* 39.4 (2004): 1–14.

Koplow, David A. *Smallpox: The Fight to Eradicate a Global Scourge.* Berkeley, CA: University of California Press, 2003.

Kostelnick, Charles. "Melting-Pot Ideology, Modernist Aesthetics, and the Emergence of Graphical Conventions: The Statistical Atlases of the United States, 1874–1925." *Defining Visual Rhetorics.* Edited by Charles A. Hill and Marguerite Helmers. Mahwah, NJ: Lawrence Erlbaum Associates, 2004. 215–42.

____. "Supra-Textual Design: The Visual Rhetoric of Whole Documents." *Technical Communication Quarterly* 5.1 (1996): 9–33.

____. "The Visual Rhetoric of Data Displays: The Conundrum of Clarity." *IEEE Transactions on Professional Communication* 51.1 (2008): 116–30.

Kostelnick, Charles, and Michael Hassett. *Shaping Information: The Rhetoric of Visual Conventions.* Carbondale, IL: Southern Illinois University Press, 2003.

Kravis, Irving B. "Comparative Studies of National Incomes and Prices." *Journal of Economic Literature* 22.1 (1984): 1–39.

Kress, Gunther R., and Theo van Leeuwen. *Reading Images: The Grammar of Visual Design.* London: Routledge, 1996.

Kruskal, J.B. *Multidimensional Scaling.* ASA Statistical Graphics Video Lending Library, 1962. Web. http://stat-graphics.org/movies/multidimensional-scaling.html

Kruskal, W. "Visions of Maps and Graphs." *Proceedings of the International Symposium on Computer-Assisted Cartography, Auto-Carto II.* Edited by J. Kavaliunas. Washington, DC: U.S. Bureau of the Census, 1975. 27–36.

Lancaster, Gillian A. "How Statistical Literacy, Official Statistics and Self-directed Learning Shaped Social Enquiry in the 19th and early 20th Centuries." *Statistical Journal of the IAOS* [International Association for Official Statistics] 27.3/4 (2011): 99–111.

Langren, Michael Florencio van. *La Verdadera Longitud por Mar y Tierra.* Antwerp, 1644.

Lima, Manuel. *Visual Complexity.* Web. http://www.visualcomplexity.com/vc/about.cfm

Lipkus, Isaac M., and J.G. Hollands. "The Visual Communication of Risk." *Journal of the National Cancer Institute Monographs* 25 (1999): 149–63.

Lupton, Ellen. "Reading Isotype." *Design Issues* 3.2 (1986): 47–58.

MacDonald, Lyn. *Somme.* London: Penguin, 1993.

Macdonald-Ross, Michael. "How Numbers Are Shown: A Review of Research on the Presentation of Quantitative Data in Texts." *Audio-Visual Communication Review* 25.4 (1977): 359–409.

MacEachern, Alan M. *How Maps Work: Representation, Visualization and Design.* New York: Guildford Press, 1995.

Magnello, M. Eileen. "Victorian Vital and Mathematical Statistics." *BSHM Bulletin* 21 (2006): 219–29.

Makdisi, Saree. "Worldly Romanticism." *Nineteenth-Century Literature* 35.4 (2011): 429–32.

Marey, Etienne-Jules. *La Méthode Graphique dans les Sciences Expérimentales et Principalement en Physiologie et en Médecine.* Paris: G. Masson, 1885.

Marois, Michael B., and Rodney Yap. "Californian's $609,000 Check Shows True Retirement Cost." *Bloomberg.com.* 14 December 2012. Web. http://www.bloomberg.com/news/2012-12-14/californian-s-609-000-check-shows-true-retirement-cost.html

Masefield, John. *The Old Front Line.* New York: Macmillan, 1917.

Mayo-Smith, Richmond. Review of Mulhall's *Industries and Wealth of Nations. Political Science Quarterly* 12.2 (1897): 351.

Mayr, Georg von. *Die Gesetzmässigkeit im Gesellschaftsleben: Statistische Studien.* Munich: Oldenbourg, 1877.

McDonald, John A. *Antelope: Data Analysis with Object-oriented Programming and Constraints.* ASA Statistical Graphics Video Lending Library, 1987. Web. http://stat-graphics.org/movies/antelope.html

McDonald, John A., and Steve Willis. *Use of the Grand Tour in Remote Sensing.* ASA Statistical Graphics Video Lending Library, 1987. Web. http://stat-graphics.org/movies/tour-sensing.html

McDonald, Lynn. "Florence Nightingale: Passionate Statistician." *Journal of Holistic Nursing* 28.1 (2002): 92–8.

McDonald, Lynn, ed. *Florence Nightingale on Society and Politics, Philosophy, Science, Education and Literature. Collected Works of Florence Nightingale.* Vol. 5. Waterloo, ON: Wilfred Laurier University Press, 2003.

____. *Women Theorists on Society and Politics.* Waterloo, Ontario, CA: Wilfrid Laurier University Press, 1998.

McKay, Betsy. "Bioterror Fears Prompt US to Keep Its Smallpox Cache." *Wall Street Journal* 18 January 2011. Web. http://online.wsj.com/news/articles/SB10001424052748704029704576088032149613692

MeasuringWorth.com. (n.d.). Web. http://www.measuringwortgh.com/ukcompare/result.php

Medical Society of the State of Pennsylvania. *Vaccination: A Message from the Medical Society of the State of Pennsylvania.* Harrisburg, PA: Medical Society of the State of Pennsylvania, 1908.

Meitzen, Alfred, and Roland P. Falkner. "History, Theory, and Technique of Statistics. Part Second: Theory and Technique of Statistics." *Annals of the American Academy of Political and Social Science* 1 (1891): 101–237.

Meyer, Morton A., Frederick R. Broome, and Richard H. Schweitzer, Jr. "Color Statistical Mapping by the US Bureau of the Census." *The American Cartographer* 2 (1975): 100–117.

Mill, John Stuart. *Principles of Political Economy with Some of their Applications to Social Philosophy*. London: John W. Parker, 1848.

Miller, Carolyn R. "Genre as Social Action." *Quarterly Journal of Speech* 70.2 (1984): 151–67.

Minard, Charles-Joseph. *Des Tableaux Graphiques et des Cartes Figuratives*. Paris: E. Thunot, 1861.

Ministère des Travaux Publics. *Album de Statistique Graphique*. Paris: Imprimerie Nationale, 1881.

Ministère du Travail et de la Prévoyance Sociale. *Album Graphique de la Statistique Générale de la France*. Paris: Imprimerie Nationale, 1907.

Mitchell, S. Augustus. "Heights of the Principal Mountains in the World. Lengths of the Principal Rivers." *A New Universal Atlas*. 1846. *David Rumsey Map Collection*. Web. http://www.davidrumsey.com/luna/servlet/detail/ RUMSEY~8~1~1963~150001:Heights-Of-The-Principal-Mountains-?trs=57& mi=23&qvq=mgid%3A2151

———. *Western Hemisphere*. Philadelphia: S. Augustus Mitchell, 1872.

Monmonier, Mark. *Air Apparent: How Meteorologists Learned to Map, Predict, and Dramatize Weather*. Chicago: University of Chicago Press, 1999.

———. "Authoring Graphic Scripts: Experiences and Principles." *Cartography and Geographic Information Systems* 19 (1992): 247–60, 272.

———. *Bushmanders and Bullwinkles: How Politicians Manipulate Electronic Maps and Census Data to Win Elections*. Chicago: University of Chicago Press, 2001.

———, ed. *Cartography in the Twentieth Century*. Vol. 6 of *The History of Cartography*. Chicago: University of Chicago Press, 2015.

———. "Coping with Qualitative-Quantitative Data in Meteorological Cartography: Standardization, Ergonomics, and Facilitated Viewing." *Proceedings of the Nineteenth International Cartographic Conference, Ottawa, Canada* 1 (1999): 947–54.

———. "Geographic Brushing: Enhancing Exploratory Analysis of the Scatterplot Matrix." *Geographical Analysis* 21 (1989): 81–4.

———. *How to Lie with Maps*. 2nd ed. Chicago: University of Chicago Press, 1996.

———. "The Rise of the National Atlas." *Cartographica* 31.1 (1994): 1–15.

Mulhall, Michael G. *Balance-Sheet of the World for Ten Years, 1870–1880*. London: Edward Stanford, 1881.

———. *The Cotton Fields of Paraguay and Corrientes*. Buenos Ayres: M.G. and E.T. Mulhall, 1864.

———. *The Dictionary of Statistics*. 3rd ed. London: G. Routledge and Sons, 1892. Web. http://catalog.hathitrust.org/Record/001885954

———. *The Dictionary of Statistics*. 4th ed., rev. to November 1898. London: G. Routledge and Sons, 1899.

———. "Five Years of American Progress." *North American Review*. 169 (July/Dec. 1899): 536–44.

———. *Fifty Years of National Progress, 1837–1887*. London: G. Routledge, 1887.

_____. *Handbook of Brazil*. London: Spottiswoode and Co., 1877.

_____. *History of Prices since the Year 1850*. London: Longmans, Green & Co., 1885. Web. http://www.archive.org/stream/cu31924013930908#page/n3/mode/2up

_____. *Industries and Wealth of Nations*. London and New York: Longmans, Green & Co., 1896.

_____. *Mulhall's Dictionary of Statistics*. 1st ed. London: Routledge, 1884.

_____. *Mulhall's Dictionary of Statistics*. New ed., rev. and corrected. 2nd ed. London: G. Routledge and Sons, 1886.

_____. *National Progress in the Queen's Reign, 1837–1897*. London: G. Routledge, 1897.

_____. *The Progress of the World in Arts, Agriculture, Commerce, Manufactures, Instruction, Railways, and Public Wealth since the Beginning of the Nineteenth Century*. London: Edward Stanford, 1880.

_____. "Progress of the United States: I.—The New England States." *The North American Review* 164 (May 1897): 566–75.

_____. "Progress of the United States: II.—The Middle States." *The North American Review* 164 (June 1897): 685.

_____. "Progress of the United States: III.—The Southern States." *The North American Review* 165 (July 1897): 43.

_____. "Progress of the United States: IV.—The Prairie States." *The North American Review* 165 (August 1897): 183.

_____. "Progress of the United States: V.—The Pacific States." *The North American Review* 165 (September 1897): 310.

_____. *Rio Grande do Sul and its German colonies*. London: Longmans, Green & Co., 1873.

_____. *The River Plate Hand-book, Guide, Directory, and Almanac for 1863: Comprising the City and Province of Buenos Ayres, the other Argentine Provinces, Montevideo, etc.* Buenos Ayres: M.G. and E.T. Mulhall, 1863.

Murray, Edmundo. "Mulhall, Michael George (1836–1900)." *Dictionary of Irish Latin American Biography*. Society for Irish Latin American Studies, 2009. Web. http://www.irlandeses.org/dilab_mulhallmg.htm

National Institutes of Health (NIH). National Library of Medicine. "Variolation." *Smallpox: A Great and Terrible Scourge*. Web. https://www.nlm.nih.gov/exhibition/smallpox/sp_variolation.html

National Research Council (NRC). *Improving Risk Communication*. Washington, DC: National Academies Press, 1989.

Neurath, Otto. *International Picture Language: The First Rules of Isotype*. London: Keegan Paul, Trench, Trubner & Co., 1936.

New South Wales Minister for Health and Medical Research. "Smallpox (Variola)." *Communicable Diseases Factsheet*. 1 July 2012. Web. http://www.health.nsw.gov.au/Infectious/factsheets/Factsheets/smallpox.PDF

Newton, Carol M. "Graphics: From Alpha to Omega in Data Analysis." Wang 59–92.

Nightingale, Florence. *Florence Nightingale on Society and Politics, Philosophy, Science, Education and Literature. Collected Works of Florence Nightingale.* Vol. 5. Edited by Lynn McDonald. Waterloo, ON: Wilfred Laurier University Press, 2003.

____. "Letter to 'Pop'" [Parthenope] 24 February 1830. MS. 8991. *Nightingale Letters.* Claydon Collection 1827–39 in the Rare Materials Room. London: Wellcome Library, 1830.

____. *Notes on Hospitals.* London: Longman, Green, Longman, Roberts and Green, 1863.

Oldford, R.W., and S.C. Peters. *Data analysis networks in DINDE.* ASA Statistical Graphics Video Lending Library, 1986. Web.

Olson, Judy M. "Spectrally Encoded Two-variable Maps." *Annals of the Association of American Geographers* 71.2 (June 1981): 259–76.

Owen, Wilfred. "Mental Cases." Breen 76.

____. "Strange Meeting." Breen 62.

____. *Wilfred Owen: Selected Poetry and Prose.* Edited by Jennifer Breen. London: Routledge, 2014.

Palsky, Gilles. "The Debate on the Standardization of Statistical Maps and Diagrams (1857–1901). Elements for the History of Graphical Language." *Cybergeo: European Journal of Geography* (1999). Web. http://cybergeo. revues.org/148

____. *Des Chiffres et des Cartes: Naissance et Developpement de la Cartographie Quantitative Française au XIXe Siècle.* Paris: Ministere de l'Enseignment Supérieur et de la Recherché. Comité des Travaux Historiques et Scientifiques (CTHS), 1996.

____. "Emmanuel de Martonne and the Ethnographical Cartography of Central Europe (1917–1920)." *Imago Mundi* 54.1 (2002): 111–19.

Patriarca, Silvana. *Numbers and Nationhood: Writing Statistics in Nineteenth-century Italy.* Cambridge: Cambridge University Press, 2003.

Peale, Richard S. *The Home Library of Useful Knowledge.* Chicago: The Home Library Association, 1886.

Pearson, E.S., ed. *The History of Statistics in the 17th and 18th Centuries against the Changing Background of Intellectual, Scientific and Religious Thought.* London: Griffin & Co. Ltd, 1978.

Pearson, Karl, and G.M. Morant. "The Wilkinson Head of Oliver Cromwell and its Relationship to Busts, Masks and Painted Portraits." *Biometrika* 26.3 (1934): 1–116.

Playfair, William. *The Commercial and Political Atlas Representing, by Means of Stained Copper-plate Charts, the Exports, Imports, and General Trade of England: The National Debt, and Other Public Accounts, with Observations and Remarks.* London: J. Debrett, 1786.

____. *The Commercial and Political Atlas.* 3rd ed. London: T. Burton, 1801.

____. *The Commercial and Political Atlas: Representing, by Copper-Plate Charts, the Progress of the Commerce, Revenues, Expenditure, and Debts of England, during the Whole of the Eighteenth Century*. London: Debrett, Robinson, and Sewell, 1786. Rpted in facsimile, Howard Wainer and Ian Spence, eds. *The Commercial and Political Atlas and Statistical Breviary*. Cambridge: Cambridge University Press, 2005.

____. *Statistical Breviary; Shewing, on a Principle Entirely New, the Resources of Every State and Kingdom in Europe*. London: Wallis, 1801. Rptd. in facsimile, Howard Wainer and Ian Spence, eds. *The Commercial and Political Atlas and Statistical Breviary*. Cambridge: Cambridge University Press, 2005.

Ponce de León, Marcia S., and Christoph P.E. Zollikofer. "Neanderthal Cranial Ontology and its Implications for Late Hominid Diversity." *Nature* 412 (2 August 2001): 534–8.

Porter, Theodore M. *The Rise of Statistical Thinking, 1820–1900*. Princeton, NJ: Princeton University Press, 1986.

Priestley, Joseph. *A Chart of Biography*. London: J. Johnson, 1765.

"The Progress of the World." *Nation* 31.801 (4 November 1880): 331.

Propen, Amy. "Visual Communication and the Map: How Maps as Visual Objects Convey Meaning in Specific Contexts." *Technical Communication Quarterly* 16 (2007): 233–54.

Quetelet, Adolphe. *Sur l'Homme et le Développement de ses Facultés, or Essai de Physique Sociale*. Paris: Bachelier, 1835.

____. *The Propensity to Crime*. Brussels, 1831.

Rafferty, Anne Marie, and Rosemary Wall. "Rereading Nightingale: Notes on Hospitals." *International Journal of Nursing Studies* 47.9 (2010): 1063–5.

Rao, C.R., Edward J. Wegman, and Jeffrey L. Solka, eds. *Handbook of Statistics: Data Mining and Visualization*. North Holland, Netherlands: Elsevier, 2005.

Rapkin, John. "A Comparative View of the Principal Waterfalls, Islands, Lakes, Rivers, and Mountains, in the Eastern Hemisphere." *The Illustrated Atlas, and Modern History of the World Geographical, Political, Commercial, and Statistical*. Edited by R. Montgomery Martin. London and New York: John Tallis, 1851.

Ravenstein, E.G. "The Progress of the World in Arts, Agriculture, Commerce, Manufactures, Instruction, Railways, and Public Wealth since the Beginning of the Nineteenth Century." *The Academy* 430 (31 July 1880): 74.

Rebok, Sandra. "Alexander von Humboldt's Perceptions of Colonial Spanish America." *Dynamis* 29 (2009): 49–72. Web. http://www.raco.cat/index.php/Dynamis/article/view/136829/238995

Review of "Mulhall's *Dictionary of Statistics*." *The Athenaeum* 2932 (1884): 22.

Riddell, Ronald C. "Parameter Disposition in pre-Newtonian Planetary Theories." *Archive for History of Exact Sciences* 23 (1980): 87–157.

Robinson, Arthur H. *Early Thematic Mapping in the History of Cartography*. Chicago: University of Chicago Press, 1982.

____. "The Genealogy of the Isopleth." *Cartographic Journal* 8 (1971): 49–53.

Rohlf, F. James, and Leslie F. Marcus. "A Revolution in Morphometrics." *Tree* 8 (1993): 129–32.

Rosen, George. *A History of Public Health*. Baltimore: The Johns Hopkins University Press, 1993.

Rosenberg, Daniel, and Antony Grafton. *Cartographies of Time: A History of the Timeline*. New York: Princeton Architectural Press, 2010.

Royston, Erica. "Studies in the History of Probability and Statistics, III. A Note on the History of the Graphical Presentation of Data." *Biometrika* 43. 3/4 (December 1956): 241–7. Rpt. *Studies in the History of Statistics and Probability Theory*. Edited by E.S. Pearson and Maurice G. Kendall. London: Griffin, 1970.

Rubin, Ernest. "The Place of Statistical Methods in Modern Historiography." *American Journal of Economics and Sociology* 2.2 (January 1943): 193–210.

Rumsey, David. *David Rumsey Map Collection*. Cartography Associates. Web. http://www.davidrumsey.com/home

____. "Heights of Mountains, Lengths of Rivers." *David Rumsey Map Collection*. Web. http://www.davidrumsey.com/blog/2009/9/5/heights-of-mountains-lengths-of-rivers

Ruskin, John. "The Nature of Gothic." *The Stones of Venice*. 2nd ed. Vol. 2. London: Smith, Elder and Co., 1867. 151–231

Rusnock, Andrea. "Catching Cowpox: The Early Spread of Smallpox Vaccination, 1798–1810." *Bulletin of the History of Medicine* 83.1 (2009): 17–36.

Russel, S.P. "Synoptic Chart Illustrating the Weekly Mortality by the Two Epidemics—Cerebro-spinal Fever and Small Pox." National Institutes of Health. US National Library of Medicine. *Images from the History of Medicine*.

Russell, David R. "Rethinking Genre in School and Society: An Activity Theory Analysis." *Written Communication* 14.4 (1997): 504–54.

Ryerson, Charles C., and Allan C. Ramsay. "Quantitative Ice Accretion Information from the Automated Surface Observing System." *Journal of Applied Meteorology and Climatology* 46 (2007): 1423–47.

Said, Edward. *Orientalism*. New York: Vintage Books, 1978.

Salch, John D., and David W. Scott. "Data Exploration with the Density Grand Tour." *Statistical Graphics and Computing Newsletter* 8.1 (1997): 7–11.

Samet, Jonathan M., Robert Schnatter, and Herman Gibb. "Invited Commentary: Epidemiology and Risk Assessment." *American Journal of Epidemiology* 148.10 (1998): 929–36.

Sarkar, Deepayan. *Lattice: Multivariate Data Visualization with R*. New York: Springer, 2008.

Sassoon, Siegfried. "Declaration against the War." 1917. *The First World War Poetry Digital Archive*. Web. http://www.oucs.ox.ac.uk/ww1lit/education/tutorials/intro/sassoon/declaration.html

Sauer, Beverly. *The Rhetoric of Risk: Technical Documentation in Hazardous Environments*. Mahwah, NJ: Lawrence Erlbaum, 2003.

Schloerke, Barret, ed. "Geo Zoo." *Geometric Data*. Web. http://streaming.stat.iastate.edu/~dicook/geometric-data/

Schmid, Calvin F. *Handbook of Graphic Presentation*. New York: Ronald, 1954.

Schryer, Catherine F. "The Lab vs. the Clinic: Sites of Competing Genres." *Genre and the New Rhetoric*. Edited by Aviva Freedman and Peter Medway. London: Taylor and Francis, 1994. 105–24.

Schulten, Susan. "How a Map Divided Virginia." *New York Times*. 5 May 2011. Web. http://opinionator.blogs.nytimes.com/2011/05/05/how-a-map-divided-virginia/

_____. *Mapping the Nation: History and Cartography in Nineteenth-century America*. Chicago: University of Chicago Press, 2012.

Shell, Charlotte M., and Karen D. Dunlap. "Florence Nightingale, Dr. Ernest Codman, American College of Surgeons Hospital Standardization Commission: Four Pillars in the Foundation of Patient Safety." *Perioperative Nursing Clinics* 3.1 (2008): 19–26.

Shneiderman, Ben. "The Eyes Have It: A Task by Data Type Taxonomy for Information Visualizations." *Proceedings of the IEEE Symposium on Visual Languages, VL '96*. Washington, DC: IEEE Computer Society Press, 1996. 336–43.

Short, James F. "The Social Fabric of Risk: Toward the Social Transformation of Risk Analysis." *American Sociological Review* 49.6 (1984): 711–25.

Shryock, Richard Harrison. *The Development of Modern Medicine: An Interpretation of the Social and Scientific Factors Involved*. Madison, WI: University of Wisconsin Press, 1979.

SIMILE. Exhibit 3.0. MIT Libraries with MIT Computer Science and Artificial Intelligence Lab. Web. http://www.simile-widgets.org/exhibit3/

SIMILE. Timeline Widget. Web. http://www.simile-widgets.org/timeline

Singer, Judith D., and John B. Willett. *Applied Longitudinal Data Analysis*. Oxford: Oxford University Press, 2003.

Sloane, Charles S. *Statistical Atlas of the United States*. U.S. Bureau of the Census. Washington, DC: Government Printing Office, 1914.

_____. *Statistical Atlas of the United States*. U.S. Bureau of the Census. Washington, DC: Government Printing Office, 1925.

Smith, Cecil Woodham. *Florence Nightingale: 1820–1910*. New York: McGraw-Hill Book Company, 1951.

Smith, Laurence D., Lisa A. Best, D. Alan Stubbs, John Johnston, and Andrea Bastiani Archibald. "Scientific Graphs and the Hierarchy of the Sciences: A Latourian Survey of Inscription Practices." *Social Studies of Science* 30 (2000): 73–94.

Smith, William. *A Delineation of the Strata of England and Wales*. London: John Cary, 1815.

Snow, John. *On the Mode of Communication of Cholera*. London: 1855.

Society for Promoting Christian Knowledge. "Comparative View of the Principal Rivers in the Four Quarters of the World." *Saturday Magazine Supplement* 26. 1 (November 1832).

_____. *Map of the Principal Rivers of the World*. [London]: Chapman and Hall, 1844.

Spence, Ian. "No Humble Pie: The Origins and Usage of a Statistical Chart." *Journal of Educational and Behavioral Statistics* 30.4 (2005): 353–68.

Spence, Ian and Robert F. Garrison. "A Remarkable Scatterplot." *The American Statistician* 47.1 (1993): 12–19.

Spence, Robert. *Information Visualization*. Harlow: Addison-Wesley, 2001.

Statistischen Bureau des eidg. Departements des Innern. *Graphisch-statistischer Atlas der Schweitz*. Bern: Stämpfli & Cie, 1897. Web. https://archive.org/details/graphischstatis00unkngoog

Steele, Julie, and Noah Iliinsky, eds. *Beautiful Visualization: Looking at Data through the Eyes of Experts*. Sebastopol, CA: O'Reilly Media, 2010.

Stigler, Stephen M. "Galton Visualizing Bayesian Inference." *Chance* 24.1 (2011): 8–10.

____. *The History of Statistics: The Measurement of Uncertainty before 1900*. Cambridge: The Belknap Press of Harvard University Press, 1986.

____. *Statistics on the Table: The History of Statistical Concepts and Methods*. Cambridge: Harvard University Press, 1999.

Stockcharts.com. "MarketCarpet: S & P Sector ETFs." Treemap chart for Aug. 26–27, 2014. *Stockcharts.com*. 17 September 2014. Web. http://stockcharts.com/freecharts/carpet.html

Street, Andrew. "The Resurrection of Hospital Mortality Statistics in England." *Journal of Health Services Research and Policy* 7.2 (2002): 104–10.

Stroup, Donna F., and Ruth L. Berkelman. "History of Statistical Methods in Public Health." *Statistics in Public Health: Quantitative Approaches in Public Health Problems*. Edited by Donna F. Stroup and Steven M. Teutsch. Oxford: Oxford University Press, 1998. 1–18.

Sutherland, Ian. "John Graunt: A Tercentenary Tribute." *Journal of the Royal Statistical Society* Series A (General) 126.4 (1963): 537–56.

Swales, John M. *Genre Analysis: English in Academic and Research Settings*. Cambridge: Cambridge University Press, 1990.

Swayne, Deborah F., Dianne Cook, and Andreas Buja. "XGobi: Dynamic Graphics for Data Analysis." ASA Statistical Graphics Video Lending Library, 1991.

Swayne, Deborah F., and Sigbert Klinke. "Introduction to the Special Issue on Interactive Graphical Data Analysis: What is Interaction?" *Computational Statistics* 14.1 (1999): 1–6.

Swayne, Deborah F., Duncan Temple Lang, Andreas Buja, and Dianne Cook. "GGobi: Evolving from XGobi into an Extensible Framework for Interactive Data Visualization." *Journal of Computational Statistics and Data Analysis* 43.4 (2003): 423–44. Web. http://www.ggobi.org

Swift, Jonathan. "Thoughts on Various Subjects (Further Thoughts on Various Subjects)." *Miscellanies*. Vol. 10. London: Robert Dodley, 1745. 232–43. Web. http://jonathanswiftarchive.org.uk/browse/title/text_12_3_2.html?page=d2e70

Symanzik, Jürgen. "Interactive and Dynamic Graphics." Gentle, Härdle, and Mori 293–336.

Thacker, Steven B. "Historical Development." *Principles and Practice of Public Health Surveillance*. Edited by Steven M. Teutsch and R. Elliott Churchill. Oxford: Oxford University Press, 2000. 1–16.

Theus, Martin. Mondrian. 2005. Web. http://www.theusrus.de/Mondrian/

Theus, Martin, and Stephan R.W. Lauer. "Visualizing Loglinear Models." *Journal of Computational and Graphical Statistics* 8.3 (1999): 396–412.

Thompson, D'Arcy. *On Growth and Form*. London: Cambridge University Press, 1917.

____. *On Growth and Form*. 2nd ed. Cambridge: Cambridge University Press, 1945.

Thomson, John, and William Home Lizars. *Comparative View of the Height of the Falls of Foyers and Corba Linn*. Drawn by William Home Lizars. Engd. by William Home Lizars. Edinburgh: John Thomson, 1832. *David Rumsey Map Collection*. Web. http://www.davidrumsey.com/luna/servlet/detail/ RUMSEY~8~1~35301~1180926:A-comparative-view-of-the-lengths-o?qv q=q:comparative;lc:RUMSEY~8~1&mi=3&trs=228

Tierney, Luke. *LispStat: An Object-orientated Environment for Statistical Computing and Dynamic Graphics*. New York: Wiley, 1991.

Tilling, Laura. "Early Experimental Graphs." *The British Journal for the History of Science* 8.3 (1975): 193–213.

Trabert, Wilhelm. *Meteorologie und Klimatologie*. Leipzig and Vienna: Franz Deuticke, 1905.

Tribou, Alex. "State Employees Cash in on Lump-Sum Retirement Checks." *Bloomberg.com*. October 2014. Web. http://www.bloomberg.com/chart/ iQ3Z7z7anAhE

Tucker, Jonathan B. *Scourge: The Once and Future Threat of Smallpox*. New York: Grove Press, 2001.

Tufte, Edward R. *Envisioning Information*. Cheshire, CT: Graphics Press, 1990.

____. *The Visual Display of Quantitative Information*. Cheshire, CT: Graphics Press, 1983.

____. *Visual Explanations: Images and Quantities, Evidence and Narrative*. Cheshire, CT: Graphics Press, 1997.

Tukey, John W. "Data-based Graphics: Visual Display in the Decades to Come." *Statistical Science* 6.3 (1990): 327–59.

____. *Exploratory Data Analysis*. Reading, MA: Addison-Wesley, 1977.

Tukey, Paul A., and Andreas Buja, eds. *Computing and Graphics in Statistics*. New York: Springer-Verlag, 1991.

Unwin, Antony R., George Hawkins, Heike Hofmann, and Bernd Siegl. "Interactive Graphics for Data Sets with Missing Values—MANET." *Journal of Computational and Graphical Statistics* 5.2 (1996): 113–22.

Unwin, Antony R., Graham Wills, and John Haslett. "REGARD: Graphical Analysis of Regional Data." *ASA Proceedings of the Section on Statistical Graphics*. Alexandria, VA: American Statistical Association, 1990. 36–41.

Unwin, David. "Off the Map?" *Significance* 9.4 (2012): 45.

U.S. Bureau of the Census. *Census of Agriculture, 1969, Vol. 5, Special Reports, Part 15, Graphic Summary.* Washington, DC: U.S. Government Printing Office, 1973.

U.S. National Institutes of Health (NIH). National Library of Medicine. "Variolation." *Smallpox: A Great and Terrible Scourge.* Web. http://www.nlm. nih.gov/exhibition/smallpox/sp_variolation.html

U.S. Surgeon-General's Office. *Meteorological Register, for Twelve Years, from 1843 to 1854, Inclusive, Compiled from Observations Made by the Officers of the Medical Department of the Army, at the Military Posts of the United States.* Washington, DC: A.O.P. Nicholson, 1855.

U.S. Works Progress Administration. *Report on Progress of the Works Program.* Washington, DC: U.S. Government Printing Office, December 1937.

Utrilla, Pilar, Carlos Mazo, M. Cruz Sopena, Manuel Martínez-Bea, and Rafael Domingo. "A Paleolithic Map from 13,660 calBP: Engraved Stone Blocks from the Late Magdalenian in Abauntz Cave (Navarra, Spain)." *Journal of Human Evolution* 57.2 (2009): 99–111.

Vaccination Law of April 8, 1874. Historical Medical Library of The College of Physicians of Philadelphia.

Vande Moere, Andrew. *Information Aesthetics.* Web. http://infosthetics.com/

Velleman, Paul F. "DataDesk: An interactive package for data exploration, display, model building, and data analysis." *Wiley Interdisciplinary Reviews: Computational Statistics* 4.4 (2012): 407–14.

Venturi, Robert. *Complexity and Contradiction in Architecture.* New York: Museum of Modern Art. 1966.

Wainer, Howard. *Graphic Discovery: A Trout in the Milk and Other Visual Adventures.* Princeton, NJ: Princeton University Press, 2005.

———. "Graphical Visions from William Playfair to John Tukey." *Statistical Science* 5.6 (1990): 340–46.

Wainer, Howard, and Paul F. Velleman. "Statistical Graphics: Mapping and the Pathways of Science." *Annual Review of Psychology* 52 (2001): 305–35.

Walker, Francis A., and the United States Census Office. *Statistical Atlas of the United States Based on the Ninth Census 1870: With Contributions from Many Eminent Men of Science and Several Departments of the Government.* New York: Julius Bien, 1874. Web. Library of Congress, Geography and Map Division: Cultural Landscapes. http://www.loc.gov/item/05019329/

Wallis, Helen M., and Arthur H. Robinson. *Cartographical Innovations: An International Handbook of Mapping Terms to 1900.* Tring, UK: Map Collector Publications in association with the International Cartographic Association, 1987.

Wang, Peter C., ed. *Graphical Representation of Multivariate Data.* New York: Academic Press, 1978.

Ward, R. De C. "The Newspaper Weather Maps of the United States." *American Meteorological Journal* 11 (1894): 96–107.

Webb, Augustus D., and Michael G. Mulhall. *The New Dictionary of Statistics: A Complement to the Fourth Edition of Mulhall's Dictionary of Statistics.* London: G. Routledge and Sons, Limited, 1911.

Wegman, Edward J. "The Grand Tour in k-Dimensions." *Computing Science and Statistics* 22 (1992): 127–36.

Wickham, Hadley. "*ggplot2: Elegant Graphics for Data Analysis (Use R!).*" New York: Springer, 2009.

Wickham, Hadley, and Heike Hofmann. "Product Plots." *IEEE Transactions on Visualization and Computer Graphics* 17.12 (2011): 2223–30.

Wilkinson, Leland. *The Grammar of Graphics.* 2nd ed. Statistics and Computing. New York: Springer Science, 2005.

Wills, Graham. *Visualizing Time: Designing Graphical Representations for Statistical Data.* Statistics and Computing. New York: Springer, 2012.

Winchester, Simon. *The Map That Changed the World.* New York: Harper Perennial, 2009.

Winterbotham, Harold. "Geographical Work with the Army in France." *The Geographical Journal* 54.1 (July 1919): 12–23.

Wood, Denis. *The Power of Maps.* London: Routledge, 1992.

World Health Organization (WHO). *Archives of the Smallpox Eradication Programme: A Guide and Inventory.* 2 vols. Geneva, 1982. Web. http://www.who.int/archives/fonds_collections/bytitle/fonds_6/en/index.html

World Health Organization (WHO). *Factsheet.* 2001. Web.

World Health Organization (WHO). Sixty-fourth World Health Assembly: Daily Notes on Proceedings. 24 May 2011. Web. http://www.who.int/mediacentre/events/2011/wha64/journal/en/index8.html

Wyld, James. "Chart of the World Shewing the Religion, Population, and Civilization of Each Country. London, 1815. Rpt. in Saul Jarcho, "Some Early Demographic Maps." *Bulletin of the New York Academy of Medicine* 49.9 (September 1973): 837–44.

Yates, JoAnne. "Graphs as a Managerial Tool: A Case Study of Du Pont's Use of Graphs in the Early Twentieth Century." *Journal of Business Communication* 22.1 (January 1985): 5–33.

Young, Forrest. W., and Penny Rheingans. *Visualizing Multivariate Structure with VISUALS/Pxpl.* ASA Statistical Graphics Video Lending Library, 1990. Web. http://stat-graphics.org/movies/visuals-pxpl.html

Young, Forrest W., Pedro. M. Valero-Mora, and Michael Friendly. *Visual Statistics: Seeing Data with Dynamic Interactive Graphics.* New York: Wiley, 2006. Web. http://www.visualstats.org/

Zande, Johan van der. "Statistik and History in the German Enlightenment." *Journal of the History of Ideas* 71.3 (2010): 411–32.

Zelditch, Miriam Leah, Donald L. Swiderski, H. David Sheets, and William L. Fink. *Geometric Morphometrics for Biologists: A Primer.* Amsterdam: Elsevier, 2004.

Index